Contemporary Topics in Immunobiology

VOLUME 13

Macrophage Activation

Contemporary Topics in Immunobiology

A Continuation Order Plan is available for this series. A continuation order will bring delivery of each new volume immediately upon publication. Volumes are billed only upon actual shipment. For further information please contact the publisher.

Contemporary Topics in Immunobiology

VOLUME 13

Macrophage Activation

Edited by
Dolph O. Adams
Duke University Medical Center
Durham, North Carolina

and
Michael G. Hanna, Jr.
Litton Institute of Applied Biotechnology
Rockville, Maryland

SPRINGER SCIENCE+BUSINESS MEDIA, LLC

Library of Congress cataloged the first volume in this series as follows:

Contemporary topics in immunobiology, v. 1-
  1972–

    v. illus. 24 cm. annual.

   1. Immunology — Periodicals.

QR180.C632	574.2'9'05	79-179761
ISSN 0093-4054	rev	
Library of Congress	72 [r74c2]	

Library of Congress Card Catalog Number 79-179761

ISBN 978-1-4757-1447-0   ISBN 978-1-4757-1445-6 (eBook)

DOI 10.1007/978-1-4757-1445-6

© 1984 Springer Science+Business Media New York
Originally published by Plenum Press, New York in 1984

Contributors

Dolph O. Adams   *Departments of Pathology and Microbiology-Immunology*
                  Duke University Medical Center
                  Durham, North Carolina 27710

Robert J. Bonney *Department of Immunology and Inflammation*
                  Merck Sharp & Dohme Research Laboratories
                  Rahway, New Jersey 07065

Zanvil A. Cohn   *Department of Cellular Physiology and Immunology*
                  The Rockefeller University
                  New York, New York 10021

Philip Davies   *Department of Immunology and Inflammation*
                  Merck Sharp & Dohme Research Laboratories
                  Rahway, New Jersey 07065

Paul J. Edelson   *Division of Infectious Diseases and Immunology*
                  Department of Pediatrics
                  Cornell Medical College
                  New York, New York 10021

R. Alan B. Ezekowitz *Sir William Dunn School of Pathology*
                      Oxford University
                      Oxford OX1 3RE, England

S. Gordon   *Sir William Dunn School of Pathology*
                  Oxford University
                  Oxford OX1 3RE, England

Frank M. Griffin, Jr. *Division of Infectious Diseases*
                       Department of Medicine
                       University of Alabama in Birmingham
                       Birmingham, Alabama 35294

Stephanie L. James *Laboratory of Parasitic Diseases*
                    National Institute of Allergy and Infectious Diseases
                    National Institutes of Health
                    Bethesda, Maryland 20205

William J. Johnson *Department of Pathology*
                    Duke University Medical Center
                    Durham, North Carolina 27710

Monte S. Meltzer *Department of Immunology*
                  Walter Reed Army Institute of Research and
                      Department of Dermatology

Walter Reed Army Medical Center
Washington, D.C. 20307

David M. Mosser   *Division of Infectious Diseases and Immunology*
Department of Pediatrics
Cornell Medical College
New York, New York 10021

Henry W. Murray   *Department of Medicine*
The Cornell University Medical College
New York, New York 10021

Carol A. Nacy   *Department of Immunology*
Walter Reed Army Institute of Research
Washington, D.C. 20307

Nadia Nogueira   *Department of Cellular Physiology and Immunology*
The Rockefeller University
New York, New York, 10021

Robert J. North   *Trudeau Institute, Inc.*
Saranac Lake, New York 12983

Charles N. Oster   *Department of Immunology*
Walter Reed Army Institute of Research
Washington, D.C. 20307

Hans Schreiber   *La Rabida–University of Chicago Institute*
Department of Pathology
*   and   Committee on Immunology*
The University of Chicago
Chicago, Illinois 60649

Robert D. Schreiber   *Research Institute of Scripps Clinic*
Scripps Clinic
La Jolla, California 92037

Scott D. Somers   *Department of Pathology*
Duke University Medical Center
Durham, North Carolina 27710

Timothy A. Springer   *Laboratory of Membrane Immunochemistry*
Dana-Farber Cancer Institute
Harvard Medical School
Boston, Massachusetts 02115

Jay C. Unkeless   *Laboratory of Cellular Physiology and Immunology*
Rockefeller University
New York, New York 10021

James Urban   *La Rabida–University of Chicago Institute*
Department of Pathology
*   and   Committee on Immunology*
The University of Chicago
Chicago, Illinois 60649

Preface

The concept of macrophage activation dates from Elie Metchnikoff, who noted that mononuclear phagocytes from animals resistant to bacterial challenge had perfected their powers of phagocytosis and microbicidal destruction (Metchnikoff, 1905). Numerous workers, including Lurie, Dannenberg, Suter, North, and Mackaness, subsequently elucidated the basis of cellular resistance to facultative and obligate intracellular parasites (for review, see North, 1981). Specifically sensitized T lymphocytes were found both *in vitro* and *in vivo* by these workers to induce the accumulation of and enhanced microbicidal ability by mononuclear phagocytes. The large, angry mononuclear phagocytes, which ultimately and nonspecifically effected the microbial destruction, were subsequently termed *activated macrophages* by Mackaness (1972). During the early 1970s, Alexander and Evans (1971) and Hibbs and co-workers (1972) made the dramatic discovery that activated macrophages could also selectively and efficiently lyse tumor cells *in vitro*.

Over the past decade, macrophage activation has been intensively studied from a variety of viewpoints (for general reviews, see Pick, 1981; Adams and Marino, 1984; Hibbs *et al.*, 1980). Two generalizations emerging from these studies merit emphasis here. First, activation of murine macrophages for destruction of tumor cells results from appropriately responsive macrophages interacting sequentially with a defined cascade of signals that includes lymphokines (Hibbs *et al.*, 1977; Ruco and Meltzer, 1978; Russell *et al.*, 1977). Second, activation for either microbicidal or tumoricidal competence is accompanied by numerous and striking changes in cellular metabolism, including increased release of reactive oxygen intermediates, altered content of ectoenzymes in addition to a variety of changes that characterize macrophages taken from sites of sterile inflammation (Nathan and Root, 1977; Edelson, 1981; Cohn, 1978; Soberman and Karnovsky, 1981).

Particularly exciting insights into macrophage activation have recently been gained from studies on four aspects of this complex phenomenon: the molecular alterations that accompany activation, the mechanistic bases of recognition and killing of microbes and of tumor cells, the induction and regulation of macrophage activation, and the role played by macrophages in destruction of neoplasms *in vivo*. In this volume, investigators from 12 laboratories describe advances in each of these four areas emphasizing recent developments in their own work.

The alterations in macrophage metabolism and biochemistry that accompany activation are now appreciated to encompass extensive but quite selective changes in proteins of the macrophage plasma membrane. In Chapter 1, Springer and Unkeless first consider those alterations detectable by monoclonal antibodies and describe how this powerful probe can be used to study altered functions of macrophages. The selective remodeling of other membrane proteins including several receptors important to endocytosis and the functional consequences of this membranous revamping are then described by Ezekowitz and Gordon in Chapter 2. Finally, in Chapter 3 Griffin describes how changes in the lateral mobility of complement receptors dramatically alter their function.

Although the precise mechanisms that activated macrophages employ to perceive and destroy tumor cells and microbes are diverse, the experimental models now analyzed generally encompass two fundamental steps: (1) initial recognition of an extracellular target, which leads either to endocytosis of the target or to its attachment to the surface of the macrophages; and (2) subsequent lysis mediated by toxic substances secreted from the macrophages. Mosser and Edelson (Chapter 4) consider the numerous means macrophages employ to recognize various microbial pathogens before shunting them to diverse intracellular compartments. The now compelling evidence that reactive oxygen intermediates are major lytic effectors in several experimental systems of microbial destruction is then reviewed by Murray in Chapter 5. In Chapter 6, Nogueira and Cohn describe, in one integrated model, the mechanisms employed by macrophages to recognize and destroy trypanosomes. Finally, the mechanisms activated macrophages deploy to kill neoplastic cells spontaneously are described by Johnson, Somers, and Adams (Chapter 7) and are then closely related to the signals of the activation cascade.

Mononuclear phagocytes increasingly appear to have a central regulatory role in governing the inflammatory as well as the immune response. Multiple signals from the extracellular millieu act on macrophages to regulate the induction and maintenance of activation; in turn, the numerous secretory products of macrophages (more than 70) act on the environment. The complex induction of macrophage activation for tumoricidal function has already received extensive attention (Hibbs *et al.*, 1977; Ruco and Meltzer, 1978; Russell *et al.*, 1977). In Chapter 8, Nacy, James, Oster, and Meltzer summarize recent studies on the multiple signals necessary to induce activation for microbicidal competence. Robert D. Schreiber then considers in Chapter 9 the exciting possibility that at least one form of macrophage activating factor (MAF) has been biochemically defined as being γ-interferon. By focusing on the release of numerous substances from macrophages, Bonney and Davies consider the provocative possiblity that macrophages regulate their own function (Chapter 10).

Despite the solid evidence indicating that activated macrophages are necessary to the destruction of certain microbes *in vivo*, compelling evidence of a similar and obligate role for macrophages in the destruction of neoplasms *in vivo* has yet to emerge. Workers in numerous laboratories have employed a variety of approaches to study this question, including analysis of leukocytic effectors within tumors, definition of the effective leukocytes in passive transfer experiments, and determination of the effect of various leukocyte inhibitors *in vivo* and have described many examples in which macrophages and T-effector lymphocytes cooperate to produce tumor rejection (Russell and Gillespie, 1980; Haskell *et al.*, 1978; Den Otter, 1981; Levy and Wheelock, 1974; Nelson *et al.*, 1978; Calvelli *et al.*, 1982). James Urban and Hans Schreiber describe in Chapter 11 the novel approach of analyzing tumor variants that are resistant to attack by various populations of leukocytes. In Chapter 12, North analyzes the importance of suppressor T cells in regulating immune-mediated destruction of established, syngeneic tumors *in vivo*.

Two common threads bind these 12 chapters: Each describes studies firmly rooted in cell biology on a different aspect of activation, and each closely relates one or more of the diverse manifestations of activation to specific alterations in the cellular physiology of macrophages. If long-held hopes for therapeutic manipulation of the mononuclear phagocyte system are ever to be realized, the regulatory physiology of this major host-defense system must first be clearly defined. These chapters depict our current degree of progress to that goal.

We thank our contributors for their excellent and timely chapters, our publishers and editors for their assistance and efficiency, and particularly our secretary Ms. Lillie Knight for her unflagging dedication and organizational skills.

D. O. Adams
M.G. Hanna, Jr.

REFERENCES

Adams, D. O., and Marino, P., 1984, Activation of mononuclear phagocytes for destruction of tumor cells as a model for study of macrophage development, in: *Contemporary Hematology/Oncology*, Vol. 3 (A. S. Gordon, R. Silber, and J. LoBue, eds.), pp. 69–136, Plenum Medical Book Company, New York.

Alexander, P., and Evans, R., 1971, Endotoxin and double stranded RNA render macrophages cytotoxic, *Nature (New Biol.)* **232**:76.

Calvelli, T. A., Freedman, V. H., Silverstein, S. C., and Silagi, S., 1982, Leukocyte subpopulations elicited by a nontumorigenic variant of B16 melanoma: Their role in direct rejection of the melanoma and in prevention of tumorigenesis in Winn assays, *J. Exp. Med.* **156**:1723.

Cohn, Z. A., 1978, The activation of mononuclear phagocytes: Fact, fancy, and future, *J. Immunol.* **121**:813.

Den Otter, W., 1981, The effect of activated macrophages on tumor growth in vitro and in vivo, *Lymphokines* **3**:389.

Edelson, P. J., 1981, Macrophage plasma membrane enzymes as differentiation markers of macrophage activation, *Lymphokines* **3**:57.

Haskill, J. S., Hayry, P., and Radov, L. A., 1978, Systemic and local immunity in allograft and cancer rejection, *Contemp. Top. Immunobiol.* **8**:114.

Hibbs, J. B., Lambert, L. H., and Remington, J. S., 1972, Possible role of macrophage mediated nonspecific cytotoxicity in tumor resistance, *Nature (New Biol.)* **235**:48.

Hibbs, J. B., Taintor, R. R., Chapman, H. A., and Weinberg, J. B., 1977, Macrophage tumor killing: Influence of the local environment, *Science* **197**:279.

Hibbs, J. B., Remington, J. S., and Stewart, C. C., 1980, Modulation of immunity and host resistance by micro-organisms, *Pharmacol. Ther.* **8**:37.

Levy, M. H., and Wheelock, E. F., 1974, The role of macrophages in defense against neoplastic disease, *Adv. Cancer Res.* **20**:131.

Mackaness, G. B., 1972, The mechanism of macrophage activation, in: *Infectious Agents and Host Reactions* (S. Mudd, ed.), p. 61, W. B. Saunders, Philadelphia.

Metchnikoff, E., 1905, *Immunity to Infectious Diseases*, Cambridge University Press, New York.

Nathan, C. F., and Root, R. K., 1977, Hydrogen peroxide release from mouse peritoneal macrophages. Dependence on sequential activation and triggering, *J. Exp. Med.* **146**:1648.

Nelson, D. S., Hopper, K. E., and Nelson, M. 1978. Role of the macrophage in resistance to cancer, in: *Handbook of Cancer Immunology*, Vol. I (H. Waters, ed.) p. 107, Garland Publishing, New York.

North, R. J., 1981, An introduction to macrophage activation. *Lymphokines* **3**:1.

Pick, E. (ed.), 1981, Lymphokines in macrophage activation. *Lymphokines* **3**:1–450.

Ruco, L. P., and Meltzer, M. S., 1978, Macrophage activation for tumor cytotoxicity: Development of macrophage cytotoxic activity requires completion of a sequence of short-lived intermediary reactions, *J. Immunol.* **121**:2035.

Russell, S. W., Doe, W. F., and McIntosh, A. J., 1977, Functional characterization of a stable, noncytolytic stage of macrophage activation in tumors, *J. Exp. Med.* **146**:1151.

Russell, S. W., Gillespie, G. Y., and Pace, J. L., 1980, Evidence for Mononuclear phagocytes in solid neoplasm and appraisal of their nonspecific cytotoxic capabilities, *Contemp. Top. Immunol.* **10**:143.

Soberman, R. J., and M. L. Karnovsky, 1981, Biochemical properties of activated macrophages, *Lymphokines* **3**:11.

Contents

Chapter 3

Activation of Macrophage Complement Receptors for Phagocytosis

Frank M. Griffin, Jr.

Chapter 4

**Mechanisms of Microbial Entry and Endocytosis by Mononuclear
Phagocytes**

David M. Mosser and Paul J. Edelson

Chapter 5

**Macrophage Activation: Enhanced Oxidative and Antiprotozoal
Activity**

Henry W. Murray

Chapter 6

**Activation of Mononuclear Phagocytes for the Destruction of
Intracellular Parasites: Studies with *Trypanosoma cruzi***

Nadia Nogueira and Zanvil A. Cohn

Chapter 7

**Expression and Development of Macrophage Activation for Tumor
Cytotoxicity**

William J. Johnson, Scott D. Somers, and Dolph O. Adams

Chapter 8

**Activation of Macrophages to Kill *Rickettsiae* and *Leishmania*:
Dissociation of Intracellular Microbicidal Activities and Extracellular
Destruction of Neoplastic and Helminth Targets**

*Carol A. Nacy, Charles N. Oster, Stephanie L. James,
and Monte S. Meltzer*

Chapter 9

Identification of Gamma-Interferon as a Murine Macrophage-Activating Factor for Tumor Cytotoxicity

Robert D. Schreiber

Chapter 10

Possible Autoregulatory Functions of the Secretory Products of Mononuclear Phagocytes

Robert J. Bonney and Philip Davies

Chapter 11

The Surveillance Role of Various Leukocytes in Preventing the
Outgrowth of Potentially Malignant Cells

James Urban and Hans Schreiber

Chapter 12

Models of Adoptive T-Cell-Mediated Regression of Established Tumors

Robert J. North

Chapter 1

Analysis of Macrophage Differentiation and Function with Monoclonal Antibodies

Timothy A. Springer

Laboratory of Membrane Immunochemistry
Dana-Farber Cancer Institute
Harvard Medical School
Boston, Massachusetts 02115

and

Jay C. Unkeless

Laboratory of Cellular Physiology and Immunology
Rockefeller University
New York, New York 10021

I. INTRODUCTION

A large number of anti-mouse and anti-human macrophage/monocyte monoclonal antibodies (MAb) have recently been obtained that are proving invaluable reagents of extraordinary specificity for the study of macrophage differentiation, function, and surface antigen structure. This chapter summarizes information on such MAb reported up to February 1983 in tabular form, and then concentrates in more detail on antigens characterized by the authors and their collaborators.

In the mouse, at least five antigens have been found that can be distinguished by molecular weight, which are present on macrophages but not on lymphocytes (Table I). One of the mouse Fc receptors, present on macrophages as well as on some lymphocytes, has also been defined with MAb. Further antigens have not been defined biochemically, but appear to have distinct distributions on functional subpopulations.

Table I. Murine Macrophage Antigens Defined by Monoclonal Antibodies[a]

Antibody/antigen designation	Antigen polypeptide (MW)	Cellular distribution	Distinguishing features, functional features	Reference
Mac-1	95,000 170,000	Resident and exudative peritoneal macrophages, splenic histiocytes, granulocytes, PB monocytes, NK cells	Blocks type 3 complement receptor; absent from Langerhans, interdigitating, and lymphoid dendritic cells; M1/70 MAb binds to human PB monocytes, PMN, NK cells; 170,000-MW chain bears epitope	Springer et al. (1979), Mellman et al. (1980), Ault and Springer (1981), Ho and Springer (1982b), Holmberg et al. (1981), Beller et al. (1982)
Mac-2	32,000	TG-peritoneal exudative macrophages; weak or absent from resident and *Listeria*-exudative macrophages	Present on Langerhans and interdigitating, but not follicular dendritic cells; present on epithelial cells; absent from granulocytes	Springer (1981), Ho and Springer (1982a), Flotte et al. (1984), Haines et al. (1983)
Mac-3	110,000	Macrophages, granulocytes	Present on Langerhans and interdigitating, but not follicular dendritic cells; present on epithelial and endothelial cells	Springer (1981), Ho and Springer (1983a), Flotte et al. (1984), Haines et al. (1983)

	MW	Reactivity	Comments	References
54-2	180,000	Cultured bone marrow macrophages, TG-peritoneal exudative macrophages, mast cells	Absent from PB monocytes and resident macrophages	Leblanc et al. (1980), Katz et al. (1981)
F4/80	160,000	PB monocytes, resident and induced macrophages	Present on 8% bone marrow cells and on P815; macrophage activation reduces expression; 75,000-MW trypsin fragment bears epitope	Austyn and Gordon (1981), Hirsch et al. (1981), Ezekowitz et al. (1981)
2.4G2	47,000–70,000	Macrophages, PMN, B cells, J774, FcR-bearing T cells	Blocks binding of IgG1 and IgG2b Fc to FcRII; protease resistant	Unkeless (1979), Mellman and Unkeless (1980)
ACM.1		Peritoneal macrophages activated by pyran or *Corynebacterium parvum*; absent from resident and protease exudative peritoneal macrophages	Cytotoxic for tumoricidal macrophage effectors	Taniyama and Watanabe (1982)
M43, M57, M102, M143		Bone marrow macrophages, resident and exudative peritoneal macrophages	Define different cytotoxic macrophage subpopulations	Sun and Lohmann-Matthes (1982)

[a]*Abbreviations*: FcR, Fc receptor; MW, molecular weight; NK, natural killer; PB, peripheral blood; PMN, polymorphonuclear leukocytes; TG, thioglycollate.

A large number of human antigens have been characterized that are present on blood monocytes but that are present on none or on a small percentage of blood lymphocytes (Table II). In cases in which the antigens have been biochemically characterized, polypeptide chain molecular weights are indicated in brackets. Unfortunately, there has been little comparison of the anti-human MAb produced by different laboratories, and no doubt many of them identify the same antigens.

II. MAC-1 ANTIGEN

Mac-1 was the first antigen to be defined by MAb that is present on macrophages and not on lymphocytes (Springer et al., 1979). The Mac-1 antigen expressed on myeloid but not on lymphoid cells. More recently, several other MAb to it have been obtained (Mellman et al., 1980; Springer et al., 1982; Sanchez-Madrid et al., 1983b). The Mac-1 antigen contains two subunits, an α-subunit of 170,000 and a β-subunit of 95,000 relative molecular weight (M_r). The α- and β-subunits are not linked by disulfide bonds, but are tightly non-covalently associated in an $\alpha_1\beta_1$ complex (Kürzinger et al., 1982; Kürzinger and Springer, 1982). Antibodies specific for the Mac-1 antigen bind to the α-subunit (Sanchez-Madrid et al., 1983b). The LFA-1 antigen on lymphocytes has a different α-subunit associated with the same β-subunit, and thus antibodies to the β-subunit have a broader pattern of cellular reactivity. Both the α- and β-subunits are glycosylated and have surface exposure (Kürzinger and Springer, 1982). After digestion of intact cells with trypsin and papain, the α- and β-chains are cleaved, but remain bound to the cell surface and remain associated as shown by immunoprecipitation with both anti-α- and anti-β-MAb (P. Simon and T. A. Springer, unpublished observations). The subunits thus appear tightly associated with each other and with the cell membrane. Biosynthesis experiments have shown that the α- and β-subunits are derived from separate precursors of 161,000 and 87,000 M_r, respectively (Ho and Springer, 1983b). Assembly into the $\alpha_1\beta_1$ complex appears to precede processing of the subunits to their mature molecular weight, which presumably involves changes in glycosylation.

Mac-1 appears to be a universal macrophage marker (Ho and Springer, 1982b). It is expressed on >95% of peritoneal resident macrophages and macrophages elicited by thioglycollate, lipopolysaccharide (LPS), peptone, Listeria monocytogenes, and concanavalin A (Con A). The average amount of Mac-1 expressed per cell varies by no more than 2-fold among those different populations. Correlating with their larger size, thioglycollate-elicited macrophages express the highest number of anti-Mac-1 MAb binding sites per cell, 1.6×10^5. Mac-1 is present on splenic macrophages in the red pulp and in the marginal zones surrounding the periarteriolar lymphoid sheath (Ho and Springer, 1982b) as well as on lymph node medullary cord macrophages (Flotte et al., 1983). Histiocytes in the lamina propria of the intestine and alveolar macrophages in the lung are positive. Kupffer cells, which are distinct from macrophages but are in the mononuclear phagocyte lineage, are negative (Flotte et al., 1983).

The Mac-1 antigen is also present on exudate granulocytes and granulocytic precursors in the bone marrow (Springer et al., 1979) and on natural killer (NK) cells. Immunofluorescent cell sorter experiments have shown that cells with natural killing activity obtained from the nylon wool nonadherent fraction of peritoneal exudates are Mac-1$^+$ (Holmberg et al., 1981). Furthermore, these Mac-1$^+$ cells have the morphology of large granular lymphocytes. The M1/70 rat anti-mouse MAb cross-reacts with human Mac-1, which has the same distribution as murine Mac-1, i.e., on monocytes, granulocytes, and null or large granular lymphocytes, which have NK and antibody-dependent cytotoxic activity (Ault and Springer, 1981). Human Mac-1 is equivalent to the OKM1 and Mo1 antigens (Todd et al., 1982). The human granulocyte–monocyte precursor cell (CFUgm) appears to be Mac-1$^-$ (Smith et al., 1984), Mo1$^-$ (Griffin et al., 1982). Thus, granulocytes and monocyte/macrophages become Mac-1$^+$ after divergence from their common stem cell. NK cells are renewed from the bone marrow (Kiessling et al., 1977), but their stem cell has not been identified. It appears that there are at least two phenotypically distinct types of NK cells. NK cells (LGL) isolated directly from blood are mostly Mac-1$^+$ and OKM1$^+$ and mostly lacking in antigens characteristic of T cells (Ault and Springer, 1981; Ortaldo et al., 1981; Breard et al., 1981; Zarling et al., 1981), while alloactivated or "cultured" NK cells express some T-cell markers and may be either Mac1/OKM1 positive (Krensky et al., 1982) or negative (Brooks et al., 1982; Sheehy et al., 1983; Hercend et al., 1983).

Mac-1 distribution on tumor cells in the monocyte/macrophage lineage parallels that on normal cells (Ralph et al., 1983a). The immature myelomonocytic leukemia cell line M1 is Mac-1$^-$, but becomes strongly Mac-1$^+$ after induction with phorbol ester or lymphokines. Induced cells stop dividing and acquire functional properties and surface markers such as Fc and complement receptors characteristic of mature macrophages. In a study of eight independent macrophage lines ranging in phenotype from mature (J774 and P388D$_1$) to somewhat less mature (WEHI-3), all were found to be Mac-1$^+$. Cells of other hematopoietic lineages, including the P815 mastocytoma cell, were found to be Mac-1$^-$.

Monoclonal antibody blocking studies suggest an association or identity between Mac-1 and the complement receptor type 3 (CR$_3$), which is specific for C3bi (Beller et al., 1982). The M1/70 anti-Mac-1 MAb strongly inhibits complement receptor-mediated rosetting of erythrocyte–IgM antibody–complement (E-IgM-C) complexes. Lack of inhibition by a panel of eight other antibodies, including anti-Mac-2, anti-Mac-3, and anti-H-2, and anti-panleukocyte MAbs which bind to a similar number of sites per cell as anti-Mac-1, demonstrates the

Table II. Human Monocyte/Macrophage Antigen Expression on Cells within the Myeloid Lineage[a,b]

Bone marrow

Erythroid precursors	Monocytic precursors		Granulocyte–monocyte stem cell	Granulocytic precursors Early	Late	Megakaryocyte
5F1	Mo1	My8	My7	Mo5	3G8	Mo4
	Mo2	B13.4	D5D6	My8	Mo1	MPA
	Mo5	5F1	C10H5	My7	Mo5	
	My3	1G10		1G10	Mo6	
	My4				My8	
	My7				B13.4	
					B43.4	
					1G10	

Peripheral blood

Monocyte[c–f]			Granulocyte[e]		Platelet
Mac-120[120] (1)	1D5 (12)	B13.4 (7)	3G8	AML-2-23	Mo4
Mo2[55] (2–4)	S4-7 (13)	B43.4 (7)	OKM1	1G10	63D3
Mo3 (4, 5)	63D3[200] (6, 14)	B34.3 (7)	Mo1	S5-7	5F1
Mo4 (4, 5)	OKM1[94, 155] (15)	5F1 (24)	Mac-1	MMA	MPA
61D3[23, 55] (6)	Mo1[94, 155] (2–4)	AML-2-23 (25)	B2.12	*My4*	
B44.1 (7)	Mac-1 (16)	MPA[93, 135] (26)	M522	*63D3*	
D5D6 (8)	My903 (17)	MφP-9 (9)	Mo5	*Mo6*	
C10H5 (8)	My904 (17)	MφS-1 (9)	My7	*My3*	
MφP-15 (9)	B2.12 (18)	MφS-39 (9)	My8	*MφS-39*	
MφP-7 (9)	Mo6[80] (4, 19)	*1G10* (24)	B13.4	*MφP-9*	
MφR-17 (9)	M522 (20)	*My7[160]* (21)	B43.4	*MφS-1*	
PHM2 (10)	My3 (21)	UC45[45] (27)	B34.3	*S4-7*	
PHM3 (10)	My4[55] (21)	S5-25 (13)		*S5-25*	

OKM5 (11) My8 (21) S5-7 (13)
4F2[40, 80] (22)
MMA (23)

Extravascular space macrophage[c,d]	
3G8[47–70] (28)	
Mo1	MφS-1
Mo2	MφS-39
Mo4	MφP-15
MφP-9	MφP-7
ID5	MφR-17
PHM2	PHM3

[a] Modified from Todd et al. (1983).
[b] Almost all the monoclonal antibodies have been tested on erythrocytes, monocytes, granulocytes, and lymphocytes, but many have not been tested on bone marrow cells, macrophages, and nonhematopoietic cells. A few of the antibodies such as PHM2 and 4F2 react with small percentages of resting lymphocytes or with activated lymphocytes.
[c] M_r of polypeptide chains $\times 10^{-3}$ is in brackets.
[d] Underline indicates specificity for monocytes or macrophages (among hematopoietic cells).
[e] Italics indicate weak expression.
[f] Numbers in parentheses are keyed to the following references:

1. Raff et al. (1980)
2. Todd et al. (1981)
3. Todd et al. (1982)
4. Todd et al. (1984)
5. Todd and Schlossman (1982)
6. Ugolini et al. (1980)
7. Perussia et al. (1982)
8. Linker-Israeli et al. (1981)
9. Dimitriu-Bona et al. (1983)
10. Becker et al. (1981)
11. Shen et al. (1983)
12. Kaplan and Gaudernack (1982)
13. Ferrero et al. (1983)
14. Rosenberg et al. (1981)
15. Breard et al. (1980)
16. Ault and Springer (1981)
17. Letvin et al. (1983)
18. Van Der Reijden et al. (1983)
19. Todd and Schlossman (1984)
20. Lohmeyer et al. (1981)
21. Griffin et al. (1981)
22. Haynes et al. (1981)
23. Hanjan et al. (1982)
24. Bernstein et al. (1982)
25. Ball et al. (1982)
26. Burckhardt et al. (1982)
27. Hogg (1983)
28. Fleit et al. (1982)

specificity of blockade. Inhibition occurs with as little as 1 μg/ml of anti-Mac-1 F(ab')$_2$ fragments, and the Fc receptor is unaffected. Macrophages bear receptors for both C3b (CR$_1$) and C3bi (CR$_3$). Primarily the latter receptor is measured when E are sensitized with C5-deficient serum (Ross, 1980). When E bearing only C3b or C3bi were prepared with homogeneous complement components, it was found that Mac-1 inhibited the CR$_3$ but not the CR$_1$ (Beller et al., 1982). Since the M1/70 anti-Mac-1 MAb is cross-reactive with human cells, it was also tested for its ability to inhibit complement receptors on human cells. Anti-Mac-1 inhibits the CR$_3$ but not the CR$_1$ on human granulocytes. The most likely interpretation of these findings is that Mac-1 antigen is the CR$_3$. Studies with monoclonal antibodies to the human OKM1 antigen, which appears identical to human Mac-1, lend further support to this idea (Wright et al., 1983). Staphylococcus aureus bacteria coated with a sandwich of OKM1 antibody and OKM1 antigen specifically agglutinate with C3bi-coated E. Since the CR$_3$ does not have sufficiently high affinity to bind soluble C3bi, it has not been possible to test for displacement of soluble ligand.

It is interesting to compare the distribution of Mac-1 with that reported for the CR$_3$ (reviewed in Ross, 1980; Fearon and Wong, 1983). The CR$_3$ is present on monocytes and neutrophils, as is Mac-1. It also is present on 6-10% of human blood lymphocytes (Ross and Lambris, 1982; Perlman et al., 1981), in agreement with the finding of Mac-1 on the null subpopulation of ~10% of human lymphocytes that contain NK activity (Ault and Springer, 1981). There are conflicting findings on the presence of the CR$_3$ on glomerular epithelial cells (Carlo et al., 1979; Beller et al., 1982); kidney is negative for Mac-1 by absorption and thin section staining. B-lymphoblastoid cell lines and less than one-third of tonsil lymphocytes have been reported to express the CR$_3$ (Ross and Lambris, 1982). However, B cells also express the complement receptors CR$_2$ and CR$_1$. The CR$_2$ binds C3bi in addition to C3d; since antibodies to the CR$_2$ largely inhibited rosetting with E-C3bi by B lymphoblastoid lines, it is possible that this rosetting is caused by the CR$_2$. Furthermore, although the CR$_1$ is specific for C3b at physiologic ionic strength, it can bind C3bi under the low-ionic-strength conditions sometimes used in complement adherence assays (Ross et al., 1983).

III. THE MAC-1 AND LFA-1 FAMILY

A second antigen has been discovered that is distinct from Mac-1 in cell distribution, function, and α-subunit structure, but appears to use the same β-subunit. In the course of studies on the molecular basis of T-cell function, MAb were selected for their ability to inhibit antigen-specific T-lymphocyte-mediated killing (Springer et al., 1982). Some of these MAb defined the LFA-1 antigen, which contains two polypeptide chains of M_r 180,000 and 95,000. MAb to LFA-1 block killing by inhibiting formation of the adhesion between the cytolytic T lymphocyte (CTL) and the target cell (Davignon et al., 1981; Springer et al., 1982). It appears that LFA-1 is distinct from the antigen receptor, but works together with it in contributing to the avidity of the CTL for the target cell (Springer et al., 1982). LFA-1 is present on B lymphocytes and myeloid cells as well as T lymphocytes (Kürzinger et al., 1981), suggesting that it plays a more general role in adhesion than do antigen receptors. Interestingly, both adhesion of CTL to target cells, the step in which LFA-1 participates (Springer et al., 1982), and adhesion of macrophages to C3bi-coated cells, which is mediated by the CR_3, are Mg^{+2}-dependent (Lay and Nussenzweig, 1968; Wright and Silverstein, 1982).

The Mac-1 and LFA-1 β-subunits of M_r 95,000 are identical by peptide mapping and by complete immunologic cross-reactivity (Kürzinger et al., 1982; Sanchez-Madrid et al., 1983b). Similar peptide map results were obtained for Mac-1 and an antigen probably identical to LFA-1 (Trowbridge and Omary, 1981). A MAb cross-reactive for Mac-1 and LFA-1 has been shown to bind to an epitope on their β-subunits (Sanchez-Madrid et al., 1983b). MAb, which are specific for Mac-1 or LFA-1, bind to α-chain epitopes. The α-subunits are non-cross-reactive, as shown with both monoclonal and conventional antisera, and have different tryptic peptide maps. However, sequencing of their N termini has shown 40% amino acid sequence homology (D. Teplow, W. Dreyer, and T. Springer, unpublished data), suggesting that the α-chains are related by gene duplication.

MAb binding to distinct topographic sites on Mac-1 and LFA-1 have been shown to differ in their functional effects (Sanchez-Madrid et al., 1983b). Two MAb recognizing closely related topographic determinants on the Mac-1 α-chain inhibit complement receptor activity, while a third anti-α-chain MAb directed against a topographically distinct α-determinant and an anti-β-MAb have no effect. In fact, the anti-β-MAb consistently enhances CR_3, activity. These results indicate that a functionally active site on the Mac-1 molecule, probably representing the ligand binding site, can be localized to a specific region of the α-chain. Similarly, a functionally active site on the LFA-1 molecule has been localized to the α-subunit.

A similar family of related molecules has been found on human cells (Sanchez-Madrid et al., 1983a). Human Mac-1 is homologous to mouse Mac-1, as shown by monoclonal antibody cross-reaction, identical cell distribution of the antigens (Ault and Springer, 1981), and identical association with the CR_3 (Beller et al., 1982). Human Mac-1 appears identical to the human OKM1 and Mo1 antigens in terms of cell distribution, and anti-OKM1 and anti-Mo1 MAb also block complement receptor activity (Todd et al., 1984; Wright et al., 1983; and Sanchez-Madrid et al., 1983a). Human LFA-1 is equivalent to mouse LFA-1 and shares a common β-subunit with OKM1 (Sanchez-Madrid et al., 1983a).

Furthermore, yet a third antigen with a distinct α-chain of 150,000 M_r has been found to be associated with the same β-subunit. It is found on granulocytes and monocytes (Sanchez-Madrid *et al.*, 1983*a*). Thus, an interrelated family of three different cell-surface molecules has been described which use a single type of β-subunit in association with differing α-subunits. Two of these molecules are associated with cell-adhesion functions, and it will be interesting to determine whether the third has a similar function.

IV. MAC-2 ANTIGEN

Mac-2 is a macrophage surface antigen of 32,000 M_r. It is biosynthesized by macrophages (Ho and Springer, 1982*a*), and the precursor identified by 5-min pulse labeling is the same molecular weight as the mature antigen (M. K. Ho and T. A. Springer, unpublished observations). Isoelectric focussing shows that Mac-2 is a basic polypeptide having a pI in the range of 7–8. It focuses in a position very close to that of the invariant chain of Ia, which is of 31,000 M_r (Jones *et al.*, 1979). However, there are no identities between the methionyl tryptic peptides of Mac-2 and the Ia invariant chain (M. K. Ho and T. A. Springer, unpublished data). Thioglycollate-elicited macrophages bear 1.7×10^5 anti-Mac-2 MAb binding sites per cell.

Mac-2 is a macrophage subpopulation marker, an inducible component of the macrophage cell surface (Ho and Springer, 1982*a*). Among resident macrophages and five different types of elicited peritoneal macrophages studied, only thioglycollate-elicited macrophages showed strong expression by immunofluorescent flow cytometry and immunoprecipitation of ^{35}S-methionine-labeled antigen. Mac-2 is expressed equally strongly by macrophages 1 day and 4 days after elicitation with thioglycollate. Thus, cells recruited into the peritoneum after 1 day are already committed to the synthesis of Mac-2. Biosynthesis of Mac-2 by resident peritoneal macrophages and macrophages elicited by peptone, LPS, Con A, and *Listeria* is detected, but is 10- to 30-fold lower than in thioglycollate-elicited macrophages. Mac-2 is essentially undetectable on these low-expressing cells by immunofluorescence, but is detectable by the much more sensitive immunoperoxidase technique (Flotte *et al.*, 1983). Mac-2 has been found on all mature macrophage cell lines examined and is absent from lymphoid and primitive erythroid and myelomonocytic lines. In contrast to Mac-1 and Mac-3, however, Mac-2 is not expressed by the M1 cell line after induction of maturation (Ralph *et al.*, 1983*a*).

Resident peritoneal macrophages or those elicited by a variety of agents synthesize and express on their surface similar amounts of the Mac-1 antigen (Ho and Springer, 1982*b*). Mac-1 is thus a constitutive macrophage marker, whereas Mac-2 and Ia (Beller *et al.*, 1980) appear to be inducible surface components.

The induction of Mac-2 and Ia is controlled independently, since some macrophages such as thioglycollate-elicited are high in Mac-2 and low in Ia, while *Listeria*-elicited macrophages are high in Ia and low in Mac-2. Resident macrophages are low in both antigens.

Immunoperoxidase shows that all tissue macrophages, such as alveolar macrophages and macrophages in the lamina propria are Mac-2$^+$, although less strongly than thioglycollate-elicited macrophages (Flotte *et al.*, 1983). Kupffer cells are also Mac-2$^+$. Surprisingly, Mac-2 is also expressed in a highly specific pattern on certain epithelial cells. It is present on bronchial epithelium, some kidney tubules, intestinal epithelium, in the skin on keratinocytes, hair follicles, and sweat ducts, and in the brain on the choroid plexus and ependyma. It appears that Mac-2 is induced during the maturation of intestinal epithelial cells. The epithelial cells of villous intestine are renewed in crypts just below the villi. They migrate from the villum base to the tip, from which they are eventually sloughed off. Crypt epithelial cells are Mac-2$^-$, those at the base of villi stain weakly, and there is a gradient of increasing Mac-2 expression from the base to the tip of villi. When intestinal epithelial cells are stained histochemically it is found that their ability to absorb nutrients such as fats from the lumen, follows a similar distribution (Ladman *et al.*, 1963).

Thus far, no functional activity of macrophages has been found to be inhibited by anti-Mac-2 MAb, including mannose uptake, which appears to be mediated by a receptor of similar molecular weight (Townsend and Stahl, 1981). Because thioglycollate-elicited macrophages are more active phagocytically than the other macrophages studied, and because certain epithelial cells are highly active endocytically, it is possible that Mac-2 plays a role in endocytosis.

V. MAC-3 ANTIGEN

Mac-3 (Ho and Springer, 1983*b*) is a less abundant antigen present in 3.6 × 10^4 sites per thioglycollate-elicited macrophages. It is expressed in similar quantities on resident peritoneal macrophages and on macrophages elicited by a variety of agents. It is found on macrophages in a number of tissues examined by immunoperoxidase staining of thin sections (Flotte *et al.*, 1983, 1984) and on eight of eight different macrophage cell lines (Ralph *et al.*, 1983*a*). Mac-3 is a glycoprotein and appears as a somewhat diffuse band in sodium dodecyl sulfate–polyacrylamide gel electrophoresis (SDS–PAGE). Mac-3 is found on macrophages and granulocytes but on no other hematopoietic cell types. It is also found on some nonhematopoietic cell types, giving highly specific staining patterns on epithelial and endothelial cells in a variety of tissues (Flotte *et al.*, 1983). Both liver parenchymal and Kupffer cells are stained, and staining of bile canaliculi is particularly intense. Intestinal epithelial cells are stained only on their luminal border, perhaps on microvilli.

An unusual feature of Mac-3 is that its molecular weight varies from 100,000 to 170,000 depending on the type of elicited macrophage or macrophage cell line from which it is isolated (Ho and Springer, 1983a; Ralph et al., 1983a). Macrophages elicited by different agents synthesize identical Mac-3 precursors of 74,000 M_r which are processed in 15 min to the higher-molecular-weight mature forms, which vary in M_r depending on the type of macrophage population (Ho and Springer, 1983a). This shift appears to be attributable to glycosylation. Recent studies of Mac-3 carbohydrate show that high-mannose, complex, and lactosaminoglycan moieties are present (A. Mercurio, P. Robbins, and T. Springer, unpublished results). It appears that Mac-3 is glycosylated to different extents in macrophages depending on their state of differentiation. Glycosylation may have important effects on the surface properties of macrophages and important consequences for macrophage functional activity. Such variation in glycosylation may thus be an important source of macrophage heterogeneity.

VI. LANGERHANS CELLS, DENDRITIC CELLS, AND MACROPHAGES

The requirement for accessory cells in the induction of antigen-specific T-lymphocyte responses is well documented. However, the relationship between different types of accessory cells has been unclear. Ia antigen-bearing macrophages are potent antigen-presenting cells (Unanue, 1981) and are probably particularly important at inflammatory sites. Langerhans cells are the antigen-presenting cells of the skin. They mediate the induction of contact sensitivity *in vivo* and the induction of antigen-specific T-lymphocyte responses *in vitro* (Silberberg-Sinakin et al., 1976; Stingl et al., 1978). Interdigitating dendritic cells are found in the T-dependent areas of lymphoid tissues. Their long dendritic processes, which resemble those of Langerhans cells, form extensive contacts with adjacent T lymphocytes in thymus, spleen, and lymph node (Thorbecke et al., 1980; Tew et al., 1982). This anatomic association and the morphologic resemblance to Langerhans cells including the sharing of the unique tennis racquet-shaped Birbeck granule suggest that interdigitating cells are important in antigen presentation and in regulating T-cell responses. Follicular dendritic cells, found in intimate contact with B lymphocytes in the corona of lymphoid follicles, differ in morphology from Langerhans and interdigitating dendritic cells. They take up antigen–antibody–complement complexes and retain them longer *in vivo* than any other cells (Thorbecke et al., 1980; Humphrey and Grennan, 1984). The lymphoid dendritic cell of Steinman and Cohn (1973) has been isolated in suspension, has a dendritic morphology, and is an active accessory cell for lymphocyte responses. Its precise relationship to the interdigitating and follicular dendritic cells is unclear. Macrophages and Langerhans cells are bone marrow derived, but the relationship of their precursors is unknown (Stingl et al., 1978).

The surface markers of dendritic cells and macrophages have been compared (Table III). All except follicular dendritic cells bear Ia antigen, which is important in the induction of antigen-specific T-lymphocyte responses (Germain et al., 1982). Macrophages can be distinguished from all the other cell types by their expression of the Mac-1 antigen. The lack of expression of Mac-1 on Langerhans cells is in agreement with the absence of C3bi receptors (Berman and Gigli, 1980). The lymphoid dendritic cell of Steinman and Cohn also lacks Mac-1. Interestingly, Langerhans cells and interdigitating cells express the Mac-2 and Mac-3 antigens, whereas follicular dendritic cells are negative for both antigens. The common Mac-$1^-2^+3^+$ phenotype of Langerhans cells and interdigitating dendritic cells supports the ideas, based on morphologic similarites, that these may be ontogenetically and functionally related cells localized in different anatomic

Table III. Properties of Murine Dendritic Cells and Macrophages[a,b]

Property/antigen	Macrophage[f]	Langerhans cells[f]	Interdigitating dendritic cell[f]	Follicular dendritic cell[f]	Lymphoid dendritic cell[f]
Ia	+/- (1)	+ (2)	+ (2)	- (3)	+ (4, 8)
Mac-1	+ (5)	- (6, 7)	- (6, 7)	- (6, 7)	- (8)
Mac-2	+ (9)	+ (6, 7)	+ (6, 7)	- (6, 7)	ND
Mac-3	+ (10)	+ (6, 7)	+ (6, 7)	- (6, 7)	ND
Ly-5/CLA	+ (5, 11)	+ (6, 7)	ND	ND	+ (8)
FcR	+ (12)	+ (6, 7, 13)	ND	+ (3)	- (8, 14)
C3b R	+ (15)	+ (16)[c]	ND	+? (3)[d]	-/+ (14)[d,e]

[a]Modified from Haines et al. (1983).
[b]The terminology is that of Tew et al. (1982). CLA, common leukocyte antigen; FcR, Fc receptor; ND, not determined.
[c]Langerhans cells bear the receptor for C3b and lack receptors for C3bi and C3d (Berman and Gigli, 1980).
[d]Results were reported for C3R, but it is not known whether C3b R, C3bi R, or both were measured.
[e]Mouse spleen dendritic cells are C3R$^-$, and human peripheral blood dendritic cells are C3R$^+$ (Van Voorhis et al., 1982).
[f]Numbers in parentheses are keyed to references:

1. Swartz et al. (1976)	9. Ho and Springer (1982a)
2. Hoffman-Fezer et al. (1978)	10. Ho and Springer (1983a)
3. Humphrey and Grennan (1984)	11. Scheid and Triglia (1979)
4. Steinman et al. (1979)	12. Berken and Benacerraf (1966)
5. Springer et al. (1979)	13. Tamaki et al. (1979)
6. Flotte et al. (1983)	14. Steinman and Cohn (1974)
7. Haines et al. (1983)	15. Fearon and Wong (1983)
8. Nussenzweig et al. (1981)	16. Burke and Gigli (1980)

sites and that these cells differ from follicular dendritic cells. It will be interesting to learn whether lymphoid dendritic cells are also Mac-2$^+$3$^+$. All cell types tested, i.e., macrophages, Langerhans cells, and lymphoid dendritic cells, share the Ly-5 or common leukocyte antigen (CLA) molecule. This marker has thus far only been found on hematopoietic cells (Scheid and Triglia, 1979; Springer, 1980; Kürzinger et al., 1981; Sarmiento et al., 1982).

VII. DEFINING MACROPHAGES BY THEIR SURFACE MARKERS

The Mac-1 and Mac-3 surface markers are acquired during differentiation from immature precursor cells, as shown in mice (Ralph et al., 1983a) and humans (Ralph et al., 1983b) with the M1 and U937 tumor line models, respectively. Studies in humans have shown that the granulocyte–monocyte colony forming unit is Mac-1$^-$ or low in Mac-1 (Griffin et al., 1982; Smith et al., 1983). Mac-1 expression is increased during differentiation of peripheral blood monocytes to peritoneal macrophages (Springer et al., 1979). Mac-1 and Mac-3 have been found on all mature macrophages and histiocytes studied and thus may be considered constitutive markers of mature macrophages. The Mac-2 antigen is found in all types of macrophages, as shown by the immunoperoxidase technique, although it should be considered an inducible marker because of the wide variation in quantitative expression. Coexpression of the Mac-1 and Mac-2 antigens, with the caveat that the sensitive immunoperoxidase technique must be used to detect Mac-2, appears to be an excellent operational definition of the macrophage in the many different anatomic sites so far investigated. Mac-1 is also found on granulocytes and NK cells and Mac-2 is also found on epithelial, Langerhans, and interdigitating cells, but the markers are found together only on macrophages.

Few surface markers are specific for what hematologists would define as a cell lineage. This is not surprising, because structures are present on cell surfaces to perform specific functions, not for the convenience of hematologists or immunologists. The cells of the immune defense system appear to have evolved a high degree of functional redundancy; e.g., cells of both the lymphoid and myeloid lineages bear Fc and C3b receptors for immune complexes and can phagocytose foreign material.

Many types of cells can act as accessory cells for the induction of antigen-specific T-lymphocyte responses, and the expression of Ia antigens is correspondingly widespread. Even skin epithelial cells become Ia$^+$ when the skin is inflamed in graft-versus-host reactions (Mason et al., 1981), and endothelial cells become Ia$^+$ in response to γ-interferon secreted by lymphocytes (Pober and Gimbrone, 1982).

Markers may be useful in an operational sense for defining lineages, but their use in predicting the relatedness of cells is limited, at best. For example, the 54-2 antigen is found on elicited and not on resident macrophages (Leblanc et al., 1980), but also marks mast cells (Katz et al., 1981). Mac-2 and Mac-3 are present on macrophages and not on lymphocytes, but are also present on epithelial, and for Mac-3, additional nonhematopoietic cells. The OKT6 antigen, originally thought to be thymocyte specific, has recently also been found on Langerhans cells (Fithian et al., 1981). The Thy-1 antigen has long been used as a marker for distinguishing T lymphocytes from other cells of the hematopoietic system; however, it was only recently discovered to be present as well on 25% of bone marrow cells, including stem cells, and on myeloid cells in bone marrow cultures (Schrader et al., 1982; Basch and Berman, 1982). In the definition of cell lineages, there appears to be no substitute for the direct study of stem cell development into mature cell types.

When the function of surface markers is known, their expression on diverse cell types takes on greater significance. The expression of Mac-1 on both macrophages and on NK/antibody-dependent cytotoxic cells is an example. On macrophages, the CR_3 (Mac-1) mediates adherence to cells or particles opsonized with the complement component C3bi. On activated macrophages, the CR_3 mediates phagocytosis (Michl et al., 1979). On resident macrophages, the CR_3 is synergistic with the FcR for phagocytosis (Bianco and Nussenzweig, 1977). What is the role of the CR_3 (Mac-1) on NK and antibody-dependent cellular cytotoxicity (ADCC) cells, which are nonphagocytic? When target cells are coated with C3bi in addition to IgG, lysis by ADCC effectors is greatly enhanced (Perlman et al., 1981), suggesting that the CR_3 synergizes the the Fc receptor (FcR) in the killing reaction. Whether the presence of C3bi on target cells would enhance natural killing has not yet been tested.

VIII. MOUSE $Fc_{\gamma 2b/\gamma 1}R$

The analysis of mouse FcR has been complicated by the apparent presence of several receptors with specificity for different subclasses of IgG. The FcR that binds mouse IgG2a ($Fc_{\gamma 2a}R$) is inactivated by trypsinization (Unkeless and Eisen, 1975; Walker, 1976), while those binding IgG2b and IgG1 ($Fc_{\gamma 2b/\gamma 1}R$) and IgG3 (Diamond and Yelton, 1981) are resistant to trypsin. Although the results of competition experiments with monomeric IgG myeloma proteins are equivocal with respect to FcR heterogeneity (Segal and Titus, 1978; Haeffner-Cavaillon et al., 1979), competition experiments using aggregated IgG and immune complexes of different subclasses indicate that there are three different FcR on macrophages (Walker, 1976; Diamond and Scharff, 1980; Diamond and Yelton, 1981). Comparable results have been obtained in rats (Boltz-Nitulescu et al., 1981). These conclusions about receptor heterogeneity are supported by the isolation of macrophage cell line variants lacking the $Fc_{\gamma2b/\gamma1}R$ and the $FcR_{\gamma3}$ receptor, respectively (Unkeless, 1979; Diamond and Yelton, 1981).

The mouse macrophage FcR specific for IgG2b and IgG1 immune complexes has been characterized using a rat monoclonal antibody, 2.4G2 (Unkeless, 1979; Mellman and Unkeless, 1980). The monoclonal antibody was isolated after the fusion of spleen cells from a rat immunized with the mouse macrophage cell lines J774 and P388D$_1$ and was identified by the ability of the culture cell supernatant to inhibit rosette formation with sheep erythrocytes (E) opsonized with monoclonal anti-E IgG2b immunoglobulin. The specificity of the monoclonal antibody 2.4G2 was examined by studying the inhibition, after preincubation of macrophages with the Fab fragment of 2.4G2, of rosette formation with opsonized erythrocytes. Only the binding of IgG2b and IgG1 immune-complex-coated E was inhibited. The binding of IgG2a-immune-complex-coated E was unaffected.

The cellular distribution of the antigen was determined by quantitative binding studies and by inhibition of rosette formation with E opsonized with rabbit IgG (EIgG). In addition to its presence on all mouse macrophages, the 2.4G2 antigen is present on monocytes, B lymphocytes, polymorphonuclear leukocytes (PMN), and several lymphoid cell lines of T-cell and null-cell origin. These results demonstrated the antigenic identity of FcR on a variety of cell types. The 2.4G2 determinant is, however, absent from mouse dendritic cells (Nussenzweig et al., 1981).

The $Fc_{\gamma2b/\gamma1}R$ was purified by affinity chromatography on 2.4G2 Fab-Sepharose 4B (Mellman and Unkeless, 1980). Nonionic detergent lysates of J774 tumors or cultured J774 cells were absorbed on an affinity column, which was then washed with Nonidet® P-40 SDS-mixed micelles, followed by 0.5% sodium deoxycholate. The bound protein was then eluted with 0.5% sodium deoxycholate adjusted to pH 11.5 with triethylamine, following a procedure developed for purification of Ia antigens (McMaster and Williams, 1979). The protocol resulted in the isolation of 0.01% of the protein in the initial lysate after clearance of nuclei, and the recovery of 2.4G2 antigen was 57%, an overall purification of $>$ 5000-fold in one step.

The $Fc_{\gamma2b/\gamma1}R$ thus isolated from the mouse macrophage line J774 consists of two poorly resolved peptides of 47,000 and 60,000 M_r. The peptides are glycosylated and can be labeled by galactose oxidase oxidation followed by reduction with $NaB[^3H_4]$. In two-dimensional isoelectric focusing SDS-PAGE, the $Fc_{\gamma2b/\gamma1}R$ exhibits the typical decrease in M_r from acidic to basic species. The isoelectric point of the purified receptor is broad, with a pI of 4.7–5.8. This was later confirmed by Lane and Cooper (1982), who isolated FcR by affinity chromatography using both 2.4G2 IgG and affinity chromatography on IgG2b-Sepharose. Lane and Cooper (1982) also observed small differences in M_r and isoelectric point between Fc-binding proteins isolated from IgG2a-Sepharose compared with proteins eluted from IgG2b-Sepharose, suggesting that the two receptors have structural differences. Others have isolated similar Fc-binding proteins (Loube et al., 1978; Loube and Dorrington, 1980; Schneider et al., 1981) from mouse macrophages or macrophage cell lines. However, differences between proteins isolated on IgG2a versus IgG2b or human IgG1-Sepharose were not detected in these studies.

The 2.4G2 antigen, and the activity of $Fc_{\gamma 2b/\gamma 1}R$ is trypsin resistant and can be solubilized from the plasma membrane only by detergents, indicating it is an integral membrane protein. However, trypsin treatment does result in a decrease in the amount of the higher M_r peptide, and a concomitant increase in the amount of the lower M_r peptide. This result suggests that the two peptides isolated from J774 cells may be related by a post-translational proteolytic event. The similarity between the typtic and chymotryptic maps of the two peptides is consistent with that interpretation (I. Mellman and J. Unkeless, unpublished data), although the possibility that two closely related peptides are translated from different messages is not ruled out. $Fc_{\gamma 2b/\gamma 1}R$ was immunoprecipitated from a variety of cell lines after surface iodination, and a significant variation in M_r was observed, with the largest species from the B-cell line WEHI-231 and the smallest from thioglycollate-elicited peritoneal macrophages. The biochemical basis for this variation and any functional correlates of these differences are unknown.

Of particular interest was the observation that the purified $Fc_{\gamma 2b/\gamma 1}R$ retained binding specificity for IgG consistent with identification of the protein as an FcR. Although there was no binding of the purified protein to $F(ab')_2$ immune-complex-coated surfaces, the specificity of binding to mouse IgG subclasses was partially lost. In the absence of detergent, the receptor bound to IgG2b-, IgG1-, and IgG2a-coated Sephadex beads, but not to IgG3-coated Sephadex beads. However, in the presence of detergent, purified and labeled $Fc_{\gamma 2b/\gamma 1}R$ bound best to IgG2b aggregates, less well to IgG1 aggregates, and not at all to either IgG2a or rabbit IgG (I. Mellman and J. Unkeless, unpublished data). The purified FcR in the absence of detergent formed aggregates of large size (S value: 15). We attribute the previously observed lack of specificity to the magnification of a low avidity of binding resulting from the multivalent nature of the receptor in the absence of detergent.

The interaction of the FcR on mouse macrophages with immune complexes results in the triggering of the cell's defense mechanisms, which range from phagocytosis of the offending particle to release of hydrolytic enzymes, superoxide, prostaglandins, and leukotrienes. The nature of the signal transmitted to the cell by the FcR was studied using the lipophilic tetraphenylphosphonium cation (TPP^+) as a probe of macrophage membrane potential (Lichtshtein et al., 1979). The effect of exposure of J774 macrophagelike cells to immune complexes and the anti-$Fc_{\gamma 2b/\gamma 1}R$ monoclonal 2.4G2 as well as other monoclonal antibodies was tested (Young et al., 1983a). The resting potential of cells of the J774 macrophage cell line determined from TPP^+ equilibration data was -15 mV.

Extensively cross-linked immune complexes or 2.4G2 Fab coupled to Sephadex beads presented to J774 cells resulted in a prompt depolarization that lasted 15-20 min. Soluble immune complexes or the bivalent 2.4G2 IgG resulted in a transient depolarization, followed by a hyperpolarization that was blocked by prior incubation with ouabain. The depolarization was due to an influx of Na^+, since replacement of Na^+ by choline, which did not affect the membrane potential, abolished the depolarization in response to immune complexes. Other monoclonal antibodies 2D2C, 2E2A, and 1.21J, all of which recognize major antigenic determinants on J774 cells (Mellman *et al.*, 1980; Nussenzweig *et al.*, 1981; Muller *et al.*, 1983) stimulated a ouabain-blockable hyperpolarization.

These results are compatible with the thesis that the FcR functions as a ligand-dependent ion channel. To investigate this possibility, the ion flux into plasma membrane vesicles isolated from J774 cells was examined by TPP^+ uptake after dilution of vesicles with entrapped cations into isotonic sucrose containing ligands and labeled TPP^+ (Young *et al.*, 1983*b*). In the presence of immune complexes or 2.4G2 IgG there was a prompt and substantial uptake of TPP^+ over that observed in the absence of ligand. Experiments in which Na^+-loaded vesicles were diluted into K^+ and vice versa showed that the ion flux was not specific with regard to these two cations. Ca^{2+} was poorly transported relative to monovalent cations. To demonstrate that these conductance changes were not caused by simple binding to the vesicles of monoclonal antibodies, the same series of monoclonal antibodies was tested in the membrane vesicle system. These reagents had no effect on the uptake of TPP^+, demonstrating the specificity of the permeability changes observed.

The results of TPP^+ uptake triggered by immune complexes or 2.4G2 suggest a role for the $Fc_{\gamma2b/\gamma1}R$ in the conductance changes, but did not rule out the possibility that the channel is formed by interaction of the receptor with another plasma membrane protein(s). To address this possibility, the purified FcR was reconstituted into phospholipid vesicles by detergent dialysis from octylglucoside, and the uptake of TPP^+ into the vesicles was measured in the presence or absence of 2.4G2 IgG (Young *et al.*, 1983*b*). Relative to the control, a substantial amount of TPP^+ was taken up in the presence of 2.4G2 IgG, demonstrating that the conductance change seen in the plasma membrane vesicles and intact J774 cells is attributable to the presence of the FcR, and not to other plasma membrane proteins.

The conductance changes observed could be caused by a nonspecific change in monovalent cation permeability rather than the formation of ion channels. To study permeability changes on a microscopic rather than a macroscopic level, as in the TPP^+ uptake experiments. $Fc_{\gamma2b/\gamma1}R$ was reconstituted into planar bilayers by the method of Montal and Muller (1972), and the ionic current flowing through the membrane was measured after the addition of appropriate ligands. The receptor was reconstituted in a lipid monolayer on one side of a two-chambered apparatus by rapid dilution of a solution containing phospholipid, octyl-glucoside, and purified $Fc_{\gamma 2b/\gamma 1}R$. The other chamber did not contain FcR. When the level of the buffer in both chambers was raised to span the annulus separating the two chambers, an asymmetric bilayer was thus formed. Addition of 2.4G2 IgG, or immune complexes, but not normal rabbit IgG to the *cis* chamber, in which the FcR was initially reconstituted, resulted in a large increase in membrane conductane when a potential was imposed between the two compartments. This conductance increase decayed with time, and addition of more ligand to the *trans* chamber had no effect on conductance. When the amount of FcR used for the reconstruction was sufficiently diluted, and the salt concentration was raised to 1 M to increase the amount of current, off-on conductance jumps were observed compatible with single channels opening and closing. The conductance of these events was 60 ± 5 pS (pico Siemens unit) and the current–voltage plot for single channels showed a linear relationship.

One of the powerful applications of immunologic reagents is to perturb the normal working of biologic systems in order to dissect the functional significance of various epitopes that the antibodies may recognize. 2.4G2 IgG has been shown, not surprisingly, to interfere with ADCC mediated by macrophages elicited with bacillus Calmette-Guérin (BCG) (Nathan *et al.*, 1980). Perhaps of more interest is the study of West Nile virus, which has been advanced as a model for Dengue hemorrhagic fever. The infectivity of West Nile virus, a flavivirus that can replicate in cells of the $P388D_1$ mouse macrophage cell line, is increased 100-fold by subneutralizing amounts of IgG. This increase in infectivity is attributable to the Fc domain of the IgG and is reversed almost totally by the addition of 2.4G2 IgG (Peiris *et al.*, 1981); 2.4G2 IgG has also been reported to act as a B-cell mitogen and to stimulate a polyclonal antibody response (Lamers *et al.*, 1982). We have, however, been unable to confirm these observations (E. Pure and J. Unkeless, unpublished results).

Another area in which monoclonal antibodies have provided useful probes is in the study of induction and regulation. Hamburg *et al.* (1980) demonstrated that, although there was enhancement of macrophage phagocytosis of IgG-opsonized E after treatment with type I interferon, there was no increase in the amount of 2.4G2 bound to the induced cells. Echoing these results, Ezekowitz *et al.* (1983) report that IgG2a binding, but not IgG2b binding, is selectively enhanced after stimulation of mouse macrophages by BCG. Yoshie *et al.* (1982) found evidence for increased levels of both $Fc_{\gamma 2a}R$ and $Fc_{\gamma 2b/\gamma 1}R$ after induction with α- and β-interferon of the mouse macrophage cell line RAW 309 Cr.1.

It is not clear whether the different mouse Fc receptors have specialized physiologic functions. Ralph *et al.* (1980) found that all IgG subclasses in mouse mediate phagocytosis and lysis of IgG-coated E. However, Ezekowitz *et al.* 1983) found that IgG2a complexes stimulate the oxidative burst of BCG-activated macrophages more efficiently. Further suggestion that $Fc_{\gamma 2a}R$ may be of particular interest in tumoricidal/microbicidal activity comes from Matthews *et al.* (1981), who found that macrophage cytotoxicity and *in vivo* protection against the 775 murine adenocarcinoma cell line were mediated by mouse IgG2a antibody. Supporting the hypothesis that different receptors have different functions, Nitta and Suzuki (1982) found differences in cyclic nucleotide responses after adherence of IgG2a- and IgG2b-sensitized E. Clearly further work is needed in this area.

IX. HUMAN Fc RECEPTOR

The analysis of human Fc receptors has largely focused on binding to monocytes of different subclasses of human IgG and has resulted in a rank order of cytophilicity in which IgG1 = IgG3 > IgG4 ≫ IgG2 (reviewed in Dickler, 1976; Unkeless *et al.*, 1981). Competitive binding experiments between different human IgG subclasses for binding to the U937 monocytic cell line failed to reveal any heterogeneity in Fc binding sites (Anderson and Abraham, 1980). However, Messner and Jelinek (1970), and Huber *et al.* (1969) reported lack of binding of some anti-Rh_0 sera to neutrophils, suggesting a possible difference between neutrophil and monocyte receptors. These results were recently confirmed by Kurlander and Batker (1982), who demonstrated that human IgG1 oligomers bound with 100- to 1000-fold higher avidity to monocytes than to neutrophils. The neutrophil receptor is thus a relatively low-avidity receptor ($Fc_\gamma R_{10}$) that is probably triggered only by immune complexes, as compared with the receptor on monocytes ($Fc_\gamma R_{hi}$), which binds IgG1 monomer with a $K_a > 10^8 M^{-1}$.

Analysis of human Fc receptors using monoclonal antibodies has provided solid evidence for human Fc receptor heterogeneity. Fleit *et al.* (1982) have isolated $Fc_\gamma R_{10}$ monoclonal antibody, 3G8, by screening hybridoma supernatants for inhibition of neutrophil rosetting with IgG-sensitized erythrocytes. The Fab fragment of 3G8 retained its potent inhibitory capacity against neutrophil Fc receptor. Immunoprecipitation of labeled neutrophils by 3G8 Fab-Sepharose revealed two poorly resolved peptides, of 53,000 and 66,000 M_r, which resemble the Fc receptor from mouse macrophages immunoprecipitated with 2.4G2 Fab-Sepharose. A monoclonal antibody, B73.1, with comparable sepcificity against $Fc_\gamma R_{10}$ has been isolated by Perussia *et al.* (1983). B73.1 IgG stains human NK cells brightly, and, like 3G8, does not react with blood monocytes, V937, or HL-60 cell lines. Affinity chromatography on IgG-Sepharose of ^{125}I-surface-labeled detergent lysates from human mononuclear cells and neutrophils resulted in a broadly migrating protein of 52,000–64,000 M_r (Kulczycki *et al.*, 1981). Comparable experiments using lysates of the U937 cell line resulted in molecules of 72,000 and 40,000–43,000 M_r (Anderson, 1982).

Although there appears to be structural similarity between the human Fc receptor recognized by 3G8 and the mouse $Fc_{\gamma2b/\gamma1}R$, the cellular distribution of the two antigens is very different. The 3G8 antigen is present on all neutrophils and eosinophils, on 15% of peripheral blood B cells, and on 6% of E-rosetting cells. However, the 3G8 antigen is absent from peripheral blood monocytes and from the promyelocytic HL-60 and monocytic U937 human cell lines, which have high-avidity receptors for human IgG1 (Anderson and Abraham, 1980; Crabtree, 1980). Although absent from monocytes, 60% of macrophages isolated from resected lung tissue bear the 3G8 determinant, as determined by immunofluorescence staining. This result, plus the observation that the antigen appears on monocytes cultered *in vitro* for 7 days, suggests that the low-avidity Fc receptor for IgG ($Fc_\gamma R_{10}$) is either an inducible protein or a marker of a particular stage in the monocyte-macrophage differentiative pathway.

We have studied the induction of $Fc_\gamma R_{10}$ on HL-60 cells after treatment with retinoic acid or dimethyl sulfoxide (DMSO) and on chronic myelogenous leukemia (CML) cells as models for the expression of $Fc_\gamma R_{10}$ during differentiation. Immunofluorescent staining of bone marrow cells for $Fc_\gamma R_{10}$ and counterstaining for nuclear morphology with *p*-phenylenediamine, a free radical scavenger used to block fluorescence bleaching (Johnson and Nogueira Araujo, 1981), revealed staining on cells at the metamyelocyte or later stages, but not on less differentiated forms. In agreement with these observations, the uninduced HL-60 cell line, which has the morphologic appearance of cells at the promyelocyte stage of differentiation, did not express the 3G8 antigen. After induction with DMSO or retinoic acid, however, both of which have been shown to drive HL-60 to more mature myeloid stages (Collins *et al.*, 1978), 5–40% of the cells synthesize $Fc_\gamma R_{10}$ (Fleit *et al.*, 1984). Finally, although the more mature cells in the peripheral circulation of patients with chronic myelogenous leukemia (CML) have the same number of 3G8 binding sites and bind the same amount of IgG in immune complexes as do peripheral neutrophils, immature cells from CML patients are completely negative for 3G8 antigen.

X. OTHER MOUSE MACROPHAGE ANTIGENS AND STUDIES ON MEMBRANE RECYCLING

Immunizations of rats with mouse macrophage cell lines have resulted in the generation of monoclonals directed against antigens which, although not unique to macrophages, have been extremely useful in the analysis of membrane flow and recycling of membrane proteins. Monoclonal antibodies used in this way include 2D2C, which immunoprecipitates a glycoprotein of 90,000 M_r and recognizes an alloantigen present on DBA/2, Balb/c, and CBA, but not A, B10, B10.D2, or AKR mice (Nussenzweig et al., 1981); 1.21J, which recognizes Mac-1, thought to be the CR_3 receptor (Beller et al., 1982); 2E2A, which recognizes a protein of 82,000 M_r; F4/80, which immunoprecipitates a macrophage-specific glycoprotein of 150,000 M_r (Austyn and Gordon, 1981); 2F44, which recognizes a protein of 42,000 M_r; 25-1, which recognizes H-2D^d; and 2.6, which immunoprecipitates a protein of 20,000 M_r (Mellman et al., 1980). These proteins together constitute about 25% of the total plasma membrane protein subject to iodination by lactoperoxidase and glucose oxidase.

The relative distribution of these proteins in the plasma membrane was compared with the distribution in vesicles, which were labeled after pinocytosis of lactoperoxidase (Mellman et al., 1980). In most cases, the relative distribution of proteins labeled on the plasma membrane was the same as the distribution of proteins in the labeled vesicles, arguing against exclusion of these proteins in pinocytic vesicles. However, one protein, recognized by monoclonal antibody 2.6, was preferentially represented in the labeled pinosome proteins relative to the plasma membrane (Mellman et al., 1980). Muller et al. (1983) have examined the protein composition of endocytic vacuoles formed by macrophage phagocytosis of Latex particles and find the plasma membrane proteins are present in the same relative amounts in phagosomes, with the notable exception of the antigen precipitated by 2.6, which was present at 7-fold the level found on the plasma membrane. The 2.6 antigen was present on macrophages, absent from lymphocytes, and present in large amounts on dendritic cells, platelets, and granulocytes (Nussenzweig et al., 1981). The function of the molecule is unknown, but it is tempting to speculate that it is involved in the specialized phagocytic and/or secretory functions carried out by these cell types.

Patients with circulating immune complexes often have a defect in the rate of clearance of IgG-coated erythrocytes (Frank et al., 1983). This defect may be secondary to internalization and clearance of Fc receptors from the surface of the phagocytic cells. Using rabbit antisera specific for the mouse $Fc_{\gamma 2b/\gamma 1}R$, prepared by immunization with protein purified by affinity chromatography on 2.4G2 Sepharose, Mellman et al. (1983) studied the rate of degradation of $Fc_{\gamma 2b/\gamma 1}R$ after phagocytosis of erythorocyte ghosts coated with IgG and found a significantly increased rate of degradation ($t_{1/2}$: <2 hr) relative to the free receptor of ($t_{1/2}$: 10 hr). The rates of turnover of other membrane proteins examined were not affected by phagocytosis of the opsonized ghosts. After ingestion of the opsonized erythrocyte ghosts, there was a small (10%) transient decrease in the binding to the plasma membrane of monoclonal antibodies 2D2C, 1.21J, and an antibody specific for H-2D^d, but a large ($>60\%$) decrease in 2.4G2 (anti-$Fc_{\gamma 2b/\gamma 1}R$) binding, which remained depressed over the next 24 hr. Thus, ligand can profoundly affect the subsequent turnover and the degradation of $Fc_{\gamma 2b/\gamma 1} R$ while not altering the turnover and recycling of other plasma membrane proteins.

ACKNOWLEDGMENTS

We are indebted to numerous colleagues, without whom much of this work could not have been done. Research was supported by USPHS grants CA31798, CA31799, CA30198, AI-14603, Council for Tobacco Research grant 1307, and American Cancer Society Junior Faculty and Faculty Research Awards.

XI. REFERENCES

Anderson, C. L., 1982, Isolation of the receptor for IgG from a human monocyte cell line (U937) and from human peripheral blood monocytes, *J. Exp. Med.* **156**:1794–1805.

Anderson, C. L., and Abraham, G. N., 1980, Characterization of the Fc receptor for IgG on a human macrophage cell line, U937, *J. Immunol.* **125**:2735–2741.

Ault, K. A., and Springer, T. A., 1981, Cross reaction of a rat-anti-mouse phagocyte-specific monoclonal antibody (anti-Mac-1) with human monocytes and natural killer cells, *J. Immunol.* **126**:359–364.

Austyn, J. M., and Gordon, S., 1981, F4/80, a monoclonal antibody directed specifically against the mouse macrophage, *Eur. J. Immunol.* **11**:805–815.

Ball, E. D., Graziano, R. F., Shen, L., and Fanger, M. W., 1982, Monoclonal antibodies to novel myeloid antigens reveal human neutrophil heterogeneity, *Proc. Natl. Acad. Sci. USA* **79**:5374–5378.

Basch, R. S., and Berman, J. W., 1982, Thy-1 determinants are present on many murine hematopoietic cells other than T cells, *Eur. J. Immunol.* **12**:359–364.

Becker, G. J., Hancock, W. W., Kraft, N., Lanyon, H. C., and Atkins, R. C., 1981, Monoclonal antibodies to human macrophage and leucocyte common antigens, *Pathology* **13**:669–680.

Beller, D. I., Kiely, J. M., and Unanue, E. R., 1980, Regulation of macrophage populations. I. Preferential induction of Ia-rich peritoneal exudates by immunologic stimuli, *J. Immunol.* **124**-1426–1432.

Beller, D. I., Springer, T. A., and Schreiber, R. D., 1982, Anti-Mac-1 selectively inhibits the mouse and human type three complement receptor, *J. Exp. Med.* **156**:1000–1009.

Berken, A., and Benacerraf, B., 1966, Properties of antibodies cytophilic for macrophages, *J. Exp. Med.* **123**:119–144.

Berman, B., and Gigli, I., 1980, Complement receptors on guinea pig epidermal Langerhans cells, *J. Immunol.* **124**:685–690.

Bernstein, I. D., Andrews, R. G., Cohen, S. F., and McMaster, B. E., 1982, Normal and malignant human myelocytic and monocytic cells identified by monoclonal antibodies, *J. Immunol.* **128**:876–881.

Bianco, C., and Nussenzweig, V., 1977, Complement receptors, *Contemp. Top. Mol. Immunol.* **6**:145–176.

Boltz-Nitulescu, G., Bazin, H., and Spiegelberg, H. L., 1981, Specificity of Fc receptors for IgG2a, IgG1/IgG2b, and IgE on rat macrophages, *J. Exp. Med.* **154**:374–384.

Breard, J., Reinherz, E. L., Kung, P. C., Goldstein, G., and Schlossman, S. F., 1980, A monoclonal antibody reactive with human peripheral blood monocytes, *J. Immunol.* **124**:1943–1948.

Breard, J., Reinherz, E., O'Brien, C., and Schlossman, S. F., 1981, Delineation of an effector population responsible for natural killing and antibody-dependent cellular cytotoxicity in man, *Clin. Immunol. Immunopathol.* **18**:145–150.

Brooks, C. G., Kuribayashi, K., Sale, G. E., and Henney, C. S., 1982, Characterization of five cloned murine cell lines showing high cytolytic activity against YAC-1 cells, *J. Immunol.* **128**:2326–2335.

Burckhardt, J. J., Anderson, W. H. K., Kearney, J. F., and Cooper, M. D., 1982, Human blood monocytes and platelets share a cell surface component, *Blood* **60**:767–771.

Burke, K., and Gigli, I., 1980, Receptors for complement on Langerhans cells, *J. Invest. Dermatol.* **75**:46–51.

Carlo, J. R., Ruddy, S., Studer, E. J., and Conrad, D. H., 1979, Complement receptor binding of C3b-coated cells treated with C3b inactivator β-1H globulin and tyrpsin, *J. Immunol.* **123**:523–528.

Collins, S. J., Ruscetti, F. W., Gallagher, R. E., and Gallo, R. C., 1978, Terminal differentiation of human promyelocytic leukemia cells induced by dimethyl sulfoxide and other polar compounds, *Proc. Natl. Acad. Sci. USA* **75**:2458–2462.

Crabtree, G. R., 1980, Fc receptors of a human promyelocytic leukemic cell line: Evidence for two types of receptors defined by binding of the staphylococcal protein A–IgG1 complex, *J. Immunol.* **127**:448–453.

Davignon, D., Martz, E., Reynolds, T., Kürzinger, K., and Springer, T. A., 1981, Monoclonal antibody to a novel lymphocyte function-associated antigen (LFA-1): Mechanism of blocking of T lymphocyte-mediated killing and effects on other T and B lymphocyte functions, *J. Immunol.* **126**:590–595.

Diamond, B., and Scharff, M. D., 1980, IgG1 and IgG2b share the Fc receptor on mouse macrophages, *J. Immunol.* **125**:631–633.

Diamond, B., and Yelton, D. E., 1981, A new Fc receptor on mouse macrophages binding IgG3, *J. Exp. Med.* **153**:514–519.

Dickler, H. B., 1976, Lymphocyte receptors for immunoglobulin, in: *Advances in Immunology*, Vol. 24 (F. J. Dixon, and H. G. Kunkel, eds.), pp. 167–214, Academic Press, New York.

Dimitriu-Bona, A., Burmester, G. R., Waters, S. J., and Winchester, R. J., 1983, Human mononuclear phagocyte differentiation antigens. 1. Patterns of antigenic expression on the surface of human monocytes and macrophages defined by monoclonal antibodies, *J. Immunol.* **130**:145–152.

Ezekowitz, R. A. B., Austyn, J., Stahl, P. D., and Gordon, S., 1981, Surface properties of bacillus Calmette-Guérin-activated mouse macrophages. Reduced expression of mannose-specific endocytosis, Fc receptors, and antigen F4/80 accompanies induction of Ia, *J. Exp. Med.* **154**:60–76.

Ezekowitz, R. A. B., Bampton, M., and Gordon, S., 1983, Macrophage activation selectively enhances expression of Fc receptors for IgG2a, *J. Exp. Med.* **157**:807–812.

Fearon, D. T., and Wong, W. W., 1983, Complement ligand-receptor interactions that mediate biological responses, *Annu. Rev. Immunol.* **1**:243–271.

Ferrero, D., Pessano, S., Pagliardi, G. L., and Rovera, G., 1983, Induction of differentiation of human myeloid leukemias: Surface changes probed with monoclonal antibodies, *Blood* **61**:171–179.

Fithian, E., Kung, P., Goldstein, G., Rubenfeld, M. Fenoglio, C., and Edelson, R., 1981, Reactivity of Langerhans cells with hybridoma antibody, *Proc. Natl. Acad. Sci. USA* **78**:2541–2544.

Fleit, H. B., Wright, S. D., and Unkeless, J. C., 1982, Human neutrophil Fc-gamma receptor distribution and structure, *Proc. Natl. Acad. Sci. USA* 79:3275-3279.

Fleit, H. B., Wright, S. D., Dune, C. J., Valinsky, J. E., and Unkeless, J. C., 1984, Ontogeny of Fc receptors and complement receptors (CR3) during human myeloid differentiation, *J. Clin. Invest.* (in press).

Flotte, T., Springer, T. A., and Thorbecke, G. J. 1983a, Dendritic cell and macrophage staining by monoclonal antibodies in tissue sections and epidermal sheets, *Am J. Pathol.* 111:112-124.

Flotte, T. J., Haines, K. A., Peckman, K., Springer, T. A., Gigli, I., and Thorbecke, J., 1984, The relation of Langerhans cells to other dendritic cells and macrophages, In *Mononuclear Phagocyte Biology* (A. Volkman, ed.), Academic Press, New York (in press).

Frank, M. M., Lawley, T. J., Hamburger, M. I., and Brown, E. J., 1983, Immunoglobulin G Fc receptor-mediated clearance in autoimmune diseases, *Ann. Intern. Med.* 98:206-218.

Germain, R. N., Bhattacharya, A., Dorf, M. E., and Springer, T. A., 1982, A single monoclonal anti-Ia antibody inhibits antigen-specific T cell proliferation controlled by distinct Ir genes mapping in different H-2 I subregions, *J. Immunol.* 128:1409-1413.

Griffin, J. D., Ritz, J., Nadler, L. M., and Schlossman, S. F., 1981, Expression of myeloid differentiation antigens on normal and malignant myeloid cells, *J. Clin. Invest.* 69:932-941.

Griffin, J. D., Beveridge, R. P., and Schlossman, S. F., 1982, Isolation of myeloid progenitor cells from peripheral blood of chronic myelogenous leukemia patients, *Blood* 60:30-37.

Haeffner-Cavaillon, N., Klein, M., and Dorrington, K. J., 1979, Studies on the Fc gamma receptor of the murine macrophage-like cell line P388D$_1$ 1. The binding of homologous and heterologous immunoglobulin G, *J. Immunol.* 123:1905-1913.

Haines, K. A., Flotte, T. J., Springer, T. A., Gigli, I., and Thorbecke, G. J., 1983, Staining of Langerhans cells with monoclonal antibodies to macrophages and lymphoid cells, *Proc. Natl. Acad. Sci. USA* 80:3448-3451.

Hanjan, S. N. S., Kearney, J. F., and Cooper, M. D., 1982, A monoclonal antibody (MMA) that identifies a differentiation antigen on human myelomonocytic cells, *Clin. Immunol. Immunopathol.* 23:172-188.

Haynes, B. F., Hemler, M. E., Mann, D. L., Eisenbarth, G. S., Shelhamer, J., Mostowski, H. S., Thomas, C. A., Strominger, J. L., and Fauci, A. S., 1981, Characterization of a monoclonal antibody (4F2) that binds to human monocytes and to a subset of activated lymphocytes, *J. Immunol.* 126:1409-1414.

Hercend, T., Reinherz, E. L., Meuer, S., Schlossman, S. F., and Ritz, J., 1983, Phenotypic and functional heterogeneity of human cloned natural killer cell lines, *Nature* 301:158-160.

Hirsch, S., Austyn, J. M., and Gordon, S., 1981, Expression of the macrophage-specific antigen F4/80 during differentiation of mouse bone marrow cells in culture, *J. Exp. Med.* 154:713-725.

Ho, M. K., and Springer, T. A., 1983a, Tissue distribution, structural characterization specific antigen defined by monoclonal antibody, *J. Immunol.* 128:1221-1228.

Ho, M. K., and Springer, T. A., 1982b, Mac-1 antigen: Quantitative expression in macrophage populations and tissues, and immunofluorescent localization in spleen, *J. Immunol.* 128:2281-2286.

Ho, M. K., and Springer, T. A., 1983a, Tissue of distribution, structural characterization and biosynthesis of Mac-3, a macrophage surface glycoprotein exhibiting molecular weight heterogeneity, *J. Biol. Chem.* 258:636-642.

Ho, M. K., and Springer, T. A. 1983b, Biosynthesis and assembly of the alpha and beta subunits of Mac-1, a macrophage glycoprotein associated with complement receptor function, *J. Biol. Chem.* 258:2766-2769.

Hoffman-Fezer, G., Gotze, D., Rodt, H., and Thierfelder, S., 1978, Immunohistochemical localization of xenogeneic antibodies against Ia[k] lymphocytes on B cells and reticular cells, *Immunogenetics* 6:367-375.

Hogg, N., 1983, Human monocytes are associated with the formation of fibrin, *J. Exp. Med.* 157:473-485.

Holmberg, L. A., Springer, T. A., and Ault, K. A., 1981, Natural killer activity in the peritoneal exudates of mice infected with *Listeria monocytogenes:* Characterization of the natural killer cells by using a monoclonal rat anti-murine macrophage antibody (M1/70), *J. Immunol.* 127:1792-1799.

Huber, H., Douglas, S. D., and Fudenberg, H. H., 1969, The IgG receptor: An immunological marker for the characterization of mononuclear cells, *Immunology* 17:7-21.

Humphrey, J. H., and Grennan, D., 1984, Isolation and properties of spleen follicular dendritic cells, *Adv. Exp. Med.,* (in press).

Johnson, G. D., and Nogueira Araujo, G. M., 1981, A simple method of reducing the fading of immunofluorescence during microscopy, *J. Immunol. Methods* 43:349-350.

Jones, P. P., Murphy, D. B., Hewgill, D., and McDevitt, H. O., 1979, Detection of a common polypeptide chain in I-A and I-E subregion immunoprecipitates, *Molec. Immunol.* 16:51-60.

Kaplan, G., and Gaudernack, G., 1982, *In vitro* differentiation of human monocytes. Differences in monocyte phenotypes induced by cultivation on glass or on collagen, *J. Exp. Med.* 156:1101-1114.

Katz, H. R., LeBlanc, P. A., and Russell, S. W., 1981, An antigenic determinant shared by mononuclear phagocytes and mast cells, as defined by monoclonal antibody, *J. Reticuloendothel. Soc.* 30:439-443.

Kiessling, R., Hochman, P. S., Haller, O., Shearer, G. M., Wigzell, H., and Cudkowicz, G., 1977, Evidence for a similar or common mechanism for natural killer cell activity and resistance to hemopoietic grafts, *Eur. J. Immunol.* 7:655-663.

Krensky, A. M., Ault, K. A., Reiss, C. S., Strominger, J. L., and Burakoff, S. J., 1982, Generation of long-term human cytolytic cell lines with persistent natural killer activity, *J. Immunol.* 129:1748-1751.

Kulcyzycki, A., Jr., Solanki, L., and Cohen, L., 1981, Isolation and partial characterization of Fc gamma-binding proteins of human leukocytes, *J. Clin. Invest.* 68:1158-1165.

Kurlander, R. J., and Batker, J., 1982, The binding of human immunoglobulin G1 monomer and small, covalently cross-linked polymers of immunoglobulin G1 to human peripheral blood monocytes and polymorphonuclear leukocytes, *J. Clin. Invest.* 69:1-8.

Kürzinger, K., and Springer, T. A., 1982, Purification and structural characterization of LFA-1, a lymphocyte function-associated antigen, and Mac-1, a related macrophage differentiation antigen, *J. Biol. Chem.* 257:12412-12418.

Kürzinger, K., Reynolds, T., Germain, R. N., Davignon, D., Martz, E., and Springer, T. A., 1981, A novel lymphocyte function-associated antigen (LFA-1): Cellular distribution, quantitative expression, and structure, *J. Immunol.* 127:596-602.

Kürzinger, K., Ho, M. K., and Springer, T. A., 1982, Structural homology of a macrophage differentiation antigen and an antigen involved in T-cell mediated killing, *Nature* 296:668-670.

Ladman, A. J., Padykula, H. A., and Strauss, E. W., 1963, A morphological study of fat transport in the normal human jejunum, *Am. J. Anat.* 112:389-394.

Lamers, M. C., Heckford, S. E., and Dickler, H. B., 1982, Monoclonal anti-Fc IgG receptor antibodies trigger B lymphocyte function, *Nature* 298:178-180.

Lane, B. C., and Cooper, S. M., 1982, Fc receptors of mouse cell lines. 1. Distinct proteins mediate the IgG subclass-specific Fc binding activities of macrophages, *J. Immunol.* 128:1819–1824.

Lay, W. H., and Nussenzweig, V., 1968, Receptors for complement on leukocytes, *J. Exp. Med.* 128:991–1009.

Leblanc, P. A., Katz, H. R., and Russell, S. W., 1980, A discrete population of mononuclear phagocytes detected by monoclonal antibody, *Infect. Immun.* 8:520–525.

Letvin, N. L., Todd, R. F. III, Palley, L. S., Schlossman, S. F., and Griffin, J. D., 1983, Conservation of myeloid surface antigens on primate granulocytes, *Blood* 61:408–410.

Lichtshtein, D., Kaback, H. R., and Blume, A. J., 1979, Use of a lipophilic cation for determination of membrane potential in neuroblastoma-glioma hybrid cell suspension, *Proc. Natl. Acad. Sci. USA* 76:650–654.

Linker-Israeli, M., Billing, R. J., Foon, K. A., and Terasaki, P. I., 1981, Monoclonal antibodies reactive with acute myelogenous leukemia cells, *J. Immunol.* 127:2473–2477.

Lohmeyer, J., Rieber, P., Feucht, H., Johnson, J., Hadam, M., and Riethmüller, G., 1981, A subset of human natural killer cells isolated and characterized by monoclonal antibodies, *Eur. J. Immunol.* 11:997–1001.

Loube, S. R., McNabb, T. C., and Dorrington, K. J., 1978, Isolation of an Fc gamma-binding proteins from detergent lysates and spent culture fluid of a macrophage-like cell line (P388D$_1$), *J. Immunol.* 125:970–975.

Loube, S. R., McNabb, T. C., and Dorrington, K. J., 1978, Isolation of an Fc gamma-binding protein from the cell membrane of a macrophage-like cell line (P388D$_1$) after detergent solubilization, *J. Immunol.* 120:709–715.

McMaster, W. R., and Williams, A. F., 1979, Identification of Ia glycoproteins in rat thymus and purification from rat spleen, *Eur. J. Immunol.* 9:426–433.

Mason, D. W., Dallman, M., and Barclay, A. N., 1981, Graft-versus-host disease induces expression of Ia antigen in rat epidermal cells and gut epithelium, *Nature* 293:150–151.

Matthews, T. J., Collins, J. J., Roloson, G. J., Thiel, H. J., and Bolognesi, D. P., 1981, Immunologic control of the ascites form of murine adenocarcinoma 755. IV. Characterization of the protective antibody in hyperimmune serum, *J. Immunol.* 126:2332–2336.

Mellman, I. S., and Unkeless, J. C., 1980, Purification of a functional mouse Fc receptor through the use of a monoclonal antibody, *J. Exp. Med.* 152:1048–1069.

Mellman, I. S., Steinman, R. M., Unkeless, J. C., and Cohn, Z. A., 1980, Selective iodination and polypeptide composition of pinocytic vesicles, *J. Cell. Biol.* 86:712–722.

Mellman, I. S., Plutner, H., Steinman, R. M., Unkeless, J. C., and Cohn, Z. A., 1983, Internalization and degradation of macrophage Fc receptors during receptor-mediated phagocytosis, *J. Cell. Biol.* 96:887–895.

Messner, R. P., and Jelinek, J., 1970, Receptors for human gamma G globulin on human neutrophils, *J. Clin. Invest.* 49:265–271.

Michl, J., Pieczonka, M. M., Unkeless, J. C., and Silverstein, S. C., 1979, Effects of immobilized immune complexes on Fc- and complement-receptor function in resident and thioglycollate-elicited mouse peritoneal macrophage, *J. Exp. Med.* 150:607–621.

Montal, M., and Miller, P., 1972, Formation of bimolecular membranes from lipid monolayers and a study of their electrical properties, *Proc. Natl. Acad. Sci. USA* 69:3561–3566.

Muller, W. A., Steinman, R. M., and Cohn, Z. A., 1983, Membrane proteins of the vacuolar system. III. Further studies on the composition and recycling of endocytic vacuole membrane in cultured macrophages, *J. Cell Biol.* 96:29–36.

Nathan, C., Brukner, L., Kaplan, G., Unkeless, J., and Cohn, Z., 1980, Role of activated macrophages in antibody-dependent lysis of tumor cells, *J. Exp. Med.* 152:183–197.

Nitta, T., and Suzuki, T., 1982, Biochemical signals transmitted by Fc gamma receptors: Triggering mechanisms of the increased synthesis of adenosine-3',5'-cyclic monophosphate mediated by $Fc_{\gamma 2a^-}$ and $Fc_{\gamma 2a^-}$ receptors of a murine macrophage-like cell line (P388D₁), *J. Immunol.* **129**:2708-2714.

Nussenzweig, M. C., Steinman, R. M., Unkeless, J. C., Witmer, M. D., Gutchinov, B., and Cohn, Z. A., 1981, Studies of the cell surface of mouse dendritic cells and other leukocytes, *J. Exp. Med.* **154**:168-187.

Ortaldo, J. R., Sharrow, S. O., Timonen, T., and Herberman, R. B., 1981, Determination of surface antigens on highly purified human NK cells by flow cytometry with monoclonal antibodies, *J. Immunol.* **127**:2401-2409.

Peiris, J. S. M., Gordon, S., Unkeless, J. C., and Porterfield, J. S., 1981, Monoclonal anti-Fc receptor IgG blocks antibody enhancement of viral replication in macrophages, *Nature* **289**:189-191.

Perlman, H., Perlman, P., Schreiber, R. D., and Muller-Eberhard, H. J., 1981, Interaction of target cell-bound C3bi and C3d with human lymphocyte receptors: Enhancement of antibody-mediated cellular cytotoxicity, *J. Exp. Med.* **153**:1592-1603.

Perussia, B., Trinchieri, G., Lebman, D., Jankiewicz, J., Lange, B., and Rovera, G., 1982, Monoclonal antibodies that detect differentiation surface antigens on human myelomonocytic cells, *Blood* **59**:382-392.

Perussia, B., Acuto, O., Terhorst, C., Faust, J., Lazarus, R., Fanning, V., and Trinchieri, G., 1983, Human natural killer cells analyzed by B73.1, a monoclonal antibody blocking Fc receptor functions. II. Studies of B73.1 antibody antigen interaction on the lymphocyte membrane, *J. Immunol.* **130**:2142-2148.

Pober, J. S., and Gimbrone, M. A., Jr., 1982, Expression of Ia-like antigens by human vascular endothelial cells is inducible in vitro: Demonstration by monoclonal antibody binding and immunoprecipitation, *Proc. Natl. Acad. Sci. USA* **79**:6641-6645.

Raff, H. V., Picker, L. J., and Stobo, J. D., 1980, Macrophage heterogeneity in man. A subpopulation of HLA-DR bearing macrophages required for antigen-induced T cell activation also contains stimulators for autologous-reactive T cells, *J. Exp. Med.* **152**:581-593.

Ralph, P., Nakoinz, I., Diamond, B., and Yelton, D., 1980, All classes of murine IgG antibody mediate macrophage phagocytosis and lysis of erythrocytes, *J. Immunol.* **125**:1885-1888.

Ralph, P., Ho, M. K., Litcofsky, P. B., and Springer, T. A., 1983a, Expression and induction in vitro of macrophage differentiation antigens on murine cell lines, *J. Immunol.* **130**:108-114.

Ralph, P., Punjabi, C. J., Welte, K., Litcofsky, P. B., Ho, M. K., Moore, M. A. S., and Springer, T. A., 1983b, Lymphokine inducing "terminal differentiation" of the human monoblast leukemia line U937, *Blood* **62**:1169-1175.

Rosenberg, S. A., Ligler, F. S., Ugolini, V., and Lipsky, P. E., 1981, A monoclonal antibody that identifies human peripheral blood monocytes recognizes the accessory- cells required for mitogen-induced T lymphocyte proliferation, *J. Immunol.* **126**:1473-1477.

Ross, G. D., 1980, Analysis of the different types of leukocyte membrane complement receptors and their interaction with the complement system, *J. Immunol. Methods* **37**:197-211.

Ross, G. D., and Lambris, J. D., 1982, Identification of a C3bi-specific membrane complement receptor that is expressed on lymphocytes, monocytes, neutrophils, and erythrocytes, *J. Exp. Med.* **155**:96-110.

Ross, G. D., Newman, S. L., Lambris, J. D., Devery-Pocius, J. E., Cain, J. A., and Lachman, P. J., 1983, Generation of three different fragments of bound C3 with purified factor I or serum II. Location of binding sites in the C3 fragments for factors B and H, complement receptors, and bovine conglutinin, *J. Exp. Med.* **158**:334-352.

Sanchez-Madrid, F., Nagy, J., Robbins, E., Simon, P., and Springer, T. A., 1983a, The human lymphocyte-function associated antigen (LFA-1), the C3bi complement receptor (OKM1/Mac-1), and the p150,95 molecule: Characterization of a leukocyte differentiation antigen family with distinct alpha subunits and a common beta subunit, *J. Exp. Med.* 158:1785–1803.

Sanchez-Madrid, F., Simon, P., Thompson, S., and Springer, T. A., 1983*b*, Mapping of antigenic and functional epitopes on the alpha and beta subunits of two related glycoproteins involved in cell interactions, LFA-1 and Mac-1, *J. Exp. Med.* 158:586–602.

Sarmiento, M., Loken, M. R., Trowbridge, I., Coffman, R. L., and Fitch, F. W., 1982, High molecular weight lymphocyte surface proteins are structurally related and are expressed on different cell populations at different times during lymphocyte maturation and differentiation, *J. Immunol.* 128:1676–1684.

Scheid, M. P., and Triglia, D., 1979, Further description of the Ly-5 system, *Immunogenetics* 9:423–433.

Schneider, R. J., Atkinson, J. P., Krause, V., and Kulczycki, A., Jr., 1981, Characterization of ligand-binding activity of isolated murine Fc$_\gamma$ receptor, *J. Immunol.* 126:735–740.

Schrader, J. W., Battye, F., and Scollay, R., 1982, Expression of Thy-1 antigen is not limited to T cells in cultures of mouse hemopoietic cells, *Proc. Natl. Acad. Sci. USA* 79:4161–4165.

Segal, D. M., and Titus, J. A., 1978, The subclass specificity for the binding of murine myeloma proteins to macrophage and lymphocyte cell lines and to normal spleen cells, *J. Immunol.* 120:1395–1403.

Sheehy, M. J., Quintieri, F. B., Leung, D. Y. M., Geha, R. S., Dubey, D. P., Limmer, C. E., and Yunis, E. J., 1983, A human large granular lymphocyte clone with natural killer-like activity and T cell-like surface markers, *J. Immunol.* 130:524–526.

Shen, H. H., Talle, M. A., Goldstein, G., and Chess, L., 1983, Functional subsets of human monocytes defined by monoclonal antibodies: A distinct subset of monocytes contains the cells capable of inducing the autologous mixed lymphocyte culture, *J. Immunol.* 130:698–705.

Silberberg-Sinakin, I., Thorbecke, G. J., Baer, R. L., Rosenthal, S. A., and Berezowsky, V., 1976, Antigen-bearing Langerhans cells in skin, dermal lymphatics and in lymph nodes, *Cellular Immunol.* 25:137–151.

Smith, B. R. Springer, T. A., Rosenthal, D. S., and Ault, K. A., 1984, Distribution of the myeloid surface antigen Mac-1 on normal and leukemic human cells, *Blood* (submitted for publication).

Springer, T. A., 1980, Cell-surface differentiation in the mouse. Characterization of "jumping" and "lineage" antigens using xenogeneic rat monoclonal antibodies, in: *Monoclonal Antibodies* (R. H. Kennett, T. J. McKearn, and K. K. Bechtol, eds.), pp. 185–217, Plenum Press, New York.

Springer, T. A., 1981, Mac-1,2,3, and 4: Murine macrophage differentiation antigens identified by monoclonal antibodies, in: *Heterogeneity of Mononuclear Phagocytes* (O. Förster and M. Landy, eds.), pp. 37–46, Academic Press, New York.

Springer, T. A., Galfre, G., Secher, D. S., and Milstein, C., 1979, Mac-1: A macrophage differentiation antigen identified by monoclonal antibody, *Eur. J. Immunol.* 9:301–306.

Springer, T. A., Davignon, D., Ho, M. K., Kürzinger, K., Martz, E., and Sanchez-Madrid, F., 1982, LFA-1 and Lyt-2,3, molecules associated with T lymphocyte-mediated killing; and Mac-1, an LFA-1 homologue associated with complement receptor function, *Immunol. Rev.* 68:111–135.

Steinman, R. M., and Cohn, Z. A., 1973, Identification of a novel cell type in peripheral lymphoid organs of mice. I. Morphology, quantitation, tissue distribution, *J. Exp. Med.* 137:1142–1162.

Steinman, R. M., and Cohn, Z. A., 1974, Identification of a novel cell type in peripheral lymphoid organs of mice. II. Functional properties in vitro, *J. Exp. Med.* 139:380–397.

Steinman, R. M., Kaplan, G., Witmer, M. D., and Cohn, Z. A., 1979, Identification of a novel cell type in peripheral lymphoid organs of mice. V. Purification of spleen dendritic cells, new surface markers, and maintenance *in vitro, J. Exp. Med.* 149:1-16.

Stingl, G., Katz, S. I., Clement, L., Green, I., and Shevach, E., 1978, Immunologic functions of Ia-bearing epidermal Langerhans cells, *J. Immunol.* 121:2005-2113.

Sun, D., and Lohmann-Matthes, M. L., 1982, Functionally different subpopulations of mouse macrophages recognized by monoclonal antibodies, *Eur. J. Immunol.* 12:134-140.

Swartz, R. H., Dickler, H. B., Sachs, D. H., and Schwartz, B. D., 1976, Studies of Ia antigens on murine peritoneal macrophages, *Scand. J. Immunol.* 5:731-743.

Tamaki, K., Stingl, G., Gullino, M., Sachs, D. H., and Katz, S. I., 1979, Ia antigens in mouse skin are predominantly expressed on Langerhans cells, *J. Immunol.* 123:784-787.

Taniyama, T., and Watanabe, T., 1982, Establishment of a hybridoma secreting a monoclonal antibody specific for activated tumoricidal macrophages, *J. Exp. Med.* 156:1286-1291.

Tew, J. G., Thorbeck, G. J., and Steinman, R. M., 1982, Dendritic cells in the immune response: Characteristics and recommended nomenclature (a report from the Reticuloendothelial Society Committee on Nomenclature), *J. Reticuloendothel. Soc.* 31:371-380.

Thorbecke, G. J., Silberberg-Sinakin, E., and Flotte, T. J., 1980, Langerhans cells as macrophages in skin and lymphoid organs, *J. Invest. Dermatol.* 75:32-43.

Todd, R. F. III, and Schlossman, S. F., 1982, Analysis of antigenic determinants on human monocytes and macrophages, *Blood* 59:775-786.

Todd, R. F. III, and Schlossman, S. F., 1984, Utilization of monoclonal antibodies in the characterization of monocyte-macrophage differentiation antigens, in: *The Reticuloendothelial System: A Comprehensive Treatise*, Vol. 6: *Immunology* (J. A. Bellanti and H. B. Herscowitz, eds.), pp. 87-111, Plenum Press, New York.

Todd, R. F. III, Nadler, L. M., and Schlossman, S. F., 1981, Antigens on human monocytes identified by monoclonal antibodies, *J. Immunol.* 126:1435-1442.

Todd, R. F. III, Van Agthoven, A., Schlossman, S. F., and Terhorst, C., 1982, Structural analysis of differentiation antigens Mo1 and Mo2 on human monocytes, *Hybridoma* 1:329-337.

Todd, R. F. III, Bhan, A. K., Kabawat, S. E., and Schlossman, S. F., 1984, Human myelomonocytic differentiation antigens defined by monoclonal antibodies, in: *Human Leukocyte Markers Detected by Monoclonal Antibodies* (A. Bernard, L. Boumsell, J. Dausset, C. Milstein, and S. F. Schlossman, eds.), Springer-Verlag, Berlin (in press).

Townsend, R., and Stahl, P., 1981, Isolation and characterization of a mannose/N-acetylglucosamine/fucose-binding protein from rat liver, *Biochem. J.* 194:209-214.

Trowbridge, I. S., and Omary, M. B., 1981, Molecular complexity of leukocyte surface glycoproteins related to the macrophage differentiation antigen Mac-1, *J. Exp. Med.* 154:1517-1524.

Ugolini, V., Nunez, G., Smith, R. G., Stastny, P., and Capra, J. D., 1980, Initial characterization of monoclonal antibodies against human monocytes, *Proc. Natl. Acad. Sci. USA.* 77:6764-6768.

Unanue, E. R., 1981, The regulatory role of macrophages in antigenic stimulation. Part Two. Symbiotic relationship between lymphocytes and macrophages, *Adv. Immunol.* 31:1-136.

Unkeless, J., 1979, Characterization of a monoclonal antibody directed against mouse macrophage and lymphocyte Fc receptors, *J. Exp. Med.* 150:580-596.

Unkeless, J. C., and Eisen, H. N., 1975, Binding of monomeric immunoglobulins to Fc receptors of mouse macrophages, *J. Exp. Med.* 142:1520-1533.

Unkeless, J. C., Fleit, H., and Mellman, I. S., 1981, Structural aspects and heterogeneity of immunoglobulin Fc receptors, in: *Advances in Immunology,* Vol. 31 (F. J. Dixon and H. G. Kunkel, eds.), pp. 247-270, Academic Press, New York.

Van Der Reijden, H. J., Van Rhenen, D. J., Lansdorp, P. M., Van't Veer, M. B., Langenhuijsen, M. M. A. C., Engelfriet, C. P., and Von Dem Borne, A. E. C. K., 1983, A comparison of surface marker analysis and FAB classification in acute myeloid leukemia, *Blood* **61**:443-448.

Van Voorhis, W. C., Hair, L. S., Steinman, R. M., and Kaplan, G., 1982, Human dendritic cells. Enrichment and characterization from peripheral blood, *J. Exp. Med.* **155**:1172-1187.

Walker, W. S., 1976, Separate Fc receptors for immunoglobulins IgG2a and IgG2b on an established cell line of mouse macrophages, *J. Immunol.* **116**:911-914.

Wright, S. D, and Silverstein, S. C., 1982, Tumor-promoting phorbol esters stimulate C3b and C3b′ receptor-mediated phagocytosis in cultured human monocytes, *J. Exp. Med.* **156**:1149-1164.

Wright, S. D., Van Voorhis, W. C., and Silverstein, S. C., 1983, Identification of the C3b′-receptor on human leukocytes using a monoclonal antibody, *Fed. Proc.* **42**:1079.

Yoshie, O., Mellman, I. S., Broeze, R. J., Garcia-Blanco, M., and Lengyel, P., 1982, Interferon action: Effects of mouse alpha and beta interferons on rosette formation, phagocytosis, and surface-antigen expression of cells of the macrophage-type line RAWCr.1, *Cell. Immunol.* **73**:128-140.

Young, J. D. E., Unkeless, J. C., Kaback, H. R., and Cohn, Z. A., 1983*a*, Macrophage membrane potential changes associated with γ2b/I Fc receptor-ligand binding, *Proc. Natl. Acad. Sci. USA.* **80**:1357-1361.

Young, J. D. E., Unkeless, J. C., Kaback, H. R., and Cohn, Z. A., 1983*b*, Mouse macrophage Fc receptor for IgG γ2b/γ1 in artificial and plasma membrane vesicles functions as a ligand-dependent ionophore, *Proc. Natl. Acad. Sci. USA* **80**:1636-1640.

Zarling, J. M., Clouse, K. A., Biddison, W. E., and Kung, P. C., 1981, Phenotypes of human natural killer cell populations detected with monoclonal antibodies, *J. Immunol.* **127**:2575-2580.

Chapter 2

Alterations of Surface Properties by Macrophage Activation: Expression of Receptors for Fc and Mannose-Terminal Glycoproteins and Differentiation Antigens

R. Alan B. Ezekowitz and S. Gordon

Sir William Dunn School of Pathology
Oxford University
Oxford OX1 3RE, England.

I. INTRODUCTION

The original definition of the activated macrophage described the altered function of these cells in antimicrobial activity and cell-mediated immunity (Mackaness, 1964), yet when the role of the macrophage in mammalian physiology and pathology is considered it is apparent that this definition may be too restricted. Inflammation, tissue remodeling, degradation, and turnover of normal body constituents and antitumor resistance could be more general expressions of enhanced macrophage functions. With these different roles in mind, it is necessary to consider the properties that distinguish acquired states of activation from the nonactivated or resident state.

There has been a long and intense interest in identifying biochemical and surface correlates of an altered functional state. Two plasma membrane ecto-enzymes are modified by inflammation. $5'$-Nucleotidase is particularly rich in resident cells and is lost progressively in elicited and bacillus Calmette–Guérin (BCG)-activated macrophages (Edelson and Cohn, 1976). In contrast, alkaline phosphodiesterase I increases in activated cells 2-3-fold (Morahan and Edelson, 1978). Other externally disposed plasma membrane polypeptides are markedly altered as well. Employing the lactoperoxidase-catalyzed iodination procedure, Yin *et al.* (1980) reported that thioglycollate-elicited macrophages demonstrate a different pattern of labeled polypeptides as compared with resident cells.

Recently, several specific receptors and antigens were defined on macrophages, including a lectinlike receptor that mediates endocytosis of mannose or fucose-terminated glycoproteins (MFR) (Stahl *et al.*, 1978), Fc receptors (Kossard and Nelson, 1968), which bind and internalize certain classes of immunoglobulin; the C3bi receptor detected by monoclonal antibody Mac-1/70 (Beller *et al.*, 1982); and F4/80, an antigenic marker for mature mouse macrophages (Austyn and Gordon, 1981). In addition, macrophages can be induced to express Ia antigens by various infectious and other stimuli (Scher *et al.*, 1980; Steinman *et al.*, 1980; Steeg *et al.*, 1980).

In this chapter we review work done on our laboratory over the past 3 years. We have made use of several specific ligands and monoclonal antibodies (Table I) to examine the effects of cell activation on plasma membrane determinants. The following approach was used:

1. The surface properties and secretion products of mouse peritoneal macrophages obtained after infection with live BCG were defined. A surface phenotype that distinguishes BCG-peritoneal macrophages (BCG-PM) from thioglycollate-elicited and other nonactivated macrophages is described (Ezekowitz *et al.*, 1981).

2. This surface phenotype is not unique to BCG-activated peritoneal macrophages; it was found on macrophages activated by other agents *in vivo* (Grosskinsky *et al.*, 1983) and could be induced in nonactivated Mϕ by lymphokine treatment *in vitro* (Ezekowitz and Gordon, 1982; Ezekowitz *et al.*, 1983).

II. SURFACE PROPERTIES OF BCG-ACTIVATED PERITONEAL MACROPHAGES

Peritoneal macrophages acquired from animals after infection with live BCG spread rapidly in culture, secrete plasminogen activator (PA) (Gordon and Cohn, 1978) and high levels of H_2O_2 when stimulated further (Nathan and Root, 1977), and display an enhanced ability to kill microorganisms (Mackaness, 1964) and tumor cells (Old *et al.*, 1961). Although macrophages obtained from mice after intraperitoneal injection of thioglycollate broth (TPM) also show enhanced spreading and secretion of plasminogen activator and superoxide anion as compared with resident cells from untreated animals, i.e., resident peritoneal macrophages (RPM), cells elicited by such an inflammatory stimulus do not exhibit microbicidal or tumoricidal activity and do not release substantial levels of H_2O_2 (Nathan *et al.*, 1979).

BCG-activated PM express inhanced Ia antigens and receptors for IgG2a immunoglobulins, but reduced receptors for mannose-terminated glycoconjugates,

Table I. Properties of Surface Molecules and Antigens Used in This Study[a,b]

Antibody/receptor	Antigen	Nature of antigen (MW)	Cell distribution	Features
F4/80 (1)	F4/80	160,000	Mouse peritoneal macrophages, Mϕ cell lines, blood monocytes, promonocytes; no other cell types	Trypsin-resistant polypeptide trypsin sensitive, 75,000-MW fragment bears Ag
2.4 G2 (2)	IgG1/IgG2b Fc receptor	47,000–60,000	Mϕ, Mϕ cell lines, B, T, and null lymphocyte lines, PMN	Trypsin resistant
Mac1/70 (3)	C3bi receptor	190,000, 105,000	Mϕ, PMN, NK cells (4)	LFA-1 homologue (5)
OX6 (6)	Ia antigens	32,000, 28,000 chains	Activated Mϕ, dendritic cells (7) B lymphocytes, activated T cells, epithelial cells (8)	Coded by I-region genes
Mannose/fucose receptor (9)	—	30,000	Peritoneal Mϕs, hepatic endothelium (10), alveolar Mϕ, monocytes, bone marrow macrophages, Mϕ hybrids	Receptor-mediated endocytosis
IgG2a Fc receptor (11)	—	60,000–70,000	Mϕ, Mϕ cell lines	Trypsin sensitive

[a] Abbreviations: Mϕ, macrophage; NK, natural killer; PMN, polymorphonuclear leukocyte.
[b] Numbers in parentheses are keyed to references:

1. Austyn and Gordon (1981)
2. Unkeless (1979)
3. Springer et al. (1979)
4. Ault and Springer (1981)
5. Kurzinger et al. (1982)
6. McMaster (1979)
7. Steinman et al. (1979)
8. Mason et al. (1981)
9. Stahl et al. (1979)
10. Hubbard et al. (1979)
11. Lane and Cooper (1982)

IgG2b receptors, and macrophage-specific (Mϕ-) antigen F4/80, whereas the expression of Mac-1 shows little change. Decreased receptor-mediated endocytosis is associated with active secretion of plasminogen activator and the ability to secrete O_2^- and H_2O_2 after surface stimulation (Ezekowitz et al., 1981).

On examination by phase-contrast microscopy, it was noted that BCG-PM differed morphologically from TPM and RPM. Electron micrographs of cells fixed in suspension showed that BCG-PM had a smooth surface compared to ruffled membrane folds of RPM, which, after 2 hr adherence did not show the flattened, spread-out appearance of BCG-PM (R. A. B. Ezekowitz, unpublished results).

A. Mannose-Specific Endocytosis

Mononuclear phagoctyes bind and internalize a variety of mannose terminal glycoproteins and glycoconjugates by a specific plasma membrane receptor (Stahl et al., 1978, 1980; Stahl and Gordon, 1982). Binding and uptake of radiolabeled mannose-bovine serum albumin (BSA) or β-glucuronidase by macrophages can be specifically prevented by mannose-rich yeast mannan. Figure 1 shows mannose-specific binding and uptake of [^{125}I] mannose-BSA by BCG-PM and TPM. TPM showed a high level of receptor activity, approaching saturation under conditions comparable to those reported earlier with rat alveolar macrophages (Stahl et al., 1978), both binding and uptake of ligand by BCG-PM were markedly depressed; i.e., 6% and 25% of TPM activity, respectively. The low level of activity by BCG-PM was still saturable and completely inhibited by mannan and was independent of the concentration of ligand or the duration of incubation. Mannose-specific degradation by BCG-PM was reduced to 20% of that displayed by TPM. Similar results were obtained with a different ligand, [^{125}I] -β-glucuronidase and diminished binding of [^{125}I] mannose-BSA was also observed with total BCG-peritoneal cells assayed in suspension, without fractionation by adherence. These results were highly reproducible.

Autoradiographic studies showed that all thioglycollate-elicited cells with macrophage morphology were labeled with [^{125}I] mannose - BSA ($>$10 grains/cell) and that BCG-PMs were less heavily, but uniformly labeled.

Control experiments established that the decrease in receptor function was not a result of nonspecific cellular injury. Phase-contrast mocroscopy indicated that the cells were viable and that $>$95% excluded trypan blue. The defect in mannose-specific endocytosis by BCG-PM persisted after 5 days of cultivation, both adherent and nonadherent, on a gelatine-coated surface and still represented 25% of uptake of TPM or RPM, which displayed similar activity. In contrast, all three populations displayed similar levels of constitutive lysozyme secretory activity (Gordon et al., 1974). The continued depression of endocytosis by BCG-PM was apparently stable and autonomous.

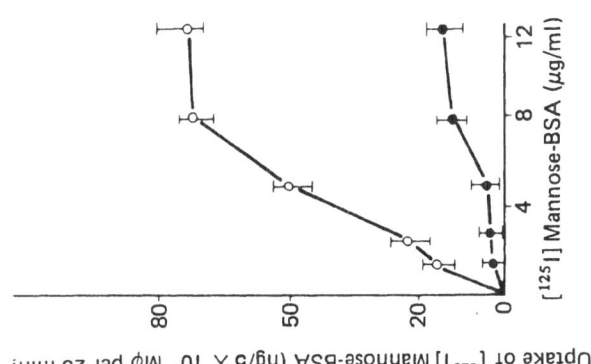

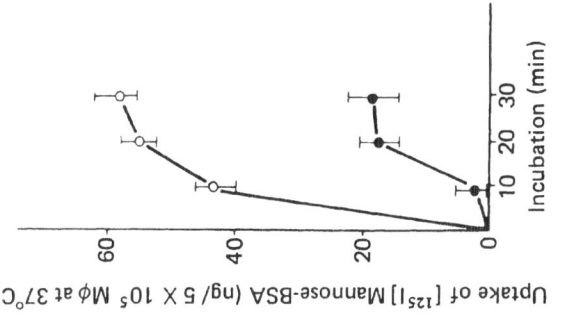

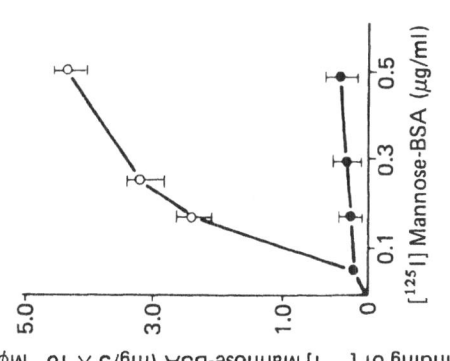

Figure 1. Specific binding and uptake of [^{125}I] mannose-BSA by BCG-PM and TPM; 5×10^5 Mϕ were cultivated for 4 hr before assay [^{125}I]-mannose-BAS (3×10^6 cpm/μg) was added with or without 1.25 mg mannan per well. Results show average SD of pooled results of five independent experiments. (○) TPM; (●) BCG-PM, BCG, bacillus Calmette-Guérin; BSA, bovine serum albumin; cpm, counts per minute; Mϕ, macrophage; PM, peritoneal macropage; TPM, thiogycollate peritoneal macrophage. (From Ezekowitz et al., 1982.)

B. Role of H_2O_2 and Plasminogen Activator Secretion

It is known from previous studies that BCG-PM Secrete H_2O_2, O_2^- (Nathan and Root, 1977), and PA (Gordon and Cohn 1978). Although TPM also secrete PA and O_2^-, H_2O_2 seems to be consumed by an ingredient of thioglycollate broth (Nathan and Cohn 1980; Spitalny, 1981). Figure 2 shows that BCG-PM with reduced specific uptake of $[^{125}I]$ mannose-BSA produced similar levels of PA as compared with TPM, and responded to phorbol myristic acetate (PMA) by releasing H_2O_2 at levels comparable with those reported by others. TPM did not produce H_2O_2 under these conditions, but both populations released high levels of O_2^- after PMA treatment. Because of the possible deleterious effect of these secretion products on cell membranes, experiments were performed in an effort to evaluate the role of secretion products in mannose-receptor function.

BCG-activated peritoneal cells were harvested and adherent macrophages

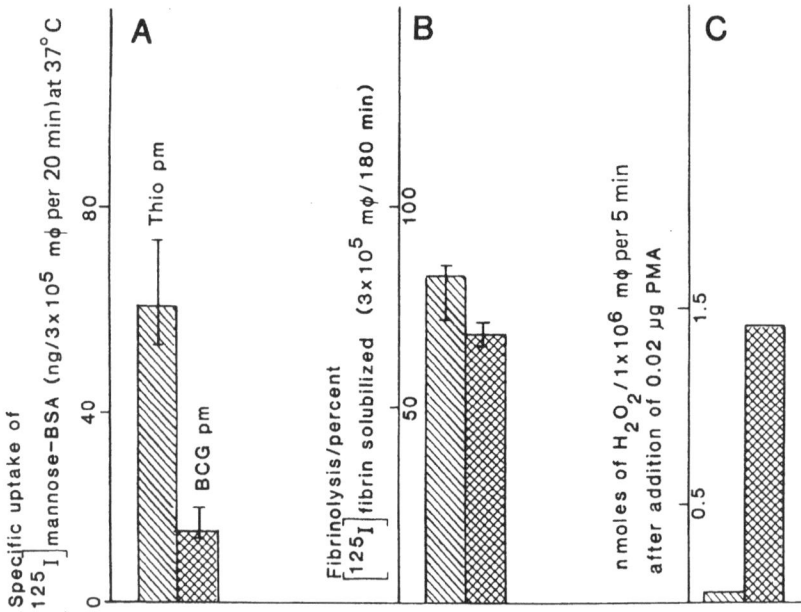

Figure 2. Receptor-mediated endocytosis (A) and secretion of plasminogen activator (B) and H_2O_2 (C) by BCG-PM and TPM. (A, B) Matched experiments; (C) independent. 0.2 nmoles H_2O_2 produced by BCG cells before PMA treatment. 2000 IU catalase + 25 μg/ml superoxide dismutase inhibited all H_2O_2 production. BCG-PM produced 301 nmoles and TPM 175 nmoles superoxide/90 min per mg cell protein after treatment with PMA, respectively. PMA, phorbol myristic acetate. Other abbreviations as in Fig. 1. (From Ezekowitz et al., 1981.)

prepared and assayed in the continuous presence of a cocktail of inhibitors, including catalase and superoxide dismutase at concentrations that inhibited H_2O_2 and O_2^- effectively. Neither this action nor the further addition of indomethacin was found to restore MFR activity. In other experiments, sublethal concentrations of $0.1M$ reagent H_2O_2 added for 30 min had no effect on uptake of $[^{125}I]$-mannose-BSA by TPM or BCG-PM. The effect of proteolysis was examined by culturing cells in serum-free conditions or by cocultivating TPM and BCG-PM. The presence of BCG-PM had no effect on TPM receptor activity as compared with unmixed controls.

It seems unlikely that diffusible secretory products could account for reduced receptor expression on BCG-PM. These studies did not rule out an intracellular site of action of these products, unique enzymes produced only by BCG-PM, or pathways not inhibited by this limited range of inhibitors.

C. Quantitative Assays of Other Surface Markers

We next asked whether BCG-activatied macrophages expressed reduced levels of other receptors for endocytosis and of other macrophage surface antigens as well. We examined expression of Fc receptors, which bind and internalize immune complexes, of Mφ-specific antigen F4/80 and Mac-1, respectively. Antibody bound to macrophages was detected by $[^{125}I]$-RAR [radiolabeled Fab fragment of a rabbit anti-rat $F(ab)^2$], with both first and second-stage antibodies at saturation. Similar assays were performed to quantitate Ia antigens, which had been reported on BCG-PM (Nussinzweig et al., 1980). The mouse monoclonal antibody used, OX6, cross-reacts with the polymorphic Ia determinants 17 or 18 (McMaster and Williams, 1979), which are present in CBA(H-2^k) but not in Balb/c(H-2^d) mice. All antigens were therefore examined in matched groups of animals of both strains. Preliminary studies confirmed that PM obtained after BCG infection of these strains spread rapidly and displayed reduced mannose-receptor activity similar to that observed with macrophages from Swiss mice and C57/BL mice. The IgG2a receptor was studied by direct binding assay using trace-labeled myeloma proteins (Ezekowitz et al., 1983).

BCG-activated macrophages from CBA strain mice assayed after 4 hr adherence and washing showed a striking 3-fold increase in expression of Ia antigen compared with TPM and RPM from the same strain. Binding of antibody was <1% in the Balb/c strain control and was therefore attributable to the presence of Ia antigen on CBA cells and not to Fc receptor (FcR) or cytophilic antibody. The enhanced expression of Ia antigen was stable after 3 days of cultivation in DMEM (Dulbecco's modified Eagles medium) + 10% FBS.

In contrast, BCG-PM showed reduced expression of F4/80 and of the trypsin-resistant IgG2b Fc receptor, with no difference between strains. F4/80 expression was greatest on the resident macrophages (71,000 molecules of rabbit anti-rat

F(ab)2' (RAR bound) reduced on TPM (28,000-40,000) and markedly reduced (18,000-20,000) in BCG-PM, in agreement with other results (Austyn and Gordon, 1981). It was found that 2.4G2 expression (IgG2b receptor) on BCG-PM was clearly reduced compared with TPM (46,000 versus 140,000), but that RPM expressed rather less antigen (35,000-80,000) relative to TPM. IgG2a receptor expression, however, was greatly enhanced on BCG-PM compared with RPM (160,000 versus 70,000), whereas TPM expressed 120,000-140,000 molecules/ macrophage. The expression of Mac-1 (C3bi receptor) was relatively stalbe in all populations studied with only minor changes (<10%) after infection (130,000-160,000 molecules/cell).

Autoradiographic studies of BCG-PM showed that >80% of adherent cells expressed reduced levels of Ag F4/80. A similar pattern was observed with 2.4G2. Although the level of Ag F4/80 is reduced by BCG activation, the Ag can still be readily detected on Mφ by immunocytochemistry of isolated cells in culture or BCG-granulomata *in vivo* (D. Hume, unpublished observations). In contrast, the expression of Ia antigens increased from 16% in TPM to 64% in BCG-PM. Most cells showed heavy labeling, resembled the cells labeled by F4/80, and were therefore macrophages. No morphologic difference was noted between Ia-positive and Ia-negative macrophages.

We nest asked whether the activation phenotype is unique to BCG infection and whether macrophages that resemble BCG-PM functionally express a similar surface phenotype.

III. SURFACE PROPERTIES OF MACROPHAGES ACTIVATED OR ELICITED BY VARIOUS AGENTS

Peritoneal macrophages harvested after intraperitoneal injection of heat-killed, endotoxin-free *Corynebacterium parvum*, endotoxin (LPS) proteose peptone, and N-acetylmuramyl-L-alanyl-D-isoglutamate $2H_2O$ (MDP) were studied. The properties of macrophages harvested after these treatments are shown in Table II. *Corynebacterium parvum*-activated PM are activated by all conventional criteria and express a surface phenotype similar to that of BCG-PM. MDP-elicited macrophages display properties intermediate between resident and BCG-PM in that there is induction of Ia expression, some decrease in MFR activity (50% of RPM), and, as reported by Pabst and Johnston (1980), inhanced secretion of superoxide, whereas fibrinolytic activity was not increased. Proteose peptone and LPS-elicited PM resemble resident PM in their surface properties and, besides a transient tumoricidal activity shown by LPS-PM (reviewed by Meltzer, 1981), both populations lack other criteria of cell activation without an additional stimulus, e.g., phagocytosis (Gordon *et al.*, 1975).

The role of macrophage activation in *Trypanosoma brucei* infection was

Table II. Properties of Macrophages Activated or Elicited by Various Agents[a,b]

Properties	Resident	BCG or Corynebacterium parvum	Trypanosoma brucei Day 9	Trypanosoma brucei Day 15	TPM	Endotoxin	MDP	Proteose peptone
Spreading	+ (7)	++++ (1)	++++	++++	++	++ (7)	++ (1)	++
Plasminogen activator	± (8)	++++ (1)	++++	++++	+++	++ (8)	− (2)	+
$H_2O_2^-$	±	++++ (2)	++++	+	++	±	++ (3)	−
Tumoricidal activity	−	++++ (3)	ND	+	++	± (4)	++	−
Enhanced antimicrobial activity		++++ (4)	ND	ND	−		++ (13)	−
5′-Nucleotidase	+++	+ (5)	ND	ND	++	±	ND	ND
Alkaline phosphodiesterase	+	+++ (5)	ND	ND	+	+ (11)	ND	ND
Surface markers								
F4/80	+++	+	+	+++	++	+++	++	+++
Ia	−	+++	+++	+++	+	+	++	−
Mac-1	+++	+++	++	++	+++	+++	+++	+++
FcR IgG2a	+	+++	+++	++	+++	ND	ND	ND
FcR IgG2b	++	+	+	+++	+++	+	++	++
MFR	+++	+++	+++	+++	+++	+++	++	−

[a] Abbreviations: BCG, bacillus Calmette–Guérin; FCR, Fc receptor; MDP, muramyl-dipeptide; MFR, mannose–fucosyl receptor; TPM, thioglycollate peritoneal macrophages; ND, not determined.
[b] Numbers in parentheses are keyed to the following references:

1. Ezekowitz and Gordon (1982)
2. Nathan et al. (1980)
3. Woodruff and Boak (1966)
4. Adlam et al. (1972)
5. Edelson (1975)
6. Grosskinsky et al. (1983)
7. Rabinovitch et al. (1977)
8. Gordon et al. (1974)
9. Johnston et al. (1978)
10. Ruco and Meltzer (1978)
11. Edelson et al. (1976)
12. Pabst and Johnston (1980)
13. Parant et al. (1978)

also studied. Although an extracellular pathogen, *T. brucei* induces the same changes in macrophages seen with BCG infection, and the level of macrophage activation fluctuates during the course of the infection. Macrophages harvested after control of the first wave of parasitemia are activated by surface, endocytic, and secretory criteria. At the beginning of the second fatal wave of parasitemia, however, the macrophage population appears to be partially deactivated, since down-regulated markers (MFR, F4/80) revert to normal, although the cells still express Ia antigens and secrete plasminogen activator (Grosskinsky *et al.*, 1983). A similar pattern of macrophage activation was observed in nude mice, implying that mature T-cell function may not be essential for macrophage activation by this organism and that macrophages activated by independent pathways share common properties.

Although not exhaustive, these studies confirmed that the activation phenotype was not unique to BCG infection, but was confined to agents with known ability to induce sustained enhanced antimicrobial or cytocidal activity.

IV. REGULATION OF MACROPHAGE ACTIVATION

The changes in the plasma membrane observed by cell activation can be explained by two hypotheses, viz., that the altered cell population represents an earlier stage or alternative pathway of macrophage differentiation or that the mature macrophage is able to modulate its surface phenotype.

A. Role of Sensitized Lymphocytes and Antigen: Adoptive Transfer *in Vivo*

To examine the role of specifically sensitized lymphocytes in modulating the surface properties of macrophage, nylon wool-enriched BCG-primed peritoneal cells were first injected intraperitoneally into uninfected syngeneic mice, with or without purified protein derivative (PPD). Peritoneal macrophages harvested 48 hr after injection of sensitized lymphocytes with PPD showed a similar pattern of surface changes and secretion as BCG infection (Table III). Specific uptake of $[^{125}I]$ mannose-BSA was reduced by 55–60% and Ag F4/80 to a lesser extent, by 35–50%, whereas Ia Ag and secretion of PA and O_2^- were enhanced 2- to 4-fold relative to mock-injected or untreated controls. Injection of sensitized lymphocytes without PPD failed to induce any change in the PM wheras PPD alone contributed up to one-third to the increase in Ia. PPD alone decreased mannose-specific endocytosis by 40% but did not diminish Ag F4/80. The effective range of lymphocytes was 1–10×10^5 per animal, the minimal effective concentration of PPD 25–50 μg.

We concluded that adoptive transfer of sensitized lymphocytes with specific

Table III. Macrophage Properties after Infection with BCG or Injection of BCG-Primed Lymphocytes Plus PPD into the Peritoneal Cavity[a,b]

Treatment	MFR (ng uptake of [^{125}I] mannose-BSA/ 5 × 10^5 Mφ per 30 min at 37°C)	Antigen expression (mol/Mφ × 10^{-4})				Release of[c] O_2^- (nmol/mg per 60 min)	Fibrinolysis (percent solubilized/ 180 min per 2 × 10^5 Mφ)
		F4/80	OX6	2.4G2	Mac-1		
BCG infection	25 ± 8	7 ± 2	20 ± 4	8 ± 1	50 ± 12	180 ± 20	55 ± 10
Thioglycollate broth	90 ± 12	36 ± 6	8 ± 2	23 ± 6	60 ± 15	80 ± 30	66 ± 5
RPM	80 ± 10	51 ± 8	4 ± 3	14 ± 5	56 ± 10	26 ± 8	12 ± 5
BCG-sensitized lymphocytes and PPD	38 ± 4	25 ± 5	15 ± 1	8 ± 4	42 ± 18	150 ± 10	60 ± 5
Mock transfer, medium alone	83 ± 12	38 ± 2	8 ± 4	ND	48 ± 16	34 ± 6	34 ± 3

[a]Abbreviations: BCG, bacillus Calmette-Guérin; PPD, purified protein derivative; MFR, mannose-fucose receptor; BAS, bovine serum albumin; RPM, resident peritoneal macrophage; Mφ, macrophage; ND, not determined.

[b]Mice were either infected as described or injected with 5 × 10^5 BCG-primed lymphocytes with 50 μg PPD. Peritoneal cells were harvested after 2 days, adherent Mφ assayed after 4 hr. Results show a representative experiment; mean ± SD of triplicate (Ag, antigen) or duplicate assays. Similar results were obtained in at least three independent experiments.

[c]Twenty ng PMA added for 60 min. Dimethyl sulfoxide control had no effect.

antigen was able to induce the complete activation phenotype in peritoneal macrophages from uninfected animals.

V. COCULTIVATION OF MACROPHAGES WITH LYMPHOCYTES AND ANTIGEN

To study the role of antigen and lymphocytes more directly, Mϕ from uninfected animals were next exposed in culture to BCG-sensitized lymphocytes with or without PPD for 2 days before assay. Untreated and thioglycollate-elicited macrophages were compared as targets in order to evaluate the response of resident and newly recruited cell populations. Results (not shown) demonstrated that BCG-sensitized lymphocytes plus PPD efficiently induced the activation phenotype *in vitro* and that resident and thioglycollate-elicited macrophages provided similar targets.

A. Role of Lymphokines

Addition of lymphokine-rich supernatants to cultivated peritoneal macrophages induces spreading, secretion of PA (Gordon and Cohn, 1978), the capacity to release H_2O_2 (Nathan *et al.*, 1979), enhanced Ia expression (Steinman *et al.*, 1980; Steeg *et al.*, 1980; Scher *et al.*, 1980), and the ability to kill intracellular organisms (Sharma and Remington, 1981) and tumor cell targets (Meltzer, 1981). Lymphokine was prepared by stimulating BCG-primed spleen cells with PPD and added to RPM and TPM for 2-5 days before assay of Mϕ markers. These lymphokines were able to enhance Ia 2- to 3-fold, and MFR activity was reduced by two-thirds in both targets. Ag F4/80 again fell to a lesser extent, whereas Mac-1 levels remained unchanged. The IgG2a Fc receptor was enhanced 2-fold. Lymphokine controls with PPD showed <20% of the efficacy of active supernatants. Although lymphokines induced similar antigenic and endocytic changes in TPM and RPM, TMP secreted twice as much O_2^- when challenged subsequently with PMA. Concanavalin A (Con A)-induced lymphokines gave results similar to those of immune lymphokine. However, some batches contained an inhibitory activity that was nondialyzable yet removable by concentration across a PM-10 ultrafiltration membrane.

B. Kinetics and Stability

Thioglycollate-elicited peritoneal macrophages were exposed to immune lymphokines continuously for 3 or 5 days before withdrawal. Although lymphokine stimulation affected several markers coordinately, the rate of change for each differed substantially. Reduction of MFR activity could be detected 4 hr after exposure to lymphokine and decreased progressively until a plateau level of ~30% of control activity was attained after 1 day. Reduction in F4/80 was more gradual and less extensive, to 70% of control values. Induction of Ia Ag showed a lag period of 1 day and reached a maximum at day 3. These changes depended on daily addition of fresh lymphokine and, after its removal, Ia levels returned toward control values. Recovery of MFR activity and F4/80 was incomplete compared with control TPM, in which receptor and Ag levels increased with cultivation. This was not attributable to cell loss or diminished viability.

C. Role of T Lymphocytes

Table IV shows that the alteration of macrophage Ag and MFR by BCG-sensitized lymphocytes and PPD depends on Thy-1 positive cells. After ablation with anti-Thy-1 Ab and complement, but not complement alone, macrophage targets retained high levels of MFR and F4/80 and showed no increase in Ia antigens when cocultivated with surviving lymphocytes and PPD.

D. Nude Mice

The properties of BCG-PM obtained from infected CBA *nu/nu* mice were compared with TPM from similar animals. Table V shows that BCG infection in nude mice reproduced all the changes associated with activation in normal mice. TPM from nudes resembled their normal counterparts, as did RPM from

Table IV. Alteration of Macrophage Surface Markers by BCG-Sensitized Lymphocytes and PPD Dependence on Thy-1-Positive Cells[a,b]

Treatment of lymphocytes	Antigen expression (mol/Mϕ × 10^4)			MFR (ng uptake/ 5 × 10^5 Mϕ per 30 min at 37°C)
	F4/80	OX6	Mac-1	
Anti-Thy-1 + C'	83 ± 10	8 ± 1	90 ± 5	72 ± 8
C'	37 ± 10	13 ± 2	93 ± 2	46 ± 6
None	40 ± 6	18 ± 4	88 ± 6	38 ± 10

[a]*Abbreviations*: BCG, bacillus Calmette–Guérin, PPD, purified protein derivative; Mϕ, macrophage.
[b]BCG-primed lymphocytes were treated with anti-Thy-1 plus complement (C') or C' alone and then incubated with TPM plus 50 μg PPD for 48 hr before assay. Results of one experiment done in triplicate, representative of three independent experiments; mean ± SD.

　　　　　　　　　　　　　　　　　　R. Alan B. Ezekowitz and S. Gordon

Table V. Macrophages from Nude Mice Infected with BCG Express Activation Phenotype[a,b]

| Mouse strain | MFR (ng uptake/ 5×10^5 Mφ per 30 min at 37°C) | Antigen expression | | | | PA (percent fibrinolysis/180 min per 2×10^5 Mφ) |
		F4/80[c]	OX6[c]	Mac-1[d]	Superoxide[d]	
BCG-PM *nu/nu*	27 ± 4	8 ± 2	11 ± 2	55 ± 2	172 ± 7	65 ± 3
RPM *nu/nu*	80 ± 6	46 ± 4	2 ± 2	80 ± 15	80 ± 40	10 ± 6
TPM *nu/nu*	68 ± 7	27 ± 5	5 ± 2	68 ± 20	33 ± 7	68 ± 7

[a] *Abbreviations*: BCG-PM, bacillus Calmette–Guérin-activated peritoneal macrophage, Mφ, macrophage, MRF, mannose-fucose receptor; PA, plasminogen activator.

[b] Macrophages from CBA nude mice (three mice pooled per group), harvested 8 days after intraperitoneal infection with BCG or 4 days after thioglycollate injection, were plated and assayed after 4 hr.　Results of triplicate assays except for PA, which was performed in duplicate.

[c] In (mol/Mφ) $\times 10^4$.

[d] In nmol/mg per 60 min.

nudes, except fro an exceptionally high level of O_2^- release in the latter. Supernatants from BCG-infected nude spleen cell suspensions cultivated with or without PPD had no lymphokine activity when cocultivated with macrophage targets (R. A. B. Ezekowitz, unpublished data).

All changes associated with macrophage activated by BCG can therefore be shown to depend on sensitized T cells and specific antigen. However, activation may also arise by independent pathways.

VI. Fc RECEPTOR EXPRESSION

Mouse macrophages express three distinct FcRs: a trypsin-resistant FcR for IgG2b/IgG1 (Unkeless, 1979), a trypsin-sensitive FcR for IgG2a (Unkeless and Eisen, 1975; Heusser et al., 1977), and a third FcR for aggregated IgG3 (Diamond and Yelton, 1981) (reviewed by Unkeless et al., 1981).

Since FcR on BCG-activated Mϕ are known to function in antibody-dependent cellular cytotoxicity (ADCC) (Nathan et al., 1979), and since Mϕ FcR for different isotypes are distinct, we investigated the specificity of FcR on BCG-induced and other Mϕ. Adherent BCG-PM, TPM and RPM, >90% pure, were incubated at 4°C with sheep erythrocytes (E) opsonized with subagglutinating concentrations of mouse monoclonal IgG2a or IgG2b ab (EIgG2a, EIgG2b). Table VI shows that binding of EIgG2a was enhanced on both BCG-PM and TPM, compared with RPM (approximately two-fold increase in EIgG2a bound/100 Mϕ and almost all Mϕ labeled), wheras bind of EIgG2b was selectively decreased in BCG-PM. Ingestion of EIgG2a by BCG-PM was also increased, unlike uptake of EIgG2b, and experiments with opsonized ^{51}Cr-labeled E gave similar results. The results shown in Table VI are representative of at least three experiments. Controls without EIgG showed <1% binding. Binding to live Mϕ of monomeric ^{125}I-labeled monoclonal mouse anti-DNP ab was measured at 4°C for 60 min. Binding to empty wells, to 1-cell fibroblasts, or to Mϕ in the presence of 100-fold excess of unlabeled myeloma protein was <0.5% of input. The number of sites determined by Scatchard analysis and by binding at saturation was within 5% of values shown.

Unkeless and Eisen (1975) found high-affinity binding of monomeric IgG2a and IgG2b myeloma proteins to RPM, TPM, and the Mϕ cell line P388D1 and showed that the FcR was sensitive to trypsin. To study the binding of monomeric IgG, Mϕ were incubated at 4°C with ^{125}I-labeled uncomplexed monoclonal anti-DNP ab of each subclass in the presence or absence of unlabeled Ig. The number of sites per cell was obtained both by Scatchard analysis and from specific binding at saturation. Table VI shows that assays of monomer binding were in good agreement with results of rosetting assays. BCG-PM showed a 70% increase in the number of IgG2a sites and a 30% decrease in IgG2b sites com-

Table VI. Binding of Monomeric and Complexed IgG Subclasses to Murine Peritoneal Macrophages[a]

Target	EIgG2a		[125I]IgG2a		EIgG2b		[125I]IgG2b	
	% Mφ rosetted	RBC/ 100 Mφ	Sites/ Mφ	Ka (M)$^{-1}$	% Mφ rosetted	RBC/ 100 Mφ	Sites/ Mφ	Ka (M)$^{-1}$
BCG-PM	95	665	110,000	2.2×10^8	68	360	65,000	4.0×10^7
RPM	65	320	72,000	4.1×10^8	80	480	90,000	1.2×10^8
TPM	86	602	120,000	3.1×10^8	94	720	200,000	1.2×10^8

[a] Abbreviations: BCG-PM, bacillus Calmette–Guérin-activated peritoneal macrophage; K_a, $K_{association}$; Mφ, macrophage; RPM, resident peritoneal macrophage; TPM, thioglycollate peritoneal macrophage.

pared with RPM, whereas both receptors were increased on TPM. There was little change in affinity of each FcR on different Mϕ. The volume of elution of both Ig subclasses on Sephadex G-200 indicated that the proteins used were monomeric; the isoelectric focusing profiles confirmed the isotype specificity. Further evidence for subclass specificity was provided by competition experiments with other monoclonal ab of the relevant subclasses. Moreover, after treatment with 0, 25% w/v trypsin at 37°C for 20 min, BCG-PM no longer bound IgG2a, but did bind both IgG2b and the Fab fragment of 2.4G2.

The isotype specificity of FcR can be reversed on nonactivated macrophages by immune lymphokines and the surface expression of FcR for different isotypes correlates well with receptor activity in endocytosis and secretion of O_2^-. Table VII shows that both BCG-PM and lymphokine-activated RPM release more O_2^- on the IgG2a immune complexes, comparable to that triggered by PMA. The high spontaneous release by BCG-PM plated directly on DNP-gelatin substrates is caused by adherence and spreading and is characteristic of BCG-activated Mϕ (G. Berton and S. Gordon, 1983). Nonactivated Mϕ release little O_2^- on immune complexes, with preference for the IgG2b subclass. We can conclude that the enhanced expression of IgG2a FcR by activated Mϕ correlates closely with its ability to mediate a secretory function.

Michl *et al.* (1980) showed that FcR can be selectively cleared from the surface of Mϕ by cultivation on immobilized immune complexes, a process known as modulation. Support for distinct FcR on activated Mϕ is provided by the evidence that the IgG2a and IgG2b receptors can be modulated independently. Table VII shows that BCG-PM plated on immune complexes of DNP-gelatin-anti-DNP (IgG2a) exhibited a 95% reduction in EIgG2a rosettes, compared control cells on DNP-gelatin alone. The results of the reciprocal experiments on IgG2b immune complexes and appropriate controls are shown in Table VII.

IgG2a and IgG2b mediated O_2^- release was measured by preparing fixed immune complexes in 24-well Linbro trays; 5×10^5 Mϕ BCG-PM and TPM were purified on a 17% metrizamide gradient, >98% Mϕ, and added to wells containing a reaction mixture of 80 μmM ferricytochrome C and 2 mM NaN$_3$ in Hank's balanced salt solution, pH 7.4, without phenol red. Cytochrome C reduction was measured at 550 nm; 5×10^5 RPM and lymphokine-activated RPM were assayed after 24-hr treatment with 5% v/v BCG-immune spleen cell supernatant under nonadherent conditions. Nonimmune O_2^- release was stimulated by adding 20 ng PMA to cells plated on DNP-gelatin alone. Results are expressed as nmole O_2^- released/mg cell protein/60 min. Superoxide dismutase inhibited >80% of cytochrome C reduction. Assays were performed in duplicate and the results shown are representative of five independent experiments; variation from the mean did not exceed 10%. In nodulation experiments BCG-PM were cultivated on fixed immune complexes and the binding of EIgG2a and EIgG2b measured at 4°C for 45 min.

Table VII. Effect of Immobilized Immune Complexes on Fc-Mediated O_2^- Production and on FcR Modulaton[a]

Addition to DNP-gelatin-coated plate	Superoxide release (nmole O_2^-/mg protein per 60 min produced)				Modulation (% BCG-PM with rosettes)	
	BCG-PM	RPM	RPM + lymphokine	TPM	EIgG2a	EIgG2b
None	160	12	40	26	92	60
PMA	316	67	120	98	ND	ND
IgG2a anti-DNP	310	48	118	35	6	52
IgG2b anti-DNP	225	64	55	42	54	12

[a] *Abbreviations*: DNP, dinitropherol; ND, not done. Other abbreviations as in Tables VI and VIII.

These studies established that at least two of the Mϕ FcR specificities can be regulated independently. The selective enhancement of IgG2a FcR expression during Mϕ activation *in vivo* or *in vitro* could account for the remarkable efficacy of the homologous antibody in mediating tumor destruction *in vivo* in some experimental systems (Matthews *et al.*, 1981; Herlyn and Koprowski, 1982).

VII. DISCUSSION

These studies show that activated macrophages known to display enhanced antimicrobial/anticellular activity express markedly different surface properties from those of elicited or resident cells. This selective remodeling of the membrane upon cell activation results in a coordinate change of several endocytic receptors and distinct surface antigens independent of a particular activating agent and can be induced both *in vitro* and *in vivo*. Although dependent on sensitized T cells and specific antigen, studies in 6-week-old nude mice suggest that an alternative pathway of activation may exist as well. Residual T-cell activity in some nude animals has been described (Huinig and Bevan, 1980; Ranges *et al.*, 1982), but the failure to demonstrate lymphokine activity in BCG-infected nude spleen cell cultures adds weight to the existence of other pathways of activation. Macrophages from nude animals are able to display other parameters of activation (Cheers and Walker, 1975; Cutler and Poor, 1981). Complement components could contribute to direct cell activation in both nudes and normal mice (Ghaffar, 1980).

Lymphokines obtained after antigen or mitogen stimulation are able to modulate all activation markers in peritoneal macrophages from uninfected animals. These effects are specific for lymphokines and have not been observed with endotoxin, MDP, phorbol ester, or L-cell-conditioned medium, a source of macrophage colony-stimulating factor.

The mechanisms by which lymphocyte products alter Mϕ properties are not understood. After lymphokine treatment, loss of MFR receptors and endocytic activity is rapid, $t_{1/2} \sim 16$ hr, and activity reaches a plateau 25% of control within 1 day. Residual activity remains stable subsequently and is not caused by a lymphokine-resistant subpopulation of cells. Loss of receptors could be attributable to decreased synthesis and/or to increased turnover or shedding, until a steady state is attained. Other workers have shown that different Mϕ plasma membrane constituents vary in their rate of turnover: $t_{1/2} 2t_o >$ 80 hr (Edelson and Cohn, 1976; Werb and Cohn, 1977; Mellman *et al.*, 1981; Kaplan *et al.*, 1979) depending on Mϕ stimulation, interaction with specific ligands and molecular size. The reduction in MFR activity was observed in all situations, whereas Ag F4/80 decreased to a lesser extent during adoptive transfer *in vivo* or *in vitro*, compared with infection in the animal. During infection, newly recruited Mϕ contribute to the activated cell population (Spector et al., 1970); it is known that less mature Mϕ express lower levels of surface markers, such as Ag F4/80 (Hirsch, 1981), which could explain this difference. The mechanism of the decreased expression of Ag F4/80 is not known. Remold-O'Donnell and Lewardrowski (1982) have described a cell-surface glycoprotein on guinea pig macrophages of similar molecular weight, (160,000), to that of Ag F4/80. The expression of this molecule is also reduced upon cell activation; reduced expression has been ascribed to a decrease in synthesis (Remold-O'Donnell and Lewandrowski, 1982).

Differences in kinetics of loss of mannosyl receptors and Ag F4/80 and induction of Ia antigens and IgG2a FcR are compatible with a common mechanism, e.g., one by which lymphokines selectively activate or repress a number of macrophage gene products. Preliminary experiments with macrophage activating factor from T-cell hybridomas (Schreiber et al., 1982) indicate that the coordinate surface changes may be induced by monokine(s) from clonal sources (S. Gordon, unpublished observations).

Several workers have proposed a sequential process of macrophage activation. The presence of one of the manifestations of activation described does not indicate the presence of others. Thioglycollate-elicited macrophages secrete high levels of plasminogen activator, yet do not display surface changes of activation and lack microbicidal and tumoricidal activity. These inflammatory Mϕ have enhanced endocytic capacity via all known macrophage endocytic receptors and are able to interact with several humoral proteinase cascades. These effects probably occur before those required for the acquisition of enhanced microbicidal mechanisms, since the maximal stimulation of toxic reactive metabolites requires sequential triggering of an already activated cell (Nathan and Root, 1977). Tumoricidal activity may represent the last of several predetermined events (reviewed by Meltzer, 1981). The recent description of an antigen AcM1 unique to tumoricidal macrophages (Taniyama and Watanabe, 1982) would support this idea.

Although activated Mϕ express reduced levels of MFR and IgG2b FcR, there is little change in the expression of CR3 (Mac 1) (Ezekowitz and Gordon, 1982) and of α_2-macroglobulin receptor (Imber et al., 1982). IgG2a FCR expression is similar in elicited and activated macrophages, although only the activated cell may stimulate enhanced superoxide secretion via this receptor. Recent studies show a reduction in fluid-phase pinocytosis and uptake of zymosan, C3b-coated zymosan, and Latex beads by the activated cell (R. A. B. Ezekowitz, manuscript in preparation). This decrease in endocytic activity upon cell activation is accompanied by down regulation of endogenous and exogenous arachidonic acid metabolite release via both lipoxygenase and cyclo-oxygenase pathways (Scott et al., 1982; Pawlowski et al., 1982). We believe that there is a profound alteration in the membrane upon cell activation, which results in altered responsiveness whether it be enhanced spreading and capacity to release superoxide, changes in prostaglandin metabolism, or altered expression of surface antigens and receptors.

The study of those changes first observed upon activation *in vivo* has been facilitated by the use of lymphokines on defined populations at similar stages of differentation. It is also possible to induce changes characteristic of activation in bone marrow Mφ cultivated under the appropriate conditions (R. A. B. Ezekowitz, unpublished observations). The plasma membrane therefore undergoes selective remodeling during macrophage differentiation and activation that enables the cell to change from a predominantly endocytic cell to a cell geared toward a pericellular effector function.

ACKNOWLEDGMENTS

This work was supported in part by grants from the Medical Research Council. R. A. B. E. is a Junior Research Fellow at Green College, Oxford. We would like to thank Pam Woodward and Paula Gaskel for typing the manuscript.

VIII. REFERENCES

Adlam, C., Broughton, E. S., and Scott, M. T., 1972, Enhanced resistance of mice to infection with bacteria following pretreatment with *Corynebacterium parvum*, *Nature* 235: 219–220.

Ault, K. A., and Springer, T. A., 1981, Cross-reaction of a rat-antimouse phagocyte specific monoclonal antibody (anti-Mac-1) with human monocytes and natural killer cells, *J. Immunol.* 126:359–364.

Austyn, J. A., and Gordon, S., 1981, F4/80, a monoclonal antibody directed specifically against the mouse macrophage, *Eur. J. Immunol.* 11:805–815.

Beller, D., Springer, T. A., and Schreiber, R. D., 1982, Anti-Mac-1 selectively inhibits the mouse and human by the type three complement receptor, *J. Exp. Med.* 156:1000–1009.

Berton, G. and Gordon, S., 1983, Superoxide release by peritoneal and bone-marrow derived mouse macrophages, modulation by adherence and cell activation, *Immunol.* 49:693–704.

Cheers, C., and Waller, R., 1975, Activated macrophages in congenitally athymic "nude" mice and in lethally irradiated mice, *J. Immunol.* 115:844–847.

Cummings, N. P., Pabst, M. J., and Johnston, R. B. Jr., 1980, Activation of macrophages for enhanced release of superoxide anion and greater killing of *Candida albicans* by injection of muramyl dipeptide, *J. Exp. Med.* 152:1659–1669.

Cutler, J. El and Poor, A. H., 1981, Effect of mouse phagocytes on *Candida albicans* in *in vivo* chambers, *Infect. Immun.* 31:1110–1116.

Diamond, B., and Yelton, D., 1981, A new Fc receptor on mouse macrophages binding IgG3, *J. Exp. Med.* 153:514–519.

Edelson, P. J., and Cohn, Z. A., 1976, 5' nucleotidase activity of mouse peritoneal macrophages. 1. Synthesis and degradation in resident and inflammatory populations, *J. Exp. Med.* 144:1581-1595.

Edelson, P. J., Zweibel, R., and Cohn, Z. A., 1975, Pinocytic rate of activated macrophages, *J. Exp. Med.* 142:1150-1164.

Ezekowitz, R. A. B., and Gordon, S., 1982, Down regulation of mannosyl-receptor-mediated endocytosis and antigen F4/80 in BCG-activated mouse macrophages. Role of T lymphocytes and lymphokines, *J. Exp. Med.* 155:1623-1637.

Ezekowitz, R. A. B., Austyn, J. A., Stahl, P., and Gordon, S., 1981, Surface properties of Bacillus Calmette-Guérin-activated mouse macrophages. Reduced expression of mannose-specific endocytosis, Fc receptors and antigen F4/80 accompanies induction of Ia, *J. Exp. Med.* 154:60-76.

Ezekowitz, R. A. B., Bampton, M., and Gordon, S., 1983, Macrophage activation selectively enhances expression of Fc receptors for IgG2a, *J. Exp. Med.* 157:807-812.

Ghaffar, A., 1980, The activation of macrophages by *Corynebacterium parvum*: Effects of anticomplementary agents cobra venom factor and sodium cyanate, *Res. J. Reticuloendothel. Soc.* 27:327-332.

Gordon, S., and Cohn, Z. A., 1978, Bacille Calmette-Guérin infection in the mouse. Regulation of macrophage plasminogen activator by T lymphocytes and specific antigen, *J. Exp. Med.* 147:1175-1188.

Gordon, S., and Werb, Z., 1975, Elastase secretion by stimulated macrophages. Characterization and regulation, *J. Exp. Med.* 142:361-377.

Gordon, S., Todd, J., and Cohn, Z. A., 1974, *In vitro* synthesis and secretion of lysozyme by mononuclear phagocytes, *J. Exp. Med.* 139:1228-1248.

Grosskinsky, M., Ezekowitz, R. A. B., Berton, G., Gordon, S., and Askonas, B., 1983, Macrophage activation in murine African Trypanosomiasis, *Infect. Immun.* 39:1080-1086.

Herlyn, D., and Koprowski, H., 1982, IgG2a monoclonal antibodies inhibit human tumor growth through interaction with effector cells, *Proc. Natl. Acad. Sci. USA* 79:4761-4765.

Heusser, C. H., Anderson, C. L., and Grey, H. M., 1977, Receptors for IgG: Subclass specificity of receptors on different mouse cell types and the definition of two distinct receptors on a macrophage cell line, *J. Exp. Med.* 145:1316-1327.

Hirsch, S., Austyn, J. M., and Gordon, S., 1981, Expression of macrophage specific antigen F4/80 during differentiation of mouse bone marrow cells in culture, *J. Exp. Med.* 152:713-725.

Huinig, T., and Bevan, M. J., 1980, Specificity of cytotoxic T cells from athymic mice, *J. Exp. Med.* 152:688-702.

Hume, D. A., Robinson, A. P., MacPherson, S. S., and Gordon, S., 1983, Mononuclear phagocyte system of mouse defined by immunohistochemical localization of antigen F4/80, *J. Exp. Med.* 158:1522-1536.

Imber, M. J., Pizzo, S. V., Johnson, W. J., and Adams, D. D., 1982, Selective diminution of the binding of mannose by murine macrophages in the late stages of activation, *J. Biol. Chem.* 257:5129-5135.

Johnston, R. B. Jr., Godzik, C. A., and Cohn, Z. A., 1978, Increased superoxide anion production by immunologically activated and chemically elicited macrophages, *J. Exp. Med.* 148:115-127.

Kaplan, G., Unkeless, J. C., and Cohn, Z. A., 1979, Insertion and turnover of macrophage plasma membrane proteins, *Proc. Natl. Acad. Sci. USA* 76:3824-3828.

Kossard, S., and Nelson, D. S., 1968, Studies on cytophilic antibodies IV. The effects of proteolytic enzymes (trypsin and papain) on the attachment to macrophages of cytophilic antibodies, *Aust. J. Exp. Biol. Med. Sci.* 46:63-69.

Kurzinger, K., Ho, M. K., Springer, T. A., 1982, Structural homology of a macrophage differentiation antigen and an antigen involved in T-cell-mediated killing, *Nature* 296:668–670.

Lane, B. C., and Cooper, M. S., 1982, Fc receptors of mouse macrophage cell lines. 1. Distinct proteins mediate the IgG subclass-specific Fc binding activities of macrophages, *J. Immunol.* 128:1819–1824.

Mackaness, G. B., 1964, The immunological basis of acquired cellular resistance, *J. Exp. Med.* 120:105–112.

McMaster, W. R., and Williams, A. F., 1979, Monoclonal antibodies to Ia antigens from rat thymus. Cross reactions with mouse and human and use in purification of Rat Ia glycoproteins, *Immunol. Rev.* 47:117–131.

Mason, D., Dallman, M., and Barclay, N., 1981, Graft-versus-host disease induces expression of Ia antigen in rat epidermal cells and gut epithelium, *Nature* 293:150–151.

Matthews, I. J., Collins, J. J., Roloson, C. J., Thiel, H. J., and Bolognesi, D. P., 1981, Immunological control of the ascites form of murine adenocarcinoma, *J. Immunol.* 126:2332–2336.

Mellman, I., Unkeless, J. C., Steinman, R., and Cohn, Z. A., 1981, Internalisation and fate of Fc receptors during endocytosis, *J. Cell Biol.* 91:124a.

Meltzer, M., 1981, Tumor cytotoxicity by lumphokine-activated macrophages: Development of macrophage tumoricidal activity requires a sequence of reactions, *Lymphokines* 3:319–329.

Michl, J., Pieczonka, M. M., Unkeless, J. C., and Silverstein, S. C., 1980, Effects of immobilized immune complexes on Fc- and complement-receptor function in resident and thioglycollate-elicited mouse peritoneal macrophages, *J. Exp. Med.* 150:607.

Nathan, C. F., and Cohn, Z. A., 1980, Role of oxygen-dependent mechanisms in antibody-induced lysis of tumor cells by activated macrophages, *J. Exp. Med.* 152:198–208.

Nathan, C. F., and Root, R. K., 1977, Hydrogen peroxide release from mouse peritoneal macrophages. Dependence on sequential activation and triggering, *J. Exp. Med.* 146:1648–1662.

Nathan, C. F., Silverstein, S. C., Brukner, L. H. and Cohn, Z. A., 1979, Extracellular cytolysis by activated macrophages and granulocytes. 11. Hydrogen peroxide as a mediator of cytotoxicity, *J. Exp. Med.* 149:100–113.

Nussensweig, M. C., Steinman, R. M., Gutchman, B., and Cohn, Z. A., 1980, Dendritic cells are accessory cells for the development of anti-trimetrophenyl cytotoxic T lymphocytes, *J. Exp. Med.* 152:1070–1084.

Old, L. J., Benacerraf, B., Clarke, D. A., Carswell, C. E., and Stockart, E., 1961, The role of the reticuloendothelial system in host reaction in neoplasia, *Cancer Res.* 21:1281–1291.

Pabst, M. J. and Johnson, R. B., Jr., 1980, Increased production of superoxide by macrophages exposed *in vitro* to muramyl-dipeptide or endotoxin, *J. Exp. Med.* 151:101–110.

Parant, M., Parant, F., and Chedid, L., 1978, Enhancement of the neonate's nonspecific immunity to Klebsiella infection by muramyl dipeptide, a synthetic immunoadjuvant, *Proc. Natl. Acad. Sci. USA* 75:23395–23399.

Paulowski, N. A., Scott, W. A., Andreach, M., and Cohn, Z. A., 1982, Uptake and metabolism of monohydroxy-eicosatetraenoic acids by macrophages, *J. Exp. Med.* 135:1653–1664.

Rabinovitch, M., Manejias, R. E., Russo, M., and Abbey, E. E., 1977, Increased spreading of macrophages from mice treated with interferon inducers, *Cell. Immunol.* 29:86–95.

Ranges, G. E., Gildstein, S., Boyse, E. A., and Schield, M. P., 1982, T cell development in normal and thymopoietin-treated nude mice, *J. Exp. Med.* 156:1057–1060.

Remold-O'Donnell, R. E., and Lewandrowski, K., 1982, Decrease of the major surface glycoprotein gp160 in activated macrophages. *Cell Immunol.* 70:85–93.

Stahl, P., Schlesinger, P. H., Sigardson, E., Rodman, J. S., and Lee, Y. C., 1980, Receptor mediated pinocytosis of mannose glycoconjugates by macrophages. Characterization and evidence for receptor recycling, *Cell* 19:207-211.

Stahl, P., and Gordon, S., 1982, Expression of a mannose-fucosyl receptor for endocytosis on cultured primary macrophages and their hybrids, *J. Cell Biol.* 93:49-62.

Steeg, P. S., Moore, R. N., and Oppenheim, J. J., 1980, Regulation of murine macrophage Ia antigen expression by products of activated spleen cells, *J. Exp. Med.* 152: 1734-1744.

Steinman, R. M., Nogueira, N., Witmer, M. D., Tydings, J. D., and Mellman, I. S., 1980, Lymphokine enhances the expression and synthesis of Ia antigens on cultured mouse peritoneal macrophages, *J. Exp. Med.* 152:1248-1261.

Taniyama, T. and Watanabe, T., 1982, Establishment of a hybridoma secreting a monoclonal antibody specific for activated tumorcidal macrophages, *J. Exp. Med.* 156:1286-1292.

Unkeless, J., and Eisen, H. N., 1975, Binding of monomeric immunoglobulins to Fc receptors of mouse macrophages, *J. Exp. Med.* 142:1520-1538.

Unkeless, J., 1979, Characterization of monoclonal antibody directed against mouse macrophage and lymphocyte Fc receptors, *J. Exp. Med.* 150:580-596.

Unkeless, J., Fleet, H., and Mellman, I. S., 1981, Structural agents and heterogeneity of immunoglobulin Fc receptors, *Adv. Immun.* 31:247-268.

Woodruff, M., and Boak, J. L., 1966, Inhibitory effect of injection of *Corynebacterium parvum* on the growth of tumor transplants in isogenic hosts, *Br. J. Cancer* 20:345-349.

Yin, H., Aley, S., Bianco, C., and Cohn, Z. A., 1980, Plasma membrane polypeptides of resident and activated mouse peritoneal macrophages, *Proc. Natl. Acad. Sci. USA* 77: 2188-2196.

Chapter 3

Activation of Macrophage Complement Receptors for Phagocytosis

Frank M. Griffin, Jr.

Division of Infectious Diseases
Department of Medicine
University of Alabama in Birmingham
Birmingham, Alabama 35294

I. INTRODUCTION

Engagement by immunoglobulin G (IgG) of Fc receptors of either polymorphonuclear or mononuclear phagocytes virtually always leads to ingestion of IgG-containing soluble and particulate complexes (Bianco *et al.*, 1975; Ehlenberger and Nussenzweig, 1977; F. M. Griffin *et al.*, 1975*b*, 1976; J. A. Griffin and Griffin, 1979; Mantovani, 1975; Mantovani *et al.*, 1972; Quie *et al.*, 1968; Rabinovitch, 1967; Schreiber and Frank, 1972; Shaw and Griffin, 1981). Engagement of C3b receptors* by C3b contained in soluble immune complexes likewise results in ingestion of the complexes (Van Snick and Masson, 1978). A number of studies (Bar-Shavit *et al.*, 1979; Diamond *et al.*, 1973, 1974; Gigli and Nelson, 1968; Horwitz and Silverstein, 1980; Huber *et al.*, 1968; Musson and Becker, 1977; Pearlman *et al.*, 1969; Schreiber *et al.*, 1982; Shurin and Stossel, 1978; Stossel *et al.*, 1973) have demonstrated enhancement of phagocytosis of a variety of

---

*Recent evidence indicates that phagocytic cells possess at least two distinct types of complement receptors: one that recognizes C3b and another that recognizes iC3b, the form of C3b that has been nicked by C3b inactivator (Carlo *et al.*, 1979; Dobson *et al.*, 1981; Fearon, 1980; Fearon *et al.*, 1981; Lambris and Ross, 1982; Rabellino *et al.*, 1978; Ross *et al.*, 1982; Ross and Rabellino, 1979; Schmitt *et al.*, 1981). While these receptors may very well serve functions different from each other, to date no functional differences have been described (Wright and Silverstein, 1982). For convenience, I therefore refer to all macrophage complement receptors collectively as C3b receptors.

particles, chiefly microoraganisms, when they were coated with complement, even in the absence of immunoglobulin, and for many years it was assumed that ingestion of these microbes was mediated by the complement receptors of phagocytic cells. Several recent studies, however, have reported that engagement by particle-bound C3b of complement receptors of neutrophils, monocytes, and "resting" tissue macrophages, although always promoting efficient particle binding, fails to promote ingestion of C3b-coated particles (Bianco *et al.*, 1975; Ehlenberger and Nussenzweig, 1977; F. M. Griffin, 1980, 1981; F. M. Griffin *et al.*, 1975*a,b*; F. M. Griffin and Griffin, 1980; J. A. Griffin and Griffin, 1979; Lay and Nussenzweig, 1968; Mantovani, 1975; Mantovani *et al.*, 1972; Newman and Johnston, 1979; Scribner and Fahrney, 1976; Shaw and Griffin, 1981; Ward and Enders, 1933). Studies in rabbits (Brown *et al.*, 1970) and guinea pigs Atkinson and Frank, 1974*a,b*; Schreiber and Frank, 1972) confirm the *in vitro* finding that macrophages are normally unable to phagocytize via their complement receptors. Erythrocytes coated with IgM, which became further coated with complement when injected intravenously, were rapidly cleared from these animals' circulation by the liver. However, within a few minutes they reappeared in the bloodstream and circulated with a normal lifespan. Presumably these C3b-coated erythrocytes were bound via the C3b receptors of Kupffer cells but were released when C3b inactivator and proteases cleaved erythrocyte-bound C3b to C3 fragments not recognized by macrophages.

These findings led several investigators (Ehlenberger and Nussenzweig, 1977; Mantovani, 1975; Scribner and Fahrney, 1976) to explore the possibilty that receptors for C3b on phagocytic cells serve a different function from receptors for IgG in the clearance of particulate immune complexes. These studies showed that when erythrocytes were coated with both IgG and C3b, the concentration of IgG on the erythrocyte surface required to promote ingestion was reduced nearly 100-fold, to a level insufficient to promote even attachment of erythrocytes coated with that quantity of IgG alone. Ehlenberger and Nussenzweig (1977) further demonstrated that erythrocytes coated with quantities of IgG insufficient to promote their attachment were ingested when physical and chemical means other than C3b were used to bind the erythrocytes to the cell surface. These results suggested that engagement of C3b receptors fails to generate a phagocytic signal, whereas engagement of Fc receptors produces a signal that results in activation of the cell's phagocytic machinery. These observations also led to the hypothesis that the chief role of C3b and its receptors is to promote particle binding, the chief role of IgG and its receptors is to promote particle ingestion, and the enhanced particle binding by C3b and its receptors serves to facilitate engagement of the cell's phagocytic signal-generating Fc receptors by IgG. The findings reported by Ehlenberger and Nussenzweig (1977) also provide an explanation for how complement can enhance phagocytosis of some microorganisms even in the absence of IgG. The microbes that C3b can opsonize alone are those that are nonencapsulated and ingestible, at least to a modest degree, in their native states, presumably because they display on their surfaces intrinsic ligands that can engage the appropriate cellular "receptors" and thereby generate a phagocytic signal. However, the density of these putative intrinsic ligands is presumably insufficient to promote microbial attachment to the phagocytic cell surface. The deposition of C3b on the microbe surface probably enhances its rate of binding to the phagocytic cell's surface thereby increasing the efficiency with which the intrinsic ligands of the microbe engage the appropriate phagocytic signal-generating receptors.

It therefore seems probable that, in the phagocyte's quiescent state, its complement receptors serve exclusively to promote particle binding and that enhancement of phagocytosis by complement is solely attributable to facilitation of the interaction of other, signal-generating receptors with their ligands that occurs as a result of complement receptor-mediated enhancement of particle–cell contact. Under certain conditions, however, the complement receptors of monocytes and macrophages can be activated so that they are capable of promoting ingestion directly (polymorphonuclear leukocytes have not been studied). It is my purpose in this chapter to describe the means by which activation can occur, to define the mechanisms responsible for activation in those situations in which they are known, to point out the potential that some of the experimental systems that were developed to study complement receptor activation offer for an exploration of other areas of cell biology and membrane function, and to present evidence suggesting that activation of macrophage complement receptors may be important in host defense *in vivo*.

II. MACROPHAGE COMPLEMENT RECEPTOR ACTIVATION BY INFLAMMATORY STIMULI

The first report of activation of macrophage complement receptors for phagocytosis was that of Bianco *et al.* (1975), who found that peritoneal macrophages elicited into the mouse peritoneal cavity by the injection of thioglycollate medium were able to ingest C3b-coated erythrocytes. These results suggested that inflammation in general might alter the complement receptor function of macrophages at the inflammatory site. The inflammatory stimulus could accomplish this effect either by altering the function of the resident macrophage population, by eliciting a physiologically different mononuclear phagocyte into the inflammatory site, or by eliciting monocytes into the inflammatory site and causing the elicited cells to mature or differentiate into macrophages with markedly different properties from those of the resident macrophage population.

Intraperitoneal injection of thioglycollate medium elicits the influx of a large number of monocytes, apparently by the generation of the chemoattractant C5a by the agar contained in the medium (Shaw and Griffin, 1982). This influx is not required, however, for activation of macrophage complement receptors because, although thioglycollate medium fails to elicit an influx of monocytes into the peritoneal cavities of C5-deficient mice, it nevertheless activates the complement receptors of their resident peritoneal macrophages for phagocytosis (Shaw and Griffin, 1982). The mechanism by which thioglycollate medium activates macrophage complement receptors in unknown. It is not by means of a T-cell lymphokine, because athymic mice respond identically to normal mice to thioglycollate injection (Shaw and Griffin, 1982).

Subsequent investigations have revealed that intraperitoneal injections of either glycogen (Mantovani, 1981), starch gel (Wellek et al., 1975), or very high doses of endotoxin (Nowadowski et al., 1980) elicit macrophages the complement receptors of which are capable of mediating phagocytosis. In vitro, the complement receptors of human monocytes can be activated by treatment with phorbol esters (Wright and Silverstein, 1982). And when cultured for several days in vitro, monocytes spontaneously acquire the ability to phagocytize complement-coated particles (Newman et al., 1980).

In contrast to the effects of these pharmacologic irritants, several physiologically relevant inflammatory stimuli, including bacillus Calmette–Guerin (BCG), tuberculin protein, Streptococcus pneumoniae, Escherichia coli, and physiologically achievable doses of endotoxin fail to activate macrophage complement receptors even though they elicit an intense inflammatory response when injected into the mouse peritoneal cavity (F. M. Griffin, in preparation). Since these stimuli failed to enhance macrophage complement receptor function, the concept that microbe-induced inflammation is an important means of activating macrophage complement receptors is open to serious question.

III. MACROPHAGE COMPLEMENT RECEPTOR ACTIVATION BY IMMUNOLOGIC MECHANISMS

Of more physiologic significance are recently described immunologic systems in which lymphocytes or lymphocyte products are responsible for macrophage complement receptor activation. Chronic infection with BCG or with Trypanosoma cruzi activates macrophage complement receptors, probably by immunologic mechanisms (Atkinson and Frank, 1974a; Nogueira et al., 1977). Murine lymphocytes can be triggered by endotoxin to release a lymphokine that enables resident mouse peritoneal macrophages to phagocytize complement-coated erythrocytes (Nowakowski et al., 1980). Saluk and colleagues (Tabor and Saluk, 1981; Wrigley and Saluk, 1981) have further characterized mechanisms by which lymphocytes can be triggered by endotoxin to activate macrophage complement receptors. These investigators found that both T and B lymphocytes were required and that only lymphocytes from endotoxin-injected mice could activate the complement receptors of resident mouse peritoneal macrophages. The precise cellular interactions were not identified, and whether or not a soluble mediator was involved was not determined.

The most thoroughly characterized *in vitro* system is one that we have been studying for the past several years. In this system lymphocytes and macrophages interact in a unique manner to generate a lymphokine that activates macrophage complement receptors for phagocytosis. The mediator responsible for this effect appears to be different from any previously described lymphokine and cannot be elicited by the classic mechanisms of lymphokine production, i.e., antigenic stimulation of appropriate clones of T lymphocytes (J. A. Griffin and Griffin, 1979). Thus, this lymphokine may not be a product of cell-mediated immune reactions; rather, elaboration of the product requires a unique series of cellular interactions. Macrophages must first phagocytize IgG-containing immune complexes; these cells thereby acquire the ability to trigger, by a cell contact-dependent mechanism, T lymphocytes to elaborate a product that imparts to freshly explanted macrophages the ability to phagocytize via their complement receptors (J. A. Griffin and Griffin, 1979). The product itself appears to be a small polypeptide (F. M. Griffin and Griffin, 1980).

Initial studies of the mechanism by which the lymphokine acts on macrophages revealed that it does not act by means that require macrophage protein synthesis (F. M. Griffin and Griffin, 1980). It does not simply activate the macrophage plasma membrane so that all particles bound to the cell's surface are indiscriminately phagocytized. Moreover, the product does not enhance the function of all phagocytic receptors, for neither the rate nor the capacity of either Fc receptor-mediated or nonimmunologically mediated phagocytosis was increased by treating macrophages with the lymphokine (F. M. Griffin and Griffin, 1980). These results suggested that the lymphokine interacts in either a catalytic or combinative manner with existing surface or cytoplasmic components to link the cell suface binding event with the intracellular machinery required for phagocytosis, i.e., that the lymphokine enabled complement receptor engagement to gererate a phagocytic signal.

More recently we have used the techniques of Michl *et al.* (1979) to identify the mechanism by which the lymphokine acts (F. M. Griffin and Mullinax, 1981). Michl *et al.* (1979) used complement-containing, immobilized immune complexes to study the topography of complement receptors of resident and thioglycollate-elicited mouse peritoneal macrophages. They found that complement receptor activity disappeared from the nonadherent surface of thioglycollate-elicited macrophages plated on these complexes but was retained by similarly plated resident macrophages. We prepared immobilized immune complexes by binding bovine serum albumin (BSA) to glass coverslips using poly-L-lysine as a cross-linker, treating the coverslips with antibody to BSA, and then treating the BSA-anti-BSA complexes with fresh mouse serum to deposit C3b on the complexes. We found, as had Michl *et al.* (1979), that when resident mouse peritoneal macrophages were plated on these BSA-anti-BSA-complement-coated coverslips, they retained the ability to bind complement-coated erythrocytes. Resident macrophages plated on the complexes and treated with the lymphokine for periods as brief a period as 5 min, however, were no longer able to bind complement-coated particles. Loss of complement receptor activity from the nonadherent macrophage surface was a consequence of receptor migration to, and sequestration on, the immobilized C3 ligands, because cleaving the ligands with a source of C3-destroying enzymes enabled complement receptor activity to return promptly to the nonadherent surface of lymphokine-treated macrophages (F. M. Griffin and Mullinax, 1981).

These findings strongly suggest that the complement receptors of resident mouse peritoneal macrophages are normally fixed and immobile within the macrophage plasma membrane and that lymphokine treatment frees the receptors from their plasma membrane anchors, enabling them to wander randomly within the plane of the plasma membrane. As they encounter immobilized C3b ligands on the immune complex substrate, they become trapped on the adherent macrophage surface.

Our finding that the lymphokine both mobilized macrophage complement receptors and enabled the receptors to promote phagocytosis raised the possibility that the two effects of the lymphokine might be related. Several additional findings add strong support to that view:

1. The concentration kinetics of lymphokine activation of macrophage complement receptors and of lymphokine-mediated complement receptor mobilization are indentical (F. M. Griffin et al., 1984).
2. Pharmacologic treatments of macrophages that abrogate one response to the lymphokine, complement receptor mobility, also abrogate the other response, complement receptor-mediated phagocytosis (F. M. Griffin and Mullinax, 1981; F. M. Griffin, et al., 1984).
3. Monoclonal antibodies that bind the lymphokine absorb both lymphokine activities at precisely the same antibody concentration (F. M. Griffin et al., 1984).

These results suggest that the lymphokine activates macrophage complement receptors for phagocytosis by freeing them from their plasma membrane achors. Results of several other studies (Douglas, 1976; Kaplan et al., 1978; Michl et al., 1979; Petty et al., 1980; Rabinovitch et al., 1975; Ragsdale and Arend, 1980) demonstrate that, in order for a receptor to be able to mediate phagocytosis, it must be able to move within the macrophage plasma membrane, suggesting that receptor mobility may be a general requirement for any receptor to promote phagocytosis.

Engagement of macrophage complement receptors does not normally result in phagocytosis, probably because the receptors cannot move. Receptor mobility appears to be required to convert ligand–receptor binding at the cell surface into the intracellular phagocytic response, i.e., receptor mobility appears to be an essential component of the phagocytic signal. Treatment of macrophages with the lymphokine enables their complement receptors to diffuse within the plane of the plasma membrane. When engaged by particle-bound C3b, these receptors can then either aggregate among themselves or perhaps become associated with an intramembrane second messenger, thereby initiating the phagocytic signal and linking the cell surface binding event with the intracellular phagocytic machinery, actin and its regulatory proteins (Boxer and Stossel, 1976; Hartwig and Stossel, 1975; Stossel, 1977; Stossel and Hartwig, 1975, 1976; Yin and Stossel, 1979). Changes in the physical state of actin then provide the motive force for ingestion of the ligand-coated particle.

This novel lymphokine enables us to turn a receptor function on and off *in vitro* and to convert an immobilized plasma membrane protein into a mobile state *in vitro*. We are unaware of any other system by which a receptor function can be quickly activated *in vitro*. The only other intrinsic plasma membrane proteins that are immobile are fibronectin of fibroblasts (Schlessinger *et al.*, 1977) and certain acetylcholine receptors of neurons (Axelrod *et al.*, 1976), and neither of these proteins can be converted to the mobile state *in vitro*. The lymphokine therefore offers the unique opportunity of studying both the mechanisms involved in transducing ligand–receptor binding into the activation of an intracellular function, in this case phagocytosis and also of identifying general mechanisms that govern membrane protein anchorage and mobility. In addition, the techniques developed to study receptor activation and receptor mobility offer excellent means to identify additional components of the phagocytic signal. For example, the complement receptors of human monocytes are spontaneously mobile yet unable to promote phagocytosis (Wright and Silverstein, 1982), indicating that, whereas receptor mobility is probably a necessary condition for phagocytosis, it is not sufficient, and suggesting that monocytes have a defect in signal transmission at a step other than receptor mobility. Since treatment of these cells with phorbol esters *in vitro* promptly activates their complement receptors for phagocytosis (Wright and Silverstein, 1982), this system may prove extremely valuable not only in further delineating the biochemical pathways that constitute the phagocytic signal, but also in elucidating general mechanisms by which ligand–receptor binding at the cell surface is transduced into an intracellular response.

IV. PHYSIOLOGIC RELEVANCE OF MACROPHAGE COMPLEMENT RECEPTOR ACTIVATION

When studied in tissue culture medium *in vitro*, Fc receptors of all phagocytic cells mediate particle attachment generate a phagocytic signal, and promote ingestion of IgG-coated particles (Bianco *et al.*, 1975; Ehlenberger and Nussenzweig,

1977; F. M. Griffin *et al.*, 1975*b*; F. M. Griffin *et al.*, 1976; J. A. Griffin and Griffin, 1979; Mantovani, 1975; Mantovani *et al.*, 1972; Quie *et al.*, 1968; Rabinovitch, 1967; Schreiber and Frank, 1972; Shaw and Griffin, 1981). Several lines of evidence indicate, however, that phagocytic cells may have difficulty in phagocytizing via their Fc receptors at sites of infection/inflammation *in vivo*. Uptake of IgG-coated particles by phagocytic cells of both experimental animals and patients with circulating immune complexes is impaired both *in vitro* (Hallgren *et al.*, 1978) and *in vivo* (Frank *et al.*, 1979; Haakenstad and Mannik, 1974). Macrophages that have either bound immobilized immune complexes (Douglas, 1976; Kaplan *et al.*, 1978; Michl *et al.*, 1979; Rabinovitch *et al.*, 1975; Ragsdale and Arend, 1980) or ingested soluble immune complexes (F. M. Griffin, 1980, 1981) *in vitro* are strikingly defective in their ability to phagocytize either IgG-coated erythrocytes or IgG-coated yeasts of *Cryptococcus neoformans*. Since many microorganisms shed substantial quantities of their surface antigens, immune complexes composed of these antigens and host antibody are probably plentiful at sites of infection in the immune host. Ingestion of these complexes may block the Fc receptors of the phagocytic cell and impair their ability to ingest IgG-coated microorganisms. Even though these microorganisms would be coated with C3b as well as IgG, they would not be opsonized for phagocytosis, because macrophages are unable, in their normal physiologic state, to phagocytize via their complement receptors.

Our findings suggest a means by which such a phagocytic impasse might be overcome and raise the possibility that complement receptor-mediated phagocytosis may be of paramount importance *in vivo*. Ingestion of immune complexes by macrophages, while it may block the Fc receptors of the cells, would at the same time trigger macrophages to signal T lymphocytes to elaborate the lymphokine that activates macrophage complement receptors. Macrophages would then be able to phagocytize immunologically coated microorganisms via their complement receptors. Note that this model proposes a most unusual type of receptor cooperation, in which ingestion of immune complex debris via macrophage Fc receptors initiates a series of events culminating in activation of macrophage complement receptors, and that it is the macrophage's complement receptors, not its Fc receptors, that are responsible for ingesting the offending pathogen.

For this model to have pathophysiologic relevance, several conditions must be satisfied: (1) ingestion of soluble immune complexes must cause severe impairment of macrophage Fc receptor funtion; (2) ingestion of these complexes must not impair macrophage interaction with particles by other means, especially via their complement receptors; and (3) macrophages that have ingested immune complexes must retain the ability to respond to the lymphokine and to phagocytize complement-coated particles. The results of the experiments described below satisfy these conditions.

Macrophages were fed soluble immune complexes composed either of BSA-anti-BSA IgG or of capsular polysaccharide antigen of *Cryptococcus neoformans* and anti-cryptococcal antibody. The cells were washed to remove excess complexes and then incubated with either IgG-coated erythocytes, IgG-coated crypto-cocci, complement-coated erythrocytes, complement-coated cryptococci, or zymosan particles. These particles were used as examples of particles whose inter-action with macrophages is governed by the cell's Fc receptors, complement re-ceptors, and nonimmunologic receptors, respectively. Immune complex-treated macrophages were strikingly defective in their ability to phagocytize IgG-coated particles. However, they were able to bind the complement-coated particles and to bind and ingest zymosan normally. Moreover, they retained the ability to re-spond to the lymphokine and ingest complement-coated erythrocytes and com-plement-coated cryptococci via their complement receptors (Griffin, 1980, 1981).

These results indicate that Fc and complement receptors may have very dif-ferent, but cooperative, functions *in vivo*. Fc receptors are capable of promoting phagocytosis quite efficienctly but are susceptible to blockade by immune com-plexes. Although innately incapable of promoting phagocytosis, complement re-ceptors can be activated to do so by a unique lymphokine, the first step in the generation of which is engagement of macrophage Fc receptors by immune com-plexes. In contrast to Fc receptor-mediated phagocytosis, complement receptor-mediated phagocytosis is not impaired by the ingestion of soluble immune com-plexes. Therefore, complement receptors, and not Fc receptors, may be responsible for phagocytizing opsonized microorganisms and other immunologically coated particles at inflammatory sites *in vivo*.

The results of Brown *et al.* (1980) also suggest that Fc receptor-mediated phogocytosis may be inadequate in defense against bacterial pathogens *in vivo* and that complement receptor-mediated phagocytosis may be essential. These investigators found that guinea pigs depleted of complement by cobra venom factor (CVF) were unable to clear intravenously injected, IgG-coated, encapsulated pneumococci from the circulation and therefore died with pneumococcal bacter-emia. In contrast, animals with normal complement levels, in which the pneu-mococci presumably became coated with complement, efficiently cleared the bacteremia and survived. Since the strain of *Pneumococcus* used was not suscepti-ble to complement-mediated killing, it is probable that complement was required to facilitate microbial ingestion. It may be that ingestion of either pneumococci-IgG complexes or soluble pneumococcal capsular antigen–IgG complexes by macrophages triggered the cells to signal T lymphocytes to elaborate the lympho-kine that permitted macrophages to ingest the pneumococci via their complement receptors.

V. SUMMARY

Macrophage complement receptors, while innately incapable of promoting phagocytosis, can be activated to do so by a number of inflammatory stimuli and by several immunologic mechanisms. Studies with a complement receptor-activating lymphokine reveal that activation occurs as a result of mobilization of innately immobile receptors and suggest that receptor mobility is a prerequisite for phagocytosis. Since Fc receptors are susceptible to blockade by immune complexes at inflammatory sites, phagocytosis mediated by macrophage complement receptors may be of prime importance *in vivo*.

ACKNOWLEDGMENTS

The work cited herein was supported by grants PCM 75-17106 and PCM 82-02455 from the National Science Foundation, by grant IM-173 from the American Cancer Society, Inc., by a Faculty Scholar Award from the Josiah Macy, Jr. Foundation, and by Research Career Development Award AI-00135 from the National Institutes of Health. I am very grateful to Ms. Susan Winn for assisting in the preparation of the manuscript.

VI. REFERENCES

Atkinson, J. P., and Frank, M. M., 1974a, The effect of bacillus Calmette-Guérin-induced macrophage activation on the in vivo clearance of sensitized erythrocytes, *J. Clin. Invest.* 53:1742–1749.

Atkinson, J. P., and Frank, M. M., 1974b, Studies on the in vivo effects of antibody. Interaction of IgM antibody and complement in the immune clearance and destruction of erythrocytes in man, *J. Clin. Invest.* 54:339–348.

Axelrod, D., Ravdin, P., Koppel, D. F., Schlessinger, J., Webb, W. W., Elson, E. L., and Podleski, T. R., 1976, Lateral motion of fluorescently labeled acetylcholine, *Proc. Natl. Acad, Sci. USA.* 73:2823–2827.

Bar-Shavit, Z., Raz, A., and Goldman, R., 1979, Complement and Fc receptor-mediated phagocytosis of normal and stimulated mouse peritoneal macrophages, *Eur. J. Immunol.* 9:385–391.

Bianco, C., Griffin, F. M., Jr., and Silverstein, S. C., 1975, Studies of the macrophage complement receptor. Alteration of receptor function upon macrophage activation, *J. Exp. Med.* 141:1278–1290.

Boxer, L. A., and Stossel, T. P., 1976, Interactions of actin, myosin, and an actin-binding protein of chronic myelogenous leukemia leukocytes, *J. Clin. Invest.* 57:964–976.

Brown, D. L., Lachman, P. J., and Dacie, J. V., 1970, The *in vivo* behaviour of complement-coated red cells: Studies in C6-deficient, C3-depleted and normal rabbits, *Clin. Exp. Immunol.* 7:401–422.

Brown, E. J., Hosea, S. W., and Frank, M. M., 1980, Complement-mediated reticuloendothelial clearance in pneumococcal bacteremia, *Fed. Proc.* 39:1204.

Carlo, J. R., Ruddy, S., Studer, E. J., and Conrad, D. H., 1979, Complement receptor binding of C3b-coated cells treated with C3b inactivator, β1H globulin and trypsin, *J. Immunol.* 123:523–528.

Diamond, R. D., May, J. E., Kane, M. A., Frank, M. M., and Bennett, J. E., 1973, The role of late complement components and the alternate complement pathway in experimental cryptococcosis, *Proc. Soc. Exp. Biol. Med.* 144:312–315.

Diamond, R. D., May, J. E., Kane, M. A., Frank, M. M., and Bennett, J. E., 1974, The role of the classical and alternate complement pathways in host defenses against *Cryptococcus neoformans* infection, *J. Immunol.* 122:2260–2270.

Dobson, N. J., Lambris, J. D., and Ross, G. D., 1981, Characteristics of isolated erythrocyte complement receptor type one (CR₁, C4b-C3b receptor) and CR₁-specific antibodies, *J. Immunol.* 126:693–698.

Douglas, S. D., 1976, Human monocyte spreading *in vitro*—inducers and effects on Fc and C3 receptors, *Cell. Immunol.* 21:344–349.

Ehlenberger, A. G., and Nussenzweig, V., 1977, The role of membrane receptors for C3b and C3d in phagocytosis, *J. Exp. Med.* 145:357–371.

Fearon, D. T., 1980, Identification of the membrane glycoprotein that is the C3b receptor of the human ertythrocyte, polymorphonuclear leukocyte, B lymphocyte, and monocyte, *J. Exp. Med.* 152:20–30.

Fearon, D. T., Kameko, I., and Thompson, G. G., 1981, Membrane distribution and absorptive endocytosis by C3b receptors on human polymorphonuclear leukocytes, *J. Exp. Med.* 153:1615–1628.

Frank, M. M., Hamburger, M. I., Lawley, T. J., Kimberly, R. P., and Plotz, P. H., 1979, Defective reticuloendothelial system Fc-receptor function in systemic lupus erythematosus, *N. Engl. J. Med.* 300:518–523.

Gigli, I., and Nelson, R. A., 1968, Complement dependent immune phagocytosis. I. Requirements for C'1, C'4, C'2, C'3, *Exp. Cell Res.* 51:45–67.

Griffin, F. M., Jr., 1980, Effects of soluble immune complexes on Fc receptor- and C3b receptor-mediated phagocytosis by macrophages, *J. Exp. Med.* 152:905–919.

Griffin, F. M., Jr., 1981, Roles of macrophage Fc and C3b receptors in phagocytosis of immunologically coated *Cryptococcus neoformans*, *Proc. Natl. Acad. Sci. USA* 78:3853–3857.

Griffin, F. M., Jr., and Griffin, J. A., 1980, Augmentation of macrophage complement receptor function *in vitro*. II. Characterization of the effects of a unique lymphokine upon the phagocytic capabilities of macrophages, *J. Immunol.* 125:844–849.

Griffin, F. M., Jr., and Mullinax, P. J., 1981, Augmentation of macrophage complement receptor function *in vitro*. III. C3b receptors that promote phagocytosis migrate within the plane of the macrophage plasma membrane, *J. Exp. Med.* 154:291–305.

Griffin, F. M., Jr., Bianco, C., and Silverstein, S. C., 1975a, Characterization of the macrophage receptor for complement and demonstration of its functional independence from the receptor for the Fc portion of immunoglobulin G, *J. Exp. Med.* 141:1269–1277.

Griffin, F. M., Jr., Griffin, J. A., Leiber, J. E., and Silverstein, S. C., 1975b, Studies on the mechanism of phagocytosis. I. Requirements for circumferential attachment of particle-bound ligands to specific receptors on the macrophage plasma membrane, *J. Exp. Med.* 142:1263–1282.

Griffin, F. M., Jr., Griffin, J. A., and Silverstein, S. C., 1976, Studies on the mechanism of phagocytosis. II. The interaction of macrophages with anti-immunoglobulin IgG-coated bone marrow-derived lymphocytes, *J. Exp. Med.* 144:788–809.

Griffin, F. M., Jr., Luben, R. A., and Golde, D. W., 1984, A human lymphokine activates macrophage complement receptors for phagocytosis: studies using monoclonal anti-lymphokine antibodies, *J. Leuk. Biol.* (in press).

Griffin, J. A., and Griffin, F. M., Jr., 1979, Augmentation of macrophage complement receptor function *in vitro*. I. Characterization of the cellular interactions required for the generation of a T-lymphocyte product that enhances macrophage complement receptor function, *J. Exp. Med.* 150:653–675.

Haakenstad, A. O., and Mannik, M., 1974, Saturation of the reticuloendothelial system with soluble immune complexes, *J. Immunol.* 112:1939-1948.

Hallgren, R., Hakanssen, L., and Venge, P., 1978, Kinetic studies of phagocytosis. I. The serum independent particle uptake of PMN from patients with rheumatoid arthritis and systemic lupus erythematosus, *Arthritis Rheum.* 21:107-113.

Hartwig, J. H., and Stossel, T. P., 1975, Isolation and properties of actin, myosin, and a new actin-binding protein in rabbit alveolar macrophages, *J. Biol. Chem.* 250:5696-5705.

Horwitz, M. A., and Silverstein, S. C., 1980, Influence of the *Escherichia coli* capsule on complement fixation and on phagocytosis and killing by human phagocytes, *J. Clin. Invest.* 65:82-94.

Huber, H., Polley, M. J., Linscott, W. D., Fudenberg, H. H., and Muller-Eberhard, H. J., 1968, Human monocytes: Distinct receptor sites for the third component of complement and for immunoglobulin G, *Science* 162:1281-1283.

Kaplan, G., Eskeland, T., and Seljelid, R., 1978, Difference in the effect of immobilized ligands on the Fc and C3 receptors of mouse peritoneal macrophages *in vitro*, *Scand. J. Immunol.* 7:19-24.

Lambris, J. D., and Ross, G. D., 1982, Assay of membrane complement receptors (CR_1 and CR_2) with C3b- and C3d-coated fluorescent microspheres, *J. Immunol.* 128:186-189.

Lay, W. H., and Nussenzweig, V., 1968, Receptors for complement on leukocytes, *J. Exp. Med.* 128:991-1009.

Mantovani, B., 1975, Different roles of IgG and complement receptors in phagocytosis by polymorphonuclear leukocytes, *J. Immunol.* 115:15-17.

Mantovani, B., 1981, Phagocytosis of immune complexes mediated by IgM and C3 receptors by macrophages from mice treated with glycogen, *J. Immunol.* 126:127-130.

Mantovani, B., Rabinovitch, M., and Nussenzweig, V., 1972, Phagocytosis of immune complexes by macrophages. Different roles of the macrophage receptor sites for complement (C3) and for immunoglobulin (IgG), *J. Exp. Med.* 135:780-792.

Michl, J., Pieczonka, M. M., Unkeless, J. C., and Silverstein, S. C., 1979, Effects of immobilized immune complexes on Fc- and complement-receptor function in resident and thioglycollate-elicited mouse peritoneal macrophages, *J. Exp. Med.* 150:607-621.

Musson, R. A., and Becker, E. L., 1977, The role of an activatable esterase in immune-dependent phagocytosis by human neutrophils, *J. Immunol.* 118:1354-1365.

Newman, S. L., and Johnston, R. B., Jr., 1979, Role of binding through C3b and IgG in polymorphonuclear neutrophil function: Studies with trypsin-generated C3b, *J. Immunol.* 123:1839-1846.

Newman, S. L., Musson, R. A., and Henson, P. M., 1980, Development of functional complement receptors during *in vitro* maturation of human monocytes into macrophages, *J. Immunol.* 125:2236-2244.

Nogueira, N., Gordon, S., and Cohn, Z., 1977, *Trypanosoma cruzi:* Modification of macrophage function during infection, *J. Exp. Med.* 146:157-171.

Nowakowski, N., Edelson, P. J., and Bianco, C., 1980, Activation of C3H/HeJ macrophages by endotoxin, *J. Immunol.* 125:2189-2194.

Pearlman, D. S., Ward, P. A., and Becker, E. L., 1969, The requirement of serine esterase function in complement-dependent erythrophagocytosis, *J. Exp. Med.* 130:745-764.

Petty, H. R., Smith, L. M., Fearon, D. T., and McConnell, H. M., 1980, Lateral distribution and diffusion of the C3b receptor of complement, HLA antigens, and lipid probes in peripheral blood leukocytes, *Proc. Natl. Acad. Sci. USA* 77:6587-6591.

Quie, P. G., Messner, R. P., and Williams, R. C., 1968, Phagoctyosis in subacute bacterial endocarditis. Localization of the primary opsonic site to Fc fragment, *J. Exp. Med.* 128:553-570.

Rabellino, E. M., Ross, G. D., and Polley, M. J., 1978, Membrane receptors of mouse leukocytes. I. Two types of complement receptors for different regions of C3, *J. Immunol.* 120:879–885.

Rabinovitch, M., 1967, Studies on the immunoglobulins which stimulate the ingestion of glutaraldehyde-treated red cells attached to macrophages, *J. Immunol.* 99:1115–1120.

Rabinovitch, M., Manejias, R. E., and Nussenzweig, V., 1975, Selective phagocytic paralysis induced by immobilized immune complexes, *J. Exp. Med.* 142:827–838.

Ragsdale, C. G., and Arend, W. P., 1980, Loss of Fc receptor activity after culture of human monocytes on surface-bound immune complexes. Mediation by cyclic nucleotides, *J. Exp. Med.* 151:32–44.

Ross, G. D., and Rabellino, E. M., 1979, Identification of a neutrophil and monocyte complement receptor (CR3) that is distinct from lymphocyte CR1 and CR2 and specific for a site contained within C3bi, *Fed. Proc.* 38:1467.

Ross, G. D., Lambris, J. D., Cain, J. A., and Newman, S. L., 1982, Generation of three different fragments of bound C3 with purified Factor I or serum. I. Requirements for Factor H *vs* CR_1 cofactor activity, *J. Immunol.* 129:2051–2060.

Schlessinger, J., Barak, L. S., Yamada, K. M., Pastan, I., and Webb, W. W., 1977, Mobility and distribution of a cell surface glycoprotein and its interaction with other membrane components, *Proc. Natl. Acad. Sci. USA* 74:2909–2913.

Schmitt, M., Mussel, H.-H., and Dierich, M. P., 1981, Qualitative and quantitative assessment of C3-receptor reactivities on lymphoid and phagocytic cell, *J. Immunol.* 126:2042–2047.

Schreiber, A. D., and Frank, M. M., 1972, Role of antibody and complement in the immune clearance and destruction of erythrocytes. I. In vivo effects of IgG and IgM complement-fixing sites, *J. Clin. Invest.* 51:575–582.

Schreiber, R. D., Pangburn, M. K., Bjornson, A. B., Brothers, M. A., and Muller-Eberhard, H. J., 1982, The role of C3 fragments in endocytosis and extracellular cytotoxic reactions by polymorphonuclear leukocytes, *Clin. Immunol. Immunopathol.* 23:335–357.

Schribner, D. J., and Fahrney, D., 1976, Neutrophil receptors for IgG and complement: Their roles in the attachment and ingestion phases of phagocytosis, *J. Immunol.* 116: 892–897.

Shaw, D. R., and Griffin, F. M., Jr., 1981, Phagocytosis requires repeated triggering of macrophage phagocytic receptors during particle ingestion, *Nature* 289:409–411.

Shaw, D. R. and Griffin, F. M., Jr., 1982, Thioglycollate-elicited mouse peritoneal machages are less efficient than resident macrophages in anitbody-dependent cell-mediated cytolysis, *J. Immunol.* 128:433–440.

Shurin, S. B., and Stossel, T. P., 1978, Complement (C3)-activated phagocytosis by lung macrophages, *J. Immunol.* 120:1305–1312.

Stossel, T. P., 1977, Contractile proteins in phagocytosis: An example of cell surface-to-cytoplasm communication, *Fed. Proc.* 36:2181–2184.

Stossel, T. P., and Hartwig, J. H., 1975, Interactions between actin, myosin, and a new actin-binding protein of rabbit alveolar macrophages. Macrophage myosin Mg^{2+}-adenosine triphosphatase requires a cofactor for activation by actin, *J. Biol. Chem.* 250:5706–5712.

Stossel, T. P., and Hartwig, J. H., 1976, Interactions of actin, myosin, and a new actin-binding protein of rabbit pulmonary macrophages. II. Role in cytoplasmic movement and phagocytosis, *J. Cell Biol.* 68:602–619.

Stossel, T. P., Alper, C. A., and Rosen, F. S., 1973, Serum-dependent phagocytosis of paraffin oil emulsified with bacterial lipopolysaccharide, *J. Exp. Med.* 137:690–705.

Tabor, D. R., and Saluk, P. H., 1981, The functional heterogeneity of murine-resident macrophages to a chemotactic signal and induction of C3b-receptor-mediated ingestion, *Immunol. Lett.* 3:371–376.

Van Snick, J. L., and Masson, P. L., 1978, The effect of complement on the ingestion of soluble antigen–antibody complexes and IgM aggregates by mouse peritoneal macrophages, *J. Exp. Med.* **148**:903–914.

Ward, H. K., and Enders, J. F., 1933, An analysis of the opsonic and tropic action of normal and immune sera based on experiments with the pneumococcus, *J. Exp. Med.* **57**:527–547.

Wellek, B., Hahn, H. H., and Opferkuch, W., 1975, Evidence for macrophage C3d-receptor active in phagocytosis, *J. Immunol.* **114**:1643–1645.

Wright, S. D., and Silverstein, S. C., 1982, Tumor-promoting phorbol esters stimulate C3b and C3b′ receptor-mediated phagocytosis in cultured human monocytes, *J. Exp. Med.* **156**:1149–1164.

Wrigley, D. M., and Saluk, P. H., 1981, Induction of C3b-mediated phagocytosis in macrophages by distinct populations of lipopolysaccharide-stimulated lymphocytes, *Infect. Immun.* **34**:780–786.

Yin, H. L., and Stossel, T. P., 1979, Control of cytosplasmic actin gel-sol transformation by gelsolin, a calcium-dependent regulatory protein, *Nature* **281**:583–586.

Chapter 4

Mechanisms of Microbial Entry and Endocytosis by Mononuclear Phagocytes

David M. Mosser and Paul J. Edelson

Division of Infectious Diseases and Immunology
Department of Pediatrics
Cornell Medical College
New York, New York 10021

I. INTRODUCTION

Phagocytosis is a primary component of the host defense against invading microorganisms. Macrophages, working in concert with cellular and humoral immune mechanisms, are active both in clearing organisms from the blood or tissues and in killing and degrading them once they have been ingested. In this chapter we examine the molecular recognition mechanisms involved in the interaction of microbial pathogens with the macrophages and how these interactions contribute either to eradication of the invaders or to their ultimate ability to survive in the host. To that end, we review the physiology of phagocytosis, the classes of membrane binding sites currently recognized on the macrophage, the alteration of these binding sites upon macrophage activation, and the ways in which these sites may contribute to or oppose microbial infection. We are especially interested in examining the hypothesis that the initial plasma membrane binding mechanisms play important roles in determing the ultimate intracellular fate of phagocytized organisms.

II. ENDOCYTOSIS

Endocytosis is an energy-requiring process used by all nucleated cells to internalize both fluid and particles (Silverstein *et al.*, 1977). Although it is often considered a low affinity–high capacity mechanism as compared with specific plasma membrane molecular transport systems, the existence of a wide variety of plasma membrane binding sites gives the system great specificity, in addition to enhancing rates of specific uptake by as much as four orders of magnitude (Steinman and Cohn, 1972).

Specific plasma membrane binding sites are important not only for their effects on phagocytosis, but also for the roles they may play in the regulation of chemotaxis (Snyderman and Goetze, 1981), in the production of various toxic oxygen products including superoxide and peroxide (Johnston et al., 1980), in the secretion of various cellular products, and in the production of low-molecular-weight inflammatory molecules such as prostaglandins and slow-reacting substance of anaphylaxis (SRS-A), (Scott et al., 1980). Some binding sites may mediate attachment only; others may be important for phagolysosomal fusion, for acidification of the endocytic vesicle, or for intracellular degradation.

A very important principle for analyzing the physiology of membrane binding sites was developed by Nussenzweig and his colleagues (Ellenberger and Nussenzweig, 1977) in their studies of receptor synergy. These workers studied the effects of IgG and C3 on red blood cell uptake. When IgG alone was used to promote phagocytosis, about 60×10^3 molecules had to be deposited on each red cell. However, when only 1000 molecules of C3b were present on each erythrocyte, the required average number of IgG molecules fell to 2×10^3. In suspension, this effect was even more striking. Furthermore, certain nonspecific mediators of attachment, such as dextran, could synergistically enhance the ingestion of particles coated with suboptimal amounts of IgG.

These experiments demonstrate that an individual particle can bind to a phagocyte by more than one mechanism and that cooperation between binding mechanisms can cause qualitative and quantitative alterations in the physiology of phagocytosis. This concept is particularly important for analyzing the physiologic consequences of microbial binding, as these organisms have heterogeneous surfaces that may interact with many different receptor classes on the phagocyte. As in Nussenzweig's studies, the qualitative consequences of such mixed binding may reflect the physiology of minority receptors out of proportion to their overall quantitative contribution to attachment. In addition, it is not inconceivable that various serum proteins, such as fibronectin, α_2-macroglobulin or C-reactive protein, may function analogously to dextran, either enhancing phagocytosis or altering the ensuing intracellular consequences.

As our understanding of both attachment mechanisms and the physiology of macrophages themselves has grown, it has become clearer that in characterizing a particular macrophage binding interaction it is important to define the chemical characteristics of both the ligand and the receptor involved, as well as the physiologic consequences of binding. These consequences can be different, not only for different receptor classes, but also for a single receptor class in different cells. In addition to the tissue origins of a given cell population, the de-

Table I. Changes in Some Macrophage Plasma Membrane Components
with Activation[a]

Marker	Activating stimulus	Effect	Reference
FcRI	BCG	Increase	Ezekowitz et al. (1983)
FcRII	BCG	Decrease	Ezekowitz et al. (1983)
Particulate activators of ACP R	Thioglycollate	Decrease	(D. Mosser and P. Edelson, unpublished observation)
Mannose/fucose R	BCG	Decrease	Ezekowitz et al. (1983)
5'-nucleotidase	BCG	Decrease	Edelson and Cohn (1976)
	Thioglycollate	Decrease	
LAP	LK	Increase	Morahan et al. (1980)
	Thioglycollate	Increase	
Ia	BCG	Increase	Beller et al. (1980)
	LK	Increase	Steeg et al. (1980); Steinman et al. (1980)
F4/80	BCG	Decrease	Ezekowitz et al. (1983)

[a] *Abbreviations:* ACP R, receptor for activators of the alternative complement pathway; BCG, bacillus Calmette-Guérin; C3R, complement receptor 3; LAP, leucine aminopeptidase; LK, lymphokine.

velopmental stage, culture conditions, state of activation, and presence of various lymphocyte products may all affect the physiology of the ligand–receptor interactions and the consequences for the cells of those interactions.

The plasma membrane of activated macrophages differs from that of the resident or unstimulated cell in a variety of both chemical and physiologic ways. Certain enzyme activities, such as leucine aminopeptidase, may be increased 10-fold in the activated cell (Morahan et al., 1980), while other enzymes, e.g., 5'-nucleotidase, disappear completely with activation (Edelson and Cohn, 1976). The mouse macrophage antigen F4/80 is decreased in inflammatory cells (Ezekowitz et al., 1981), but Ia antigen expression is stimulated when cells are treated with lymphokine. These changes are further summarized in Table I.

III. MECHANISMS OF ATTACHMENT TO THE MACROPHAGE PLASMA MEMBRANE: IMMUNOLOGIC LIGANDS

A. Fc Receptors

As Rabinovitch (1967b) has pointed out, a characteristic of the "professional phagocyte" is its plasma membrane mechanism specialized to promote the uptake of IgG-coated particles. This opsonization by antibody depends on a membrane component that recognizes the crystallizable or Fc portion of the IgG molecule, known as the Fc receptor (see Unkeless et al., 1981).

Studies by Unkeless (1977) and Walker (1976) on mouse macrophages and by Grey and colleagues on human cells (Grey *et al.*, 1976) have distinguished at least two different classes of Fc receptors. In the mouse, the FcRI receptor recognizes IgG2a molecules, whether as monomers or in aggregates, and the FcRII recognizes mouse IgG1 and IgG2b molecules, either as aggregates or in combination with antigen. For human cells, similar receptor classes exist that distinguish IgG1 and IgG3 molecules. An additional general category of FcR that recognizes non-IgG, including IgM, IgA, and IgE, and a third type of IgG FcR have also been identified (see Unkeless for review, 1981), but to date, have not been directly implicated in microbial attachment.

The molecular weight of the mouse FcRII has been estimated at 50,000–70,000, depending on the isolation procedure and cell population used; it has been estimated that there are $\sim 10^5$ such receptors per macrophage (Unkeless, 1980). Although the FcRII, which is the classic trypsin-insensitive Fc receptor, clearly mediates phagocytosis, it is not clear whether this is also true for the FcRI.

While virtually all particles that elicit a humoral immune response are potential targets for F_c-mediated phagocytosis, the organisms of greatest relevance are the encapsulated, pyogenic bacteria, including *Streptococcus pneumoniae*, *Hemophilus influenzae*, *Neisseria meningitidis*, and the group A and group B streptococci. All possess a capsule, the makeup of which varies from species to species, but which seems generally to have an antiphagocytic effect. *Strep. pyogenes,* for example, has two antiphagocytic surface components, an acidic fibrillar M protein (McCarty, 1980) as well as a relatively nonimmunogenic hyaluronic acid envelope, while *H. influenza* contains a polysugarphosphate capsular polysaccharide.

In all cases, host resistance to these organisms is correlated with levels of antibody found in the serum, and clinical cure is invariably manifested in high antibody titers. Antimicrobial antibody can work in a number of ways: by blocking microbial attachment to epithelial surfaces, by neutralizing endotoxins, by fixing complement, and by opsonizing a particle for phagocytic uptake. *In vitro* studies of a number of encapsulated bacteria confirm the requirement for antibody in their phagocytosis by macrophages.

The FcII receptor is probably the primary microbial attachment mechanism of phagocytic cells. It mediates internalization in resident and activated cells and triggers the respiratory burst; organisms that interact with this receptor are eventually contained in secondary lysosomes. The activated macrophage expresses quantitatively more FcRI activity than does the resident (Ezekowitz *et al.*, 1983). An excellent example of the significance of the FcRII comes from studies with the protozoan parasite, *Toxoplasma gondii*. This organism can interact directly with the macrophage plasma membrane. This interaction does not trigger the production of toxic oxygen species (Wilson *et al.*, 1980), nor do the phagosomes formed fuse with lysosomes (Jones and Hirsch, 1972). However, when similar parasites are first coated with IgG antibody, the binding triggers the respiratory burst (Hirsch *et al.*, 1974) and, perhaps as a result of this microbicidal mechanism, the fusion of phagosomes with lysosomes is no longer inhibited.

B. Complement Receptors

Bianco and Nussesnzweig (1977) operationally define complement receptors as plasma membrane structures that are able to interact specifically with a fragment of C3 or C4. As in the work on the Fc receptors, it is now clear that there is more than one class of complement receptors. The complement 1 receptor (CR1) recognizes C3b and C4b and is present on lymphocytes, monocytes, and neutrophils (Fearon, 1980), while CR2 recognizes the cleavage product of C3bINA, C3d and is probably present only on B lymphocytes (Dobson *et al.*, 1981). Recent studies (Ross and Rabellino, 1979) indicate that a third class of complement receptors, CR3, recognizes C3bi and is present on human neutrophils, monocytes and erythrocytes. The human CR1 and CR2 have been isolated and are glycoproteins of MW 205,000 (Fearon, 1980) and 72,000 (Lambris *et al.*, 1981), respectively.

Particle binding to complement receptors has been characterized by the use of C3b-coated erythrocytes (EIgMC). The most striking characteristic of the interaction of particles with complement receptors is that in the resident cell they mediate binding only, whereas in the activated macrophage they mediate binding and internalization (Bianco *et al.*, 1975).

While the primary role of the Fc receptor is to mediate particle uptake and degradation, the role of the C3 receptor is probably to promote particle binding and to augment other phagocytic interactions. This augmentation was first described using red blood cells and IgG, as described earlier, but has now been observed in a number of other systems. Particulate activators of the alternative complement pathway, such as zymosan or rabbit red blood cells, can be taken up by the macrophage in a serum-free system (see Section VII). The addition of C3 augments this uptake, markedly increasing the number of particles per macrophage (Czop and Austen, 1980). The addition of C3 also augments bacterial uptake (Horowitz and Silverstein, 1980) presumably by synergizing with other macrophage mechanisms that interact with the bacteria.

The other role that complement-mediated phagocytosis plays is undoubtedly when the nonimmune host is first challenged with an organism. Lipopolysaccharide from bacterial cell walls has been shown to fix complement both by the alternative (Morrison and Kline, 1976) and by the classic (Loos *et al.*, 1974) pathway. Liang-Takasaki and colleagues have related the virulence of various strains of *Salmonella typhimurium* to differences in their O-antigen structure, with the less virulent being more likely to fix C3 by the alternative complement pathway and more likely to be taken up by phagocytic cells (Liang-Takasaki et al., 1983). Clas and Loos (1981) recently reported that *Salmonella minnesota* can bind C1 independent of antibody and activate the classic pathway. Horowitz (1980) has shown that unencapsulated *Escherichia coli* can be phagocytized by neutrophils upon addition of normal serum, while the encapsulated form requires bacteria-specific antibody.

Finally, two intriguing suggestions about complement and their receptors have recently been proposed: (1) the addition of fluid-phase C3 or C3b *in vitro* can enhance the intracellular killing of *Staphylococcus aureus* by macrophages (Leigh et al., 1982), and (2) CR1 may perform the additional function of causing local regulation of complement activation, thereby preventing red cell sensitization (Iida and Nussenzweig, 1983).

IV. MECHANISMS OF ATTACHMENT TO THE MACROPHAGE PLASMA MEMBRANE: NONIMMUNOLOGIC LIGANDS

The biology of the classic phagocytic receptors, Fc and C3, has been investigated extensively and their importance in immune clearance phenomena is well established. Nonimmune phagocytic events have also been observed for many years (see review in Rabinovitch, 1968), but only recently have the receptors involved in these types of recognition become the object of widespread interest (Stahl et al., 1978; Mosser and Roberts, 1982). These specific interactions, which work in the absence of a serum response, may be of particular importance, both in the nonimmune host and in defense against those intracellular pathogens that fail to provoke a strong humoral response. In addition, the phagocytosis of inert hydrophobic particles has often been used as a model for phagocytosis. From a cell-biologic perspective, these three types of recognition—immunologic, nonimmunologic, and nonspecific—are quite distinct, and the interaction of a particle by each may elicit three very different sets of cellular responses.

The nonimmunologic recognition mechanisms described to date have been identified largely on the basis of studies of the *in vitro* binding of exogenous ligands and include lectinlike molecules, a recognition site for fibronectin, and a site that recognizes particulate activators of the alternative complement pathway.

Stahl and colleagues (1980) have identified a lectinlike receptor on macrophages that recognizes glycoproteins terminating in either mannose or fucose. The receptor shows saturability and specificity and has been isolated (Townsend and Stahl, 1981). It has a MW of approximately 30,000 and there are ~75,000 copies per cell. Other lectinlike binding phenomena have been described by Weir (1977) and by Muchmore and Blaese (1980). Weir (1980) suggests that a lectinlike mechanism may be involved in the phagocytosis of *Salmonella typhimurium,* *Klebsiella aerogenes,* and *Corynebacterium parvum.*

Bevilacqua and colleagues (1981) have demonstrated the presence of specific binding sites for fibronectin on macrophages, a fragment of which may have phagocytosis-enhancing capacity (Czop and Fearon, 1982) and that may particularly augment the phagocytosis of activators of the alternative complement pathway. Other receptors have been described for various serum proteins that may also adhere to bacterial surfaces, including α_2-macroglobulin (Kaplan and Nielsen, 1979) and lactoferrin (Broxmeyer, 1979).

Models of phagocytosis have been developed using polystyrene latex (Singer *et al.,* 1967), aldehyde-treated red blood cells (Rabinovitch, 1967*a*), and a number of other inert hydrophobic particles (Brogan, 1966). The relevance of these models to the *in vivo* process of phagocytosis, however, is uncertain. In general, these particles do not seem to bind to specific plasma membrane sites, and they do not elicit many of the normal cellular consequences of the phagocytic process. They do not stimulate a migration of receptors in the plane of the membrane. They do not trigger a respiratory burst or the production of prostaglandins (perhaps for that reason), although they may act as stimuli for the release of neutral proteases. Unlike other receptor-mediated processes, their uptake is not affected by 2-deoxy-D-glucose. Thus, information developed from experiments with these particles may be quite misleading when extended to other systems. It should be noted, however, that although studies of nonspecific phagocytosis may have only limited applicability, some older work in which uptake was attributed to nonspecific mechanisms may actually have involved quite specific interactions with previously unrecognized nonimmunologic binding mechanisms.

The phagocytosis of zymosan is a good example to illustrate some of the points we have made about the complexity of analyzing uptake mechanisms. Zymosan interacts with the macrophage in the absence of serum by what was formerly thought to be a nonspecific mechanism. This interaction does not trigger the respiratory burst but does mediate particle internalization. Recent studies (Sung *et al.,* 1983), however, taking into account the mannose in the yeast cell wall, suggest that the particle interacts with the mannose receptor. When exposed to normal serum, zymosan fixes complement by the alternative pathway. Zymosan opsonized in this manner is used routinely as a phagocytic stimulus that triggers the respiratory burst. From a cell-biologic perspective, these phenomena need to be unraveled as follows: nonopsonized zymosan may interact with at least two macrophage recognition mechanisms: the mannose/ fucose receptor and the serum-independent receptor for particulate activators of the alternative complement pathway (Mosser and Roberts, 1982). Neither of these two recognition mechanisms triggers the respiratory burst, but at least one can mediate internalization. In the presence of serum the particle may also interact with a third mechanism, the macrophage C3b receptor, with consequent triggering of the respiratory burst. Thus, an analysis of the consequences of adding opsonized zymosan to macrophages involves not one, but probably three combined mechanisms. Even a single consequence, such as triggering the respiratory burst, cannot be ascribed simply to a single binding mechanism, but may reflect an interaction of the C3b receptor with one or more other plasma membrane binding sites.

V. PHYSIOLOGIC CONSEQUENCES OF ATTACHMENT

We have already alluded to a number of processes that may result from an initial ligand–plasma membrane interaction. These processes include internalization, triggering of the respiratory burst, formation of secondary lysosomes as a result of the fusion of phagosomes and primary lysosomes, acidification of these phagolysosomes, and stimulation of the secretion of various macrophage products.

A. Membrane Kinetics

One of the long-standing questions about the internalization process concerns the fate of internalized receptors (Edelson and Cohn, 1978). Are they internalized randomly or does selection occur? Once internalized, are they recycled intact or are they degraded? The most thoroughly investigated model for receptor recycling during phagocytosis is the low-density lipoprotein (LDL) receptor described by Brown et al. (1983). In this system, LDL receptors, which have bound ligand, form clusters with one another and are concentrated in clathrin-coated pits for internalization into coated vesicles. Once inside the endosome, the ligand is released from the receptor, probably because of a decrease in pH (Tyco and Maxfield, 1982); the receptor is reshuttled intact to the plasma membrane for reutilization. The entire process can be completed in 12 min (Brown et al., 1982). This helps explain the ability of cells to continue to internalize receptor-bound ligands for many hours without depleting their surface receptors, even in the presence of cyclohexamide (Brown and Goldstein, 1974), a process that can be inhibited with weak bases such as chloroquine (Anderson et al., 1982).

Using the latex phagocytosis model, a number of investigators have examined the issue of preferential internalization during phagocytosis. In comparing SDS banding patterns from phagosomes to those from normal plasma membrane, Werb and colleagues (Werb and Cohn, 1972) found a general agreement between the two. Mellman and co-workers used a lactoperoxidase (LPO)-glucose oxidase system to label macrophages intracellularly and found similar results (Muller et al., 1980). Latex phagosomes have been claimed to contain the plasma membrane ectoenzyme 5′-nucleotidase in amounts proportional to the area of membrane internalized (Werb and Cohn, 1972), although in un-published experiments performed by us, 5′-nucleotidase appeared to be prefer-entially included in latex phagosomes. The phagocytosis of IgG-coated red cells seemed to cause preferential internalization of Fc receptors. Some 70% were removed from the surface, while less than 20% of other unrelated antigens were internalized. Taken together with the pinocytosis data, this finding seems to be consonant with Steinman's suggestion that proteins are often free to diffuse into the area of plasma membrane to be internalized and that there can be some selective concentration of, for example, the ligand binding site (Steinman *et al.*, 1983).

Receptor recycling during phagocytosis has also been investigated by the Rockefeller group. Muller *et al.* (1983) demonstrated recycling by feeding cells latex beads with covalently coupled LPO, which lables phagolysosomes. After 10 min at 37°C, the label could be visualized in the plasma membrane. Not all phagocytized material was reshuttled back to the membrane intact, however. Fc receptor activity remained low for 24 hr after phagocytosis of sensitized red cells, suggesting that internalized receptors were degraded rather than recycled (Mellman *et al.*, 1983). Furthermore, the normal rate of receptor degradation increased from a half-life ($t_{1/2}$) of 10 hr to <2 hr in the presence of IgG-coated red cell ghosts.

The lessons for receptor recycling to be learned from this work are that there is a great deal of movement of plasma membrane constituents from one cellular compartment to another, that this movement can probably go in both direc-tions, and that certain migrant proteins can be segregated or sorted from more stationary membrane proteins by such mechanisms as clathrin pits.

B. Intracellular Consequences

Once a particle has been internalized, the usual sequence of events is for the vesicle containing the particle to fuse with cytoplasmic granules to form what are called secondary lysosomes, or phagolysosomes. These lysosomes typically contain some 40 or more hydrolytic enzymes (Werb, 1982), including phospha-tases, proteinases, aryl sulfatases, various lipases, and glycosidases. These en-zymes can degrade ingested proteins, carbohydrates, and lipids to components of ≤MW 200 that have a sufficiently small Stokes radius to escape the lysosome and enter the cytoplasm for reutilization (Ehrenreich and Cohn, 1969).

One way to define the series of events that normally follows phgocytosis is to examine some of the intracellular parasites that have evolved the means of circumventing one or more of the phagocytes' killing mechanisms. In this chapter, we present this work selectively, emphasizing the initial events that occur on the plasma membrane.

Armstrong and D'Arcy Hart (1971) were the first workers to observe that healthy strains of *Mycobacterium tuberculosis* could inhibit the fusion of primary lysosomes to incoming phagosomes. Jones and Hirsch (1972) noted that viable *Toxoplasma gondii* are also capable of inhibiting phagolysosome formation. Some observations have been made that may relate this inhibition of phagolysosome formation to initial events taking place at the plasma membrane. Damaged or nonviable organisms are unable to inhibit phagolysosome formation. In addition, antibody-coated organisms, although still viable, are no longer able to prevent fusion. Interestingly, in D'Arcy Hart's studies of *M. tuberculosis*, despite specific antimycobacterial antibody that was particularly effective at preventing the inhibition of lysosomal fusion, even the preimmune rabbit serum control had some effect in this direction.

These observations prompt several suggestions. First, nonviable, antibody-coated, or serum-treated organisms may present surfaces that have been substantially altered or masked in important ways, so that their interaction with the phagocyte involves a different range of membrane components. Second, antibody, and even nonimmune serum, may promote an additional interaction with plasma membrane receptors. Thus, in both instances, one may imagine that an alteration in the initial microbe–plasma membrane interaction is associated with a substantial change in the intracellular fate of the organism. It should be noted, however, that in D'Arcy Hart's studies, although antibody-coated organisms could not prevent phagosome-lysosome fusion, there was little effect on their intracellular survival, suggesting that other as yet unidentified virulence mechanisms were still intact.

A model system has been described in which direct ligand–membrane interaction governs the ability of phagosomes to fuse with lysosomes (Edelson and Cohn, 1974*a,b*). In these studies, the bivalent lectin concanavalin A (Con A) was allowed to bind to the macrophage plasma membrane, and the fate of the Con A pinosomes was examined. As judged both by ultrastructural studies in which Con A pinosomes were marked either by horseradish peroxidase or by colloidal gold, and by metabolic studies of the rate of degradation of either horseradish peroxidase or [^{125}I] bovine serum albumin (BSA), pinosomes in which Con A was closely associated with the inner surface of the enclosing membrane showed little or no evidence of fusion with either primary or secondary lysosomes. However, elution of the lectin from the pinosome membrane by the competitive sugar mannose led to the restoration of normal fusion. It is certainly possible that microorganisms might produce either surface components or secreted products that may act in a way similar to Con A, and that such materials might be additional virulence factors for these organisms.

Klebanoff (1967) first showed that one mechanism of microbial killing occurs at the level of the plasma membrane as a result of the production by the phagocyte of such reactive oxygen species as superoxide, hydroxyl radical, and peroxide. The biochemistry of this reaction and the localization of the membrane-bound oxidase have been reviewed (Babior *et al.*, 1981) and are not dealt with in detail here except to make several comments: (1) particles need not be internalized to trigger the respiratory burst, as binding with subsequent receptor crosslinking is often sufficient; and (2) not all plasma membrane recognition sites trigger the respiratory burst, e.g., it has recently been argued that the C3b receptor does not stimulate O_2^- production (R. Johnston, personal communication). Intracellular parasites may trigger the respiratory burst to very different degrees. *Toxoplasma gondii* enters the resident macrophage with little demonstrable release of O_2^-. Amastigotes of *Leishmania donovani* (Murray, 1982) provoke a very modest release, while promastigotes of the same species trigger a more substantial respiratory burst (Murray, 1981).

In addition to the Klebanoff mechanism, which is clearly highly regulated, there are other non-oxygen-dependent microbicidal mechanisms the bases of which are far less well understood, but whose existence may be surmised from several observations: (1) in chronic granulomatous disease, a genetic disorder characterized by a deficient respiratory burst (Quie *et al.*, 1967), phagocytic cells retain the ability to kill a number of pathogens, including *Chlamydia* (Yong *et al.*, 1982) and *Leishmania* (Murray and Cartelli, 1983); (2) cells treated with lymphokine preparations after particle internalization may still demonstrate enhanced microbistatic or microbicidal capacity; and (3) the addition of fluid phase IgG or C3b has been described as having an effect on the intracellular survival of phagocytized *Staphylococcus aureus* (Leigh *et al.*, 1982). Mechanisms by which microbial binding may affect these intracellular processes await definition.

VI. EXPERIMENTAL CONSIDERATIONS

Because of the size and complexity of many microbes that have been studied, one of the most difficult tasks is to identify the specific molecular mechanism(s) by which a phagocyte recognizes a given target. In other systems involving soluble ligands, a number of guidelines have been presented that have helped demonstrate that binding occurs by a specific mechanism that fulfills the criteria required to define a plasma membrane receptor. Despite some special artifactual pitfalls, many of these approaches can also be applied to the study of microbial recognition. Experimental approaches to the definition of the specificity of hormone–receptor interactions have been masterfully reviewed by Cuatrecasas and Hollenberg (1976); their conclusions may also be applied to studies of microbial binding.

A. Criteria for Receptor Identification: Soluble Ligands

While no single characteristic can, in itself, identify a specific receptor, several criteria taken together can distinguish specific receptors from relatively nonspecific interactions:

1. Binding *in vitro* parallels that seen *in vivo* and elicits the same biologic response both qualitatively and quantitatively. Structural analogues or antagonists can be used that either mimic or compete for the ligand, producing a quantitative change in these cellular responses.

2. Ligand binding to tissue samples should be confined to those cell types that are capable of responding to the ligand appropriately *in vivo*.

3. Receptor activity *in vitro* should be related to the levels of ligand that the cell might normally be exposed to *in vivo*. For example, affinity should be consistent with established tissue sensitivities, while saturability should generally be demonstrated at ligand levels that do not greatly exceed *in vivo* concentrations.

4. Receptor–ligand interactions are often sensitive to various membrane perturbations. Enzymatic or chemical treatment of the receptor may affect its ability to bind ligand and thereby alter the biologic response. A specific receptor will often be characterized by its sensitivity or resistance to a battery of enzymatic or other chemical treatments.

5. Mixing experiments can often be used to help demonstrate specificity. In general, the amount of labeled ligand not displaced from the membrane by high concentrations of unlabeled ligand can be considered nonspecifically bound, while the amount displaced by low concentrations of unlabeled ligand is thought to reflect specific binding. This assumes that the ligand interacts reversibly with the receptor, which is generally true in the case of soluble ligands, provided that endocytosis of the ligand has not occurred.

Having presented these guidelines, we hasten to point out that there are exceptions to virtually every one of these criteria, although most recognized receptors for soluble ligands do demonstrate many of these characteristics.

B. Microbial Recognition and the Criteria for Receptor Identification

In trying to apply the above-listed criteria to microbial recognition, one is immediately presented with obvious difficulties.

First, the biologic response to be elicited by interactions of the particle-bound ligand with its receptor is, in this case, phagocytosis. How, then, should we classify recognition mechanisms which mediate binding but not internalization, as for example the C3b receptor in resident cells? In some instances, particle binding is sufficient to elicit other cellular responses, including the release of lysosomal enzymes (Gordon *et al.*, 1974), alterations in glucose utilization (Johnston *et al.*, 1976), and changes in intracellular cyclic adenosine monophosphate (cAMP) levels, but the relevance of these changes is often uncertain.

Second, ligand binding should be confined to those cell types capable of responding to the ligand appropriately. However, various binding sites may occur on a range of cell types. Complement receptors, for example, are present on a number of nonphagocytic cells and have actually been isolated from red blood cells (Fearon, 1980). The intracellular parasite *Trypanosoma cruzi*, which binds to the macrophage and is taken up by an active phagocytic process (Nogueira and Cohn, 1976), can also enter cells that are not specialized for phagocytosis.

Third, it is impossible to determine receptor affinity and saturability from an examination of microbial binding. Kinetic analysis requires that the amount of ligand be expressed in molecular concentration, with each particle interacting independently of the others, and also that the interactions with a receptor be reversible. Neither is the case in studies of large particle attachment. Even if the concentration of particle-bound ligand were known, large particles would still interact with the phagocyte in a manner that would cause multiple simultaneous attachments, thereby potentially decreasing the off rate of the particle by several orders of magnitude.

Finally, in order to select for specific interactions *in vitro*, the macrophage monolayer is usually rinsed briskly after particle exposure. This procedure tends to underestimate the amount of specific particle attachment because of the high shear forces generated when washing large particles that are attached to the cell surface.

The most obvious way to circumvent these problems is to isolate the microbial ligand responsible for binding and to study it in solution. Although a difficult task, it should be the goal of most binding investigations because it is probably the only way to show definitively that the receptor is capable of recognizing a well-defined series of ligands and their analogues.

However, using intact microbes, certain experiments can still be performed that address the issue of receptor-mediated phagocytosis. Particle interaction with nonprofessional phagocytes is one of the most straightforward ways to distinguish between specific and nonspecific interactions. Generally a number of different cell types should be assessed for their ability to bind the particle to be tested and compared with macrophage binding in order to establish the background nonspecific binding for a given assay. Specific interactions can then be characterized with respect to binding requirements and temperature sensitivity. If the binding is sensitive to proteolysis, recovery experiments can be performed to look at the rapidity with which the binding mechanism is restored. A very rapid recovery may indicate either a large intracellular pool of the binding molecule or a relatively nonspecific particle interaction. It should be noted that increases in receptor synthesis upon exposure to proteases may perhaps necessitate the use of metabolic inhibitors if a prolonged treatment is necessary. Cell viability and phagocytic capacity must also be documented after proteolysis. Antibody or $F(ab')^2$ fragments to the phagocytic receptor, if available, should inhibit the target interaction, as has been shown for Fc (Unkeless, 1980) and C3 (Fearon,

1980) receptors. Lastly, a time course for binding should be examined. High time-zero binding is usually indicative of a nonspecific interaction.

While specific receptor–ligand interactions are always saturable, it is difficult to demonstrate saturability with large particles. Steric hindrances may prevent molecular saturation despite a plateau in particle binding. Isolated molecules, membrane fragments, and crude homogenates of the target have been used as competitive inhibitors of binding as an indirect means of indicating receptor saturability. For example, using membrane fragments from transformed cells, Adams and Marino (Marino *et al.*, 1981) have inhibited macrophage attachment to tumor cells. Mosser and Roberts (1982) have used crude homogenates of two developmental forms of *Trypanosoma brucei* to demonstrate that binding to macrophages by these organisms is both saturable and specific to a particular developmental stage. The results of studies with such crude preparations must be interpreted with some reservations unless adequate relevant controls are available, as there are a number of substances in unpurified preparations, including proteases or other hydrolases, that may impair attachment and thereby mimic specific competition. There are other caveats to these competition experiments as well. Often monovalent fluid-phase ligands will not compete successfully with large particles despite binding by the same mechanism. Lectin specificity, for example, is often determined by inhibition studies using monosaccharides to compete for the ligand. These studies must be interpreted cautiously, as the concentrations of simple sugars necessary to compete are usually so high that toxic effects may become a factor. Stahl and colleagues (Townsend and Stahl, 1981) have partially overcome this problem by complexing sugars to BSA. These multivalent ligands are able to inhibit the binding of a soluble ligand with 1000 times more efficiency than the unconjugated monosaccharide. Nevertheless, these glycoconjugates are still often unable to compete for large particulate targets, such as tumor cells or protozoa. Rabinovich and colleagues (Montovani *et al.*, 1972) have offered an interesting alternative to this type of competition study. Surfaces are prepared to which a given ligand has been bound with poly-L-lysine. Macrophages are then plated on this surface and, after specific receptors have been given time to modulate to the underside of the cell, the ligand to be tested is added to the monolayer. These elegant studies demonstrate that Fc binding could be competitively inhibited by immobilized immune complexes, illustrating the caveat that receptors that mediate binding but not internalization (e.g., the C3b receptor in the resident macrophage) do not migrate freely in the plane of the membrane and therefore cannot be modulated (Michl *et al.*, 1979).

C. Quantitation of Parasite Binding and Uptake

In order to identify and characterize fully the mechanism whereby a particle binds to the phagocyte, one must devise a method for quantitating macrophage-particle interactions. The technique of light microscopy, which can yield qualitative answers (i.e., the particle does or does not bind) is often inadequate in dissecting out quantitative changes in receptor expression or the relative interplay of more than one receptor. Unfortunately, alternative techniques are currently limited primarily to radiobinding assays or election microscopy.

1. Radiolabeling

While radiolabeling is probably the most accurate means of determining the number of macrophage-associated organisms, this technique is not without shortcomings. A system must be designed that will be sensitive enough to permit the quantitation required. The organisms must be relatively unaffected by the label and should demonstrate growth curves indistinguishable from unlabled cells. The label must be taken up into the organism and retained in a stable form that will remain associated with the parasite throughout the duration of the experiment. Controls examining the location of label in the parasite and the amount of time before free label is released into the supernatant must therefore be performed. Intracellular parasites present another problem. If the organisms are to be labeled before their addition to the macrophage, the label chosen must be one that will not be readily reused by the macrophage upon parasite degradation. If the organism is to be labeled within the phagocyte, the label must be one that is used preferentially by the parasite and generally needs to be one that will not be greatly diluted by macrophage intracellular pools. Whenever possible, the amount of label expressed as disintegrations per minute (DPM) should be translated back to the actual number of organisms in the system. In addition to making quantitation absolute rather than relative, this approach also leads one to examine the extent of label uptake on a per-parasite basis, which is often an excellent indicator of parasite viability.

A major problem with radiolabeling methods is that they may not distinguish particle binding from uptake. Often this can be overcome by performing experiments in the cold or in the presence of metabolic inhibitors in order to estimate binding alone. If uptake is being assayed, other manipulations may be required that will either release or lyse extracellular forms. Each of these techniques presents the investigator with problems. Attachment in the cold reduces molecular migration in the plane of the plasma membrane, preventing multiple attachment sites from interacting with the parasite. Upon washing, even specifically attached parasites may therefore be released because of the high shear forces generated. Extracellular lysis again introduces the problem of free label, which may then be reused by the macrophage. One successful approach has been the use of chelators to release peripherally attached ligand (Stahl et al., 1980). Macrophages so treated, however, may tend to round up and may even be released from the substratum during washing. If the total number of macrophage-associated particles is to be quantitated (both bound and internalized), visualization must be performed each time to determine whether differences between two experimental situations are the result of altered binding or merely attributable to a change in uptake.

2. Light Microscopy

A number of investigators have quantitated parasite uptake (Zenian, 1981) and survival (Berman *et al.*, 1979) by light microscopy. A common procedure is simply to expose macrophages to parasites under specified conditions and then to wash, fix, and stain the macrophages for counting. While light microscopic data can give a good indication of what is occurring qualitatively, they are at best semiquantitative. Values derived from light microscopy can be expressed as either the percentage of macrophages infected (percentage of infectivity) or as the average number of organisms per macrophage (average infectivity). The percentage of infectivity tends to be relatively insensitive because, even at high parasite:macrophage ratios, the percentage of macrophages infected with most organisms is generally below 70%. The average infectivity tends to be more difficult to quantitate. Care must be taken to use a dilution of parasites that does not result in such extensive uptake as to make counting impossible. For example, we have found it impossible to distinguish reliably by visual methods an average infectivity of four organisms from an average of seven using light microscopy alone. Finally, it should be noted that the use of Student's t test to compare percentages of infectivity is inappropriate.

3. Alternative Methods of Quantitation

The enormous sampling error inherent in electron microscopy makes it cumbersome to use for quantitative studies. In addition, it is a relatively expensive and time-consuming approach. However, it may be a valuable adjunctive tool, particularly when examining the anatomic correlates of binding or the intracellular consequences of particle uptake.

While biochemical assays are not commonly used to quantitate phagocytosis, a logical way to demonstrate the presence of a particle in phagocytic cells would be to measure the amount of some parasite-specific enzyme in detergent lysates of the monolayer. Analogous techniques have been employed to identify the presence of herpes virus in tissue culture (Laskin *et al.*, 1983).

Reculture techniques are often used after mild detergent lysis to free intracellular forms in order to assess intracellular viability. Care must be taken in these cases to ensure that (1) quantitation is done when parasites are in an active growth phase, (2) the technique for freeing the organisms does them no damage, and (3) the amount quantitated at a given time is a true reflection of the original inoculum.

D. Macrophage Culture

Perhaps the most important issue, after defining the stage of differentiation of the macrophage (activated versus resident), is to establish the conditions of macrophage culture, as particle recognition by macrophages may vary greatly depending on such variables as the time held in culture, the substrate on which they are induced to settle, or the medium in which they are cultivated.

Macrophages are usually separated from other cell types by adherence to a glass or plastic substrate. After 1–2 hr, mouse peritoneal macrophages, for example, are adherent enough to withstand washing procedures rigorous enough to remove many of the nonadherent cells, which are primarily lymphocytes. Adherent cells are then generally held in culture overnight, with subsequent washing, to yield a monolayer that is usually 95% macrophages. It is at this time that *in vitro* investigations are generally performed. The disadvantages of using cells at earlier times are that they are usually less well spread and may be interspersed with contaminating lymphocytes. Cells maintained in culture for longer periods of time begin to undergo a series of metabolic and functional changes that may affect their ability to undertake specific phagocytosis.

In the case of peripheral blood monocyte-derived macrophages, the optimal time for *in vitro* infectivity is less precise. Indeed, fresh monocytes put into culture are continually undergoing developmental changes, progressing from the monocyte to the macrophage and then to one of two further differentiative stages: either the activated cell or the multinucleated giant cell. The path of terminal differentiation depends in part on the substrate on which it is cultured.

The precise time at which cultured cells should be considered macrophages is unclear; cells from different individuals vary considerably in the rate of expression of various stage-specific markers. Circulating monocytes put into culture and allowed to adhere for 2 hr express no detectable 5'-nucleotidase activity (Johnson *et al.*, 1977), are weakly adherent, and are only poorly spread. By 24 hr the cells express considerable levels of 5'-nucleotidase activity, and by 48 hr enzyme expression reaches a plateau. Macrophage spreading continues to increase throughout the first week in culture. By day 2, however, the shape of these cells begins to resemble that of resident tissue macrophages. Lysozyme secretion also increases through the first week in culture. Peroxidase activity, which is present in circulating monocytes, rapidly declines over days 1 or 2 in culture (Johnson *et al.*, 1977). Lastly, freshly isolated monocytes cultured on glass for $\leqslant$24 hr bind but do not ingest via the C3 receptor. However, at 3–4 days in culture, these cells are fully able to ingest via the C3 receptor (Kaplan and Gaudernack, 1982).

From these comments it should be clear that, depending on the length of time that cells have been kept in culture, very different populations of mononuclear cells may be studied. Just where, along this continuum, these cells are most like resident tissue macrophages is open to debate, although by 48–72 hr in culture the cells are relatively well spread, lack peroxidase, demonstrate 5'-nucleotidase activity, and bind; for the most part, however, they do not ingest via their C3 receptors.

Even after 1 week in culture, however, cells continue to undergo developmental changes en route to what may be their terminal differentiative stage. Macrophage cultures maintained for 2–3 weeks on glass display numerous binucleated cells, and by day 20 multinucleated giant cells are the most prominent cell type. These cells demonstrate a 50% reduction in Fc receptor activity and are no longer able to internalize by their C3 receptor. In addition to decreased phagocytic activity, cells cultivated on glass for 3 weeks show a reduced respiratory burst upon stimulation, as well as an inability to limit the intracellular growth of *Leishmania donovani* (Murray and Cartelli, 1983). In contrast to cells plated on glass, macrophages cultured on nonwettable surfaces such as collagen or Teflon differentiate over the 3-week period to a stage resembling the classic activated macrophage, as described by Mackaness (1976). The two populations can be distinguished by stage-specific monoclonal antibodies (Kaplan and Gaudernack, 1982). These cells continue to internalize via the C3 receptor, have tumorcidal activity for K562 and MOLT4 tumor targets (Andreesen *et al.*, 1983), and are bactericidal for staphylococci (Van der Meer *et al.*, 1982).

VII. SERUM-INDEPENDENT RECOGNITION OF ACTIVATORS OF THE ALTERNATIVE COMPLEMENT PATHWAY

Czop *et al.* (1978*a,b*) and Fearon (1978) first recognized that certain particulate activators of the alternative complement pathway are also able to interact directly with the plasma membrane of cultivated human monocytes. These investigators showed that both zymosan and rabbit erythrocytes are readily bound to and internalized by monocytes in the absence of added serum proteins. Moreover, by reducing the quantity of sialic acid present on sheep erythrocytes, without changing the net charge on these cells, they were also able to convert the erythrocytes to activators of the alternative pathway, at the same time transforming them into particles able to interact directly with the monocyte membrane.

A similar concordance between alternative pathway activation and direct interaction with mouse peritoneal macrophages was noted by Mosser and Roberts (1982) in their work on the midgut form of the parasite *Trypanosoma brucei*. It is of interest that work by Nogueira and colleagues on the closely related hemoflagellate *T. cruzi* followed the same pattern (Nogueira and Cohn, 1976).

We have identified a similar mechanism in our study of the intracellular parasite *Leishmania* (Mosser and Edelson, 1983). The promastigote form of *Leishmania* is susceptible to complement-mediated lysis when placed in normal serum for 20 min

Table II. Characterization of Macrophage Binding Mechanisms

Characteristics	FcRII IgG-RBC	CRI IgM-C-RBC	FN[d]	M/F[e]	Leishmania	Trypanosoma brucei	Desialated or old RBC
					Test particles		
Protease sensitivity							
Trypsin	No	Yes	Yes	Yes	Yes	Yes	Yes
Chymotrypsin	No	No	No		Yes	Yes	—
Divalent cation requirement							
Calcium	No	No	No	Yes	Yes	Yes	Yes
Magnesium	No	No[a]	Yes	No	No	No	No
Binding at 4°C	Yes	No	No	Yes	No		No
Mannan inhibition	—	—	—	Yes	No	—	No
Mediation of internalization	Yes	No[b]	No	Yes[c]	Yes	Yes	Yes

[a]Mouse peritoneal macrophage requires magnesium.
[b]But does occur (yes) with activated macrophage.
[c]Pinocytosis of soluble ligands.
[d]Fibronectin-coated particle.
[e]Mannose/fucose-derivatized albumin.

at 37°C. The requirement for Mg^{2+}, but not Ca^{2+}, and the ability of both C2-deficient and C4-deficient sera to cause lysis indicate that lysis is accomplished by the alternative complement pathway. When these parasites are added to macrophages in the absence of serum they are readily bound and ingested.

The *Leishmania* model can be used to highlight some of the issues involved in analyzing the mechanisms of microbial recognition by macrophages. We have identified two distinct mechanisms for macrophage recognition of the promastigote form *in vitro*. One depends directly on the ability of the parasite to fix C3 with consequent interaction with macrophage C3b receptors. The other is the serum-independent recognition for particulate activators of complement. Binding in the presence of serum is therefore the sum of both binding mechanisms and reflects the physiology of both receptors. By taking advantage of different binding characteristics (cation requirements and chymotrypsin sensitivity), we can separate the two mechanisms and examine parasite binding to each.

This approach has allowed us to examine the different physiologic consequences of interaction with each of the two plasma membrane receptors and, in particular, to examine the relationship between binding to one or the other of these mechanisms and microbial virulence. We have already evidence that promastigotes of the more virulent strains, as well as the amastigote form, may have evolved ways by which they can more efficiently take advantage of the serum-independent uptake mechanism and avoid both serum and cellular lytic mechanisms.

There is currently no evidence that these various particles—erythrocytes, *Trypanosoma*, and *Leishmania*—all bind to the same plasma membrane receptor. The binding characteristics of various immune and nonimmune mechanisms are compiled in Table II, highlighting both their differences and similarities. While the similarities among the final three columns are striking, it is quite possible that similar but distinct uptake mechanisms, which have very different consequences for the fate of the particle, may be involved.

Finally, it is intriguing to consider what role these serum-independent receptors may play in the clearance of other sorts of particles and in the normal

Table III. Particulate Activators of Complement That Interact with the Macrophage in the Absence of Serum

Complement activator	Reference
Rabbit red blood cells	Czop and Austen (1980)
Neuriminidase-treated sheep red blood cells	Czop and Fearon (1982)
Zymosan	Sung *et al.* (1983)
Midgut forms of *Trypanosoma brucei*	Mosser and Roberts (1982)
Epimastigotes of *T. cruzi*	Nogueira and Cohn (1976)
Promastigotes of *Leishmania*	Mosser and Edelson (1983)

economy of the body. We now know of a number of particles that are able both to activate complement by the alternative pathway and to be taken up by the macrophage in the absence of serum (Table III). A long-standing question of great interest concerns the mechanism by which macrophages effect the clearance of old red blood cells. Several workers have proposed that such clearance is related to the progressive loss of sialic acid residues as the erythrocyte circulates (Danon *et al.*, 1971). Indeed, Aminoff *et al.* (1977) have reported that disialated rat erythrocytes are preferentially ingested by cultivated rat Kupffer cells or by splenic macrophages. It is not inconceivable, then, that this serum-independent receptor developed as a mechanism for erythrocyte destruction; it may have been seized upon by various microbial pathogens for their own purposes—a situation that we may describe as a parasite in sheep's (red blood cell's) clothing.

VIII. REFERENCES

Aminoff, D., Bruegge, W., Bell, W., Sarpolis, K., and Williams, R., 1977, Role of sialic acid in survival of erythrocytes in the circulation: Interaction of neuraminidase-treated and untreated erythrocytes with spleen and liver at the cellular level, *Proc. Natl. Acad. Sci. USA* 74:1521.

Anderson, R., Brown, M., Biesigel, U., and Goldstein, J., 1982, Surface distribution and recycling of the LDL receptor as visualized with anti-receptor antibodies, *J. Cell Biol.* 93:523.

Andreesen, R., Picht, J., and Lohr, G., 1983, Primary cultures of human blood-born macrophages grown on hydrophobic teflon membranes, *J. Immunol. Methods* 56:295.

Armstrong, J., and D'Arcy Hart, P., 1971, Response of cultured macrophages to *Mycobacterium tuberculosis* with observations on fusion of lysosomes with phagosomes, *J. Exp. Med.* 134:713.

Babior, G., Rosen, R., McMurrich, B., Peters, W., and Babior, B., 1981, Arrangement of the respiratory burst oxidase in the plasma membrane of the neutrophil, *J. Clin. Invest.* 67: 1724.

Beller, D., Keily, J., and Unanue, E., 1980, Regulation of macrophage populations. I. Preferential induction of Ia rich peritoneal exudates by immunologic stimuli, *J. Immunol.* 124:1426.

Berman, J., Dwyer, D., and Wyler, D., 1979, Multiplication of *Leishmania* in human macrophages *in vitro, Infect. Immun.* 26:375.

Bevilacqua, M., Amrani, D., Mosesson, M., and Bianco, C., 1981, Receptors for cold insoluble globulin (plasma fibronectin) on human monocytes, *J. Exp. Med.* 153:42.

Bianco, C., and Nussenzweig, V., 1977, Complement receptors, in: *Contemporary Topics in Molecular Immunology*, Vol. 6 (R. Porter and G. Ada, eds.), pp. 145–176, Plenum Press, New York.

Bianco, C., Griffin, F., and Silverstein, S., 1975, Studies of the macrophage complement receptor. Alteration of receptor function upon macrophage activation, *J. Exp. Med.* 141: 1278.

Brogan, T., 1966, Mechanisms of phagocytosis in human polymorphonuclear leukocytes, *Immunology* 10:137.

Brown, M., and Goldstein, J., 1974, Binding and degradation of low density lipoproteins by cultured human fibroblasts, *J. Biol. Chem.* 249:5153.

Brown, M. S., Anderson, R., Basu, R., and Goldstein, J., 1982, Recycling of cell surface receptors: Observations from the LDL receptor system, *Cold Spring Harbor Symp. Quant. Biol.* **46**:713.

Brown, M., Anderson, R., and Goldstein, J., 1983, Recycling receptors: The round-trip itinery of migrant membrance proteins, *Cell* **32**:663.

Broxmeyer, H., 1979, Lactoferrin acts on Ia-like antigen positive subpopulations of human monocytes to inhibit production of colony stimulatory activity *in vitro*, *J. Clin. Invest.* **64**:1717.

Clas, F., and Loos, M., 1981, Antibody-independent binding of the first component of complement (C1) and its subcomponent C1q to the S and R forms of *Salmonella minnesota*, *Infect. Immun.* **31**:1138.

Cuatrecasas, P., and Hollenberg, M., 1976, Membrane receptors and hormone action, *Adv. Protein Chem.* **30**:252.

Czop, J., and Austen, K. F., 1980, Functional discrimination by human monocytes between their C3b receptors and their recognition units for particulate activators of the alternative complement pathway, *J. Immunol.* **125**:124.

Czop, J., and Fearon, D., 1982, Augmentation of phagocytosis by a specific fibronectin fragment that links particulate activators to the fibronectin adherence receptor of human monocytes, *J. Immunol.* **129**:2678.

Czop, J., Fearon, D., and Austen, K. F., 1978a, Opsonin-independent phagocytosis of activators of the alternative complement pathway by human monocytes, *J. Immunol.* **120**:1132.

Czop, J., Fearon, D., and Austen, K. F., 1978b, Membrane sialic acid on target particles modulates their phagocytosis by a trypsin sensitive mechanism on human monocytes, *Proc. Natl. Acad. Sci. USA* **75**:3831.

Danon, D., Marikovsky, Y., and Skutelsky, M., 1971, The sequestration of old red cells and extruded erythroid nuclei, in: *Red Cell Structure and Metabolism* (E. Rabot, ed.), p. 23, Academic Press, New York.

Dobson, N., Lambris, J., and Ross, G., 1981, Characteristics of isolated erythrocyte complement receptor type I (CRI, C4b–C3b receptor) and CRI-specific antibodies, *J. Immunol.* **126**:693.

Edelson, P., and Cohn, Z., 1974a, Effects of concanavalin A on mouse peritoneal macrophages. I. Stimulation of endocytic activity and inhibition of phagolysosome formation, *J. Exp. Med.* **140**:1364.

Edelson, P., and Cohn, Z., 1974b, Effects of concanavalin A on mouse peritoneal macrophages. II. Metabolism of endocytized proteins and reversibility of the effects of mannose, *J. Exp. Med.* **140**:1387.

Edelson, P., and Cohn, Z., 1976, 5′-nucleotidase activity of mouse peritoneal macrophages. I. Synthesis and degradation in resident and inflammatory populations, *J. Exp. Med.* **144**:1581.

Edelson, P., and Cohn, Z., 1978, Endocytosis: Regulation of membrane interactions, *Cell Surface Rev.* **5**:387.

Ellenberger, A., and Nussenzweig, V., 1977, The role of membrane receptors for C3b and C3d in phagocytosis, *J. Exp. Med.* **145**:357.

Ehrenreich, B., and Cohn, Z., 1969, The fate of peptides pinocytosed by the macrophage *in vitro*, *J. Exp. Med.* **129**:227.

Ezekowitz, R., Austyn, J., Stahl, P., and Gordon, S., 1981, Surface properties of bacillus Calmette-Guérin-activated macrophages. Reduced expression of mannose specific endocytosis, Fc receptors, and antigen F4/80 accompanies induction of Ia, *J. Exp. Med.* **154**:60.

Ezekowitz, R., Bampton, M., and Gordon, S., 1983, Macrophage activation selectively enhances expression of Fc receptors for IgG2a, *J. Exp. Med.* 157:807.

Fearon, D., 1978, Regulation by membrane sialic acid of B1H-dependent decay-dissociation of amplification C3 convertase of the alternative complement pathway, *Proc. Natl. Acad. Sci. USA* 75:1971.

Fearon, D., 1980, Identification of the membrane glycoprotein that is the C3b receptor of the human erythrocyte, polymorphonuclear leukocyte, B lymphocyte, and monocyte, *J. Exp. Med.* 152:20.

Fearon, D., and Collins, L., 1983, Increased expression of C3b receptors on polymorphonuclear leukocytes induced by chemotactic factors and by purification procedures, *J. Immunol.* 130:370.

Gordon, S., Unkeless, J., and Cohn, Z., 1974, Induction of macrophage plasminogen activator by endotoxin stimulation and phagocytosis, *J. Exp. Med.* 140:995.

Grey, H., Anderson, C., Heusser, C., Borthistle, B., Von Eschen, K., and Chiller, J., 1976, Structural and functional heterogeneity of Fc receptors, *Cold Spring Harbor Symp. Quant. Biol.* 41:315.

Hirsch, J., Jones, T., and Len, L., 1974, Interactions *in vitro* between *Toxoplasma gondii* and mouse cells, in: *Ciba Foundation Symposium 25*, p. 205, Elsevier, Amsterdam.

Horowitz, M., 1980, The roles of the Fc and C3 receptors in the phagocytosis and killing of bacteria by human phagocytes, *J. Reticuloendothel. Soc.* 28:17s.

Horowitz, M., and Silverstein, S., 1980, Influence of the *Escherichia coli* capsule on complement fixation and on phagocytosis and killing by human phagocytes, *J. Clin. Invest.* 65:82.

Iida, K., and Nussenzweig, V., 1983, Functional properties of membrane-associated complement receptor CRI, *J. Immunol.* 130:1876.

Johnson, W., Mei, B., and Cohn, Z., 1977, The separation, long term cultivation, and maturation of the human monocyte, *J. Exp. Med.* 146:1613.

Johnston, R., Lehmeyer, J., and Guthrie, L., 1976, Generation of superoxide anion and chemiluminescence by human monocytes during phagocytosis and on contact with surface bound immunoglobulin G, *J. Exp. Med.* 143:1551.

Johnston, R., Chadwick, D., Pabst, M., 1980, Release of superoxide anion by macrophages: Effect of *in vivo* or *in vitro* priming, in: *Mononuclear Phagocytes* (R. van Furth, ed.), pp. 1143–1159, Martinus Nijhoff, The Hague.

Jones, T., and Hirsch, J., 1972, The interaction between *Toxoplasma gondii* and mammalian cells. II. The absence of lysosomal fusion with phagocytic vacuoles containing living parasites, *J. Exp. Med.* 136:1173.

Kaplan, G., and Gaudernack, G., 1982, *In vitro* differentiation of human monocytes. Differences in monocyte phenotypes induced by cultivation on glass or on collagen, *J. Exp. Med.* 156:1101.

Kaplan, J., and Nielsen, M., 1979, Analysis of macrophage surface receptors. I. Binding of α-macroglobulin protease complexes to rabbit alveolar macrophages, *J. Bio. Chem.* 254:7323.

Kelbanoff, S., 1967, A peroxidase-mediated antimicrobial system in leukocytes, *J. Clin. Invest.* 46:1078.

Lambris, J., Dobson, N., and Roos, G., 1981, Isolation of lymphocyte complement receptor type two (the C3d receptor) and preparation of receptor-specific antibody, *Proc. Natl. Acad. Sci. USA* 78:1828.

Laskin, O., Griffin, D., Gibson, W., and Leitman, P., 1983, Antiviral drugs as probes for identification of cells containing Herpesvirus thymidine kinase, *Clin. Res.* 31:294A.

Leigh, P., van den Barselaar, T., Daha, M., and van Furth, R., 1982, Stimulation of the intracellular killing of *Staphylococcus aureus* by monocytes: regulation by immunoglobulin G and complement components C3/C3b and B/Bb, *J. Immunol.* 129:332.

Liang-Takasaki, C., Grossman, N., and Leive, L., 1983, *Salmonella* activate complement differentially via the alternative pathway depending on the structure of their lipopolysaccharide O-antigen, *J. Immunol.* 130:1867.

Loos, M., Bitter-Suermann, D., and Dierich, M., 1974, Interaction of the first (C1), the second (C2) and or fourth (C4) component of complement with different preparations of bacterial lipopolysaccharides and with lipid A, *J. Immunol.* 112:935.

Mackaness, G., 1976, Cellular immunity, in: *Mononuclear Phagocytes* (R. van Furth, ed.), Blackwell, Oxford, p. 461.

Marino, P., Whisnant, C., and Adams, D., 1981, Binding of Bacillus Calmette-Guérin-activated macrophages to tumor targets. Selective inhibition by membrane preparations from homologous and heterologous neoplastic cells, *J. Exp. Med.* 154:77.

McCarty, M., 1980, Streptococci, in: *Microbiology* (B. Davis, R. Dulbecco, H. Eisen, and H. Ginsberg, eds.), p. 607, Harper & Row, Hagerstown, Md.

Mellman, I., Plutner, H., Steinman, R., Unkeless, J., and Cohn, Z., 1983, Internalization and degradation of macrophage Fc receptor during receptor-mediated phagocytosis, *J. Cell. Biol.* 96:887.

Michl, J., Pieczonka, M. Unkeless, J., and Silverstein, S., 1979, Effects of immobilized immune complexes on Fc and complement-receptor function in resident and thioglycollate-elicited mouse peritoneal macrophages, *J. Exp. Med.* 150:607.

Montovani, B., Rabinovitch, M., and Nussenzweig, V., 1972, Phagocytosis of immune complexes by macrophages. Different roles of the macrophage receptor sites for Complement (C3) and for Immunoglobulin (IgG), *J. Exp. Med.* 135:780.

Morahan, P., Edelson, P., and Gass, K., 1980, Changes in macrophage ectoenzymes associated with anti-tumor activity, *J. Immunol.* 125:1312.

Morrison, D., and Kline, K., 1976, Activation of the classical and properdin pathways of complement by bacterial lipopolysaccharides (LPS), *J. Immunol.* 118:362.

Mosser, D., and Roberts, J., 1982, *Trypanosoma brucei:* Recognition *in vitro* of two developmental forms by murine macrophages, *Exp. Parasitol.* 54:310.

Mosser, D., and Edelson, P., 1984, Activation of the alternative complement pathway by *Leishmania* promastigotes: parasite lysis and attachment to macrophages, *J. Immunol.* 132: (in press).

Muchmore, A., and Blaese, R., 1980, Evidence that monocyte-mediated cellular recognition phenomena are mediated by receptors with specificity for simple oligosaccharides, in: *Macrophage Regulation of Immunity* (E. Unanue and A. Rosenthal, eds.), p. 505, Academic Press, New York.

Muller, W., Steinman, R., and Cohn, Z., 1980, The membrane proteins of the vacular system. 1. Analysis by a novel method of intralysosomal iodination, *J. Cell Biol.* 86:292.

Muller, W., Steinman, R., and Cohn, Z., 1983, Membrane proteins of the vacuolar system. III. Further studies on the composition and recycling of endocytic vacuole membrane in cultured macrophages, *J. Cell. Biol.* 96:29.

Murray, H., 1981, Susceptibility of *Leishmania* to oxygen intermediates and killing by normal macrophages, *J. Exp. Med.* 153:1302.

Murray, H., 1982, Cell mediated immune response in experimental visceral leishmaniasis. II. Oxygen dependent killing of intracellular *Leishmania donovani* amastogotes, *J. Immunol.* 129:351.

Murray, H., and Cartelli, D., 1983, Killing of intracellular *Leishmania donovani* by human mononuclear phagocytes, evidence for oxygen-dependent and -independent leishmanicidal activity, *J. Clin. Invest.* 72:32.

Nogueira, N., and Cohn, Z., 1976, *Trypanosoma cruzi:* Mechanism of entry and intracellular fate in mammalian cells, *J. Exp. Med.* 143:1402.

Quie, P., White, J., Holmes, B., and Good, R., 1967, *In vitro* bacteriocidal capacity of hyman polymorphonuclear leukocytes: Diminished activity in chronic granulomatous disease of childhood, *J. Clin. Invest.* 46:668.

Rabinovitch, M., 1967a, The dissociation of the attachment and ingestion phases of phagocytosis by macrophages, *Exp. Cell Res.* 46:19.

Rabinovitch, M., 1967b, "Non-professional" and "professional" phagocytosis: Particle uptake by L cells and by macrophages, *J. Cell Biol.* 35:108A.

Rabinovitch, M., 1968, Phagocytosis: The engulfment stage, *Semin. Hematol.* 5:134.

Ross, G., and Rabellino, E., 1979, Identification of a neutrophil and monocyte complement receptor (CR3) that is distinct from lymphocyte CR_1 and CR_2 and specific for a site contained within C3bi, *Fed. Proc.* 38:1467.

Scott, W., Zrike, J., Hamill, A., Kempe, J., and Cohn, Z., 1980, Regulation of arachidonic acid metabolites in macrophages, *J. Exp. Med.* 152:324.

Silverstein, S., Steinman, R., and Cohn, Z., 1977, Endocytosis, *Annu. Rev. Biochem.* 46:669.

Singer, J., Lavie, S., Aldesberg, L., Ende, E., Hoenig, E., and Tchorsh, Y., 1967, The use of radioiodinated latex particles for *in vivo* studies of phagocytosis, in: *The Reticuloendothelial System and Atherosclerosis* (N. Diluzia and R. Paoletti, eds.), p. 17, Plenum Press, New York.

Snyderman, R., and Goetze, E., 1981, Molecular and cellular mechanisms of leukocyte chemotaxis, *Science* 213:830.

Stahl, P., Rodman, J., Miller, M., and Schlesinger, P., 1978, Evidence for receptor-mediated binding of glycoproteins, glycoconjugates, and lysosomal glycosidases by alveolar macrophages, *Proc. Natl. Acad. Sci. USA* 75:1399.

Stahl, P., Schlesinger, P., Sigardson, E., Rodman, J., and Lee, Y., 1980, Receptor-mediated pinocytosis of mannose glycoconjugates by macrophages: Characterization and evidence for receptor recycling, *Cell* 19:207.

Steeg, P., Moore, P., and Oppenheim, J., 1980, Regulation of murine macrophage Ia-antigen expression by products of activated spleen cells, *J. Exp. Med.* 152:1734.

Steinman, R., and Cohn, Z., 1972, The interaction of particulate horseradish peroxidase (HRP)–anti-HRP immune complexes with mouse peritoneal macrophages *in vitro*, *J. Cell Biol.* 55:616.

Steinman, R., Nogueira, N., Witmer, M., Tydings, J., and Mellman, I., 1980, Lymphokine enhances the expression and synthesis of Ia antigens on cultured mouse peritoneal macrophages, *J. Exp. Med.* 152:1248.

Steinman, R., Mellman, I., Muller, W., and Cohn, Z., 1983, Endocytosis and the recycling of plasma membrane, *J. Cell Biol.* 96:1.

Sung, S., Nelson, R., and Silverstein, S., 1983, Yeast mannons inhibit binding and phagocytosis of zymosan by mouse peritoneal macrophages, *J. Cell Biol.* 96:160.

Townsend, R., and Stahl, P., 1981, Isolation and characterization of a mannose/N-acetylglucosamine/fucose binding protein from rat liver, *Biochem. J.* 194:209.

Tyco, B., and Maxfield, F., 1982, Rapid acidification of endocytic vesicles containing α_2-macroglobulin, *Cell* 28:643.

Unkeless, J., 1977, The presence of two Fc receptors on mouse macrophages: Evidence from a variant cell line and differential trypsin sensitivity, *J. Exp. Med.* 145:931.

Unkeless, J., 1980, Mouse macrophage Fc receptors, *J. Retriculoendothel. Soc.* 28:11-S.

Unkeless, J., Fleit, H., and Mellman, I., 1981, Structural aspects and heterogeneity of immunoglobulin Fc receptors, *Adv. Immunol.* 31:247.

Van der Meer, J., van de Geuel, J., van oud Albas, Kramps, J., van Zwet, T., Leijh, P., and van Furth, R., 1982, Characteristics of human monocytes cultured in the teflon culture bag, *Immunology* 47:617.

Walker, W., 1976, Separate Fc receptors for immunoglobulins IgG2a and IgG2b on an established cell line of mouse macrophages, *J. Immunol.* 116:911.

Weir, D., 1980, Surface carbohydrates and lectins in cellular recognition, *Immunol. Today* 1:45.

Weir, D., and Ogmundsdottir, H., 1977, Non-specific recognition mechanisms by mononuclear phagocytes, *Clin. Exp. Immunol.* 30:323.

Werb, Z., 1982, Phagocytic cells: Chemotaxis and effector functions of macrophages and granulocytes. I. Macrophages, in: *Basic and Clinical Immunology* (D. Stites, J. Stobo, H. Fudenberg, and J. Wells, eds.), p. 109, Lange, Los Altos, Calif.

Werb, Z., and Cohn, Z., 1972, Plasma membrane synthesis in macrophages following phagocytosis of polystyrene latex, *J. Biol. Chem.* 247:2439.

Wilson, C., Tsai, V., and Remington, J., 1980, Failure to trigger the oxidative metabolic burst by normal macrophages. Possible mechanism for survival of intracellular pathogens, *J. Exp. Med.* 151:328.

Yong, E., Klebanoff, S., and Kuo, C., 1982, Toxic effect of human polymorphonuclear leukocytes on *Chlamydia trachomatis, Infect. Immun.* 37:422.

Zenian, A., 1981, *Leishmania tropica:* Biochemical aspects of promastigotes attachment to macrophages *in vitro, Exp. Parasitol.* 51:175.

Chapter 5

Macrophage Activation: Enhanced Oxidative and Antiprotozoal Activity

Henry W. Murray

Department of Medicine
The Cornell University Medical College
New York, New York 10021

I. INTRODUCTION

Although only one of the array of secretory properties displayed by activated macrophages, the capacity to generate increased amounts of reactive oxygen intermediates is a consistent biochemical marker of activation with clearly relevant biologic effects (Nathan *et al.*, 1980). Thus, the ability to secrete high levels of superoxide anion (O_2^-) and hydrogen peroxide (H_2O_2) appears to contribute in an important way to the capacity to exert enhanced antimicrobial activity—a particularly key expression of the activated state (North, 1979). Current evidence suggests, for example, that (1) those mononuclear phagocytes that are capable of killing intracellular pathogens such as protozoa, fungi, and mycobacteria depend in large measure on this oxygen-dependent mechanism (Haidaris and Bonventre, 1982; Murray *et al.*, 1979; Murray and Cohn, 1980; Murray, 1981*a*, 1982*b*; Nathan *et al.*, 1979; Sasada and Johnston, 1980; Walker and Lowrie, 1981), and (2) secreted O_2^- and H_2O_2 may also be active against certain extracellular microbial targets too large to be ingested (Diamond *et al.*, 1982; Kazura *et al.*, 1981).

This chapter reviews a series of studies that have examined the capacity of the activated macrophage to display both enhanced oxidative and antimicrobial activity and summarizes evidence suggesting that these two expressions of the complex activation process are indeed closely linked. The activity of macrophages against two clinically relevant intracellular protozoal targets, *Toxoplasma gondii* and *Leishmania donovani*, is discussed in detail. Interaction of the activated macrophage with other intracellular pathogens including *Trypanosoma cruzi, Leishmania tropica,* and *Rickettsia* is considered in Chapters 6 and 8 in this volume. Although most of the information presented in this chapter has been derived from the mouse peritoneal macrophage model, observations now being made with human macrophages also support the bulk of these findings.

II. INDUCTION OF MACROPHAGE ANTIPROTOZOAL ACTIVITY: MODELS USING *T. GONDII* AND *L. DONOVANI* AS INTRACELLULAR MICROBIAL TARGETS

Unstimulated resident peritoneal macrophages from normal mice kill few if any ingested *T. gondii* and readily permit this protozoan to replicate intracellularly (Table I) (Murray *et al.*, 1979). The number of toxoplasmas per vacuole increases from one to four to five 18 hr after infection, (Fig. 1A). Inflammatory macrophages, obtained from mice after intraperitoneal (IP) injection of sterile agents such as proteose peptone, thioglycollate, or heart infusion broth (HIB), behave identically to resident cells and support *Toxoplasma* replication (Murray and Cohn, 1980). Macrophages from chronically infected *T. gondii*-immune mice also exert little toxoplasmacidal activity, but do effectively inhibit the organism's intracellular replication. In contrast, when immune mice are boosted IP with specific antigen (e.g., heat-killed toxoplasmas) 3 days before harvest, their peritoneal macrophages appear to achieve a considerably heightened state of activation and display striking microbicidal effects (Murray *et al.*, 1979). These immune-boosted (IB) macrophages kill >80% of ingested *T. gondii* and prevent replication of the few surviving parasites. Cells from mice immunized with viable or dead bacillus Calmette–Guérin (BCG) or killed *Corynebacterium parvum,* and boosted IP with homologous antigen, also exert toxoplasmacidal activity (Murray and Cohn, 1980), indicating that although macrophage activation requires antigen specificity, its expression is nonspecific.

Presumably simulating *in vivo* events in the immune animal, normal resident peritoneal macrophages can also be activated to achieve antitoxoplasma activity by exposure to immune lymphocytes plus antigen or to soluble products, i.e., lymphokines (LKs) secreted by sensitized T lymphocytes. Such activation can be induced *in vivo* by using the peritoneal cavity as a culture compartment, i.e., injecting normal mice IP with immune lymphocytes plus heat-killed toxoplasmas (HKTs) and harvesting the peritoneal macrophages 3 days later, or *in vitro* by co-cultivating normal resident cells for 18 hr with supernates of mitogen- or antigen-stimulated lymphocyte cultures (Murray and Cohn, 1980). In the latter instance, however, *in vitro* exposure to lymphokine alone is not sufficient to activate normal macrophages to display antitoxoplasma activity, and a synergistically acting inflammatory agent (e.g., proteose peptone, thioglycollate, HIB) is required to induce toxoplasmastatic activity successfully (Murray and Cohn,

Table I. Macrophage Antiprotozoal and Oxidative Activities[a,b]

Macrophage population and pretreatment	Antiprotozoal activity against:			H_2O_2 release[c] (PMA)	Percentage of cells NBT-positive[d]				
	Toxoplasma gondii	LDP	LDA		PMA	Zymosan	Toxoplasma gondii	LDP	LDA
Normal resident	None	Cidal	None	86	97	87	16	82	49
Inflammatory	None	Cidal	None	70–95	94	89	19	87	–
In vitro-activated									
+LK × 24 hr	None	Cidal	Cidal	140	92	91	20	88	91
+LK + HIB × 24 hr	Static	–	–	125	90	78	62	–	–
+LK × 72 hr	None	–	Cidal	450	94	86	67	–	–
In vivo-activated Toxoplasma gondii, BCG, Corynebacterium parvum	Static	–	None	325–580	–	–	74	–	–
+IP antigen boost	Cidal	Cidal	None	950	94	88	84	92	20–29
J774G8 cells	None	None	None	14	3	8	13	10	7
+LK × 24 hr	None	Cidal	–	35	34	60	–	65	–

[a]From Murray et al., 1979; Murray and Cohn, 1980; Murray, 1981a, b, 1982a. Macrophages were first cultivated for 24 hr before infection or assay.

[b]Abbreviations: BCG, bacillus Calmette-Guérin; HIB, heart infusion broth; IP, intraperitoneal; LDA, Leishmania donovani amastigote; LDP, Leishmania donovani promastigote; LK, lymphokine; NBT, nitroblue tetrazolium; PMA, phorbol myristate acetate.

[c]Nanomoles of H_2O_2/mg protein per 90 min after triggering with PMA (100 ng/ml).

[d]Percentage of macrophages showing clumped formazan precipitation after PMA stimulation or formazan precipitation at site of parasite or particle ingestion.

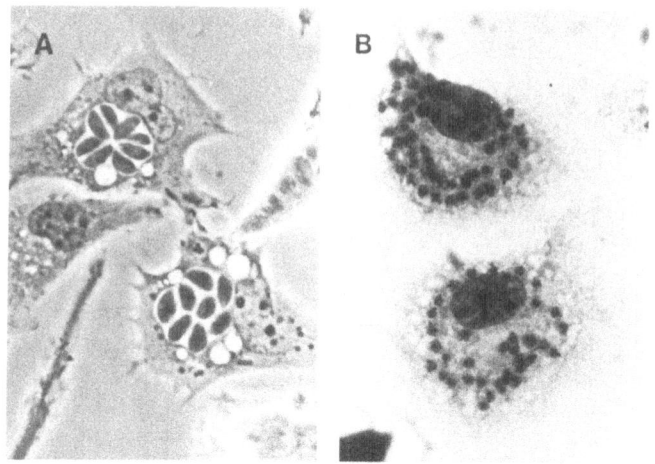

Figure 1. (A) Normal resident peritoneal macropliages supporting *Toxoplasma gondii* replication 18 hr after infection and (B) remaining heavily parasitized with L. *donovani*
amastigotes (LDAs) 72 hr after infection (×600). (Reproduced with permission from
Murray et al., 1979; Murray et al., 1982.)

1980). These inflammatory agents, which by themselves do not impart antimicrobial effects, presumably represent a rich source of endotoxin. Adding lymphokine (with or without HIB) to normal macrophages after infection only does
not confer any antitoxoplasma activity in this model.

  In contrast to their lack of effect against toxoplasmas, resident peritoneal
macrophages are able to achieve swift killing of ingested L. donovani promastigotes (LDPs). Once intracellular, 50% of these flagellates are digested within
6 hr, 80–90% are eradicated by 18 hr, and surviving LDPs fail to replicate in
these unstimulated cells (Murray, 1981a). Thus, macrophage activation is not required for effective killing of LDPs. The conditions under which the intracellular
(amastigote) form of L. donovani (LDA) is killed, however, are quite different.
Promastigotes transform to amastigotes within phagolysosomes, and the latter is
the form of the protozoan that causes persistent tissue infection. Like T. gondii
(but in direct contrast to LDPs), LDAs readily parasitize normal resident cells
(Fig. 1B). An 18-hr pretreatment with antigen- or mitogen-stimulated lymphokines (no additional inflammatory agent required), however, induces normal
macrophages to destroy 90% of ingested LDAs by 72 hr. The addition of
lymphokine after infection only does not result in amastigocidal activity for
normal resident cells (Murray, 1982a). Surprisingly, in vivo-activated macrophages from mice immunized and boosted with T. gondii, bacillus Calmette-
Guérin (BCG), or Corynebacterium pawum, cells that promptly kill toxoplasmas

and other intracellular protozoa [e.g., *T. cruzi* (Nathan *et al.*, 1979)], fail to exert any activity against LDAs and in this regard behave like normal resident macrophages.

The observations derived from these models of intracellular infection indicate, then, that with the exception of LDP, macrophage activation achieved either *in vivo* by systemic microbial stimuli with consequent T-cell stimulation or *in vitro* by pretreatment with soluble T-lymphocyte products is required in order to kill both *T. gondii* and LDAs. As judged by the results of other studies, intracellular activity against the trypomastigote form of *T. cruzi* and *L. tropica* amastigotes also requires the activated state (Nacy *et al.*, 1981; Nathan *et al.*, 1979). However, as indicated earlier and in Table I, macrophages activated by lymphokine alone can kill LDAs but not toxoplasmas, cells stimulated by lymphokine plus HIB inhibit but do not kill toxoplasmas, and *in vivo*-activated macrophages can kill *T. gondii* but not LDA. These apparent paradoxes are addressed in Section V in the discussion of the differential capacity of these organisms to trigger the oxygen-dependent respiratory burst mechanism of the activated macrophage.

III. CAPACITY OF NORMAL AND ACTIVATED MACROPHAGES TO GENERATE TOXIC OXYGEN INTERMEDIATES

Once triggered by any one of a number of plasma membrane-perturbing stimuli including soluble agents, inert or viable ingestible particles, or non-phagocytosable surfaces, the macrophage responds with a coordinated series of biochemical events initiated by oxygen consumption and oxygen reduction (reviewed by Badwey and Karnovsky, 1980; Klebanoff, 1980). Although the nature, cofactor(s), and cellular location of the macrophage's oxidase system are unknown, within moments of being triggered this mechanism's activity results in the partial reduction of molecular oxygen and the formation of O_2^-. Spontaneously or in the presence of superoxide dismutase (SOD), O_2^- dismutes to form H_2O_2 and, in the presence of trace metal ions (e.g., iron), O_2^- and H_2O_2 appear to interact to form hydroxyl radical ($OH^{\cdot}$), a highly reactive species (Rosen and Klebanoff, 1979). Singlet oxygen ($^1O_2$) may also be another product of this reaction. As judged by the reduction of ferricytochrome *c* (Johnston *et al.*, 1978) and the oxidation of scopoletin (Nathan and Root, 1977), assays that detect the extracellular release of O_2^- and H_2O_2, respectively, the macrophage's respiratory burst probably shuts down after 60-90 min irrespective of the triggering stimulus. In the peripheral blood monocyte, granular myeloperoxidase (MPO) is abundant, and in the presence of H_2O_2 and an oxidizable halide, the H_2O_2-MPO-halide sytem can appreciably augment the toxicity of H_2O_2 and enhance the phagocyte's antimicrobial activity (Klebanoff, 1980; Locksley *et al.*, 1982). The mature macrophage, however, is devoid of granular MPO, making it unlikely that the H_2O_2-MPO–halide reaction plays any meaningful role in the macrophage's antiprotozoal activity.

Resident peritoneal macrophages from normal mice certainly release O_2^- and H_2O_2 once effectively triggered by soluble agents such as phorbol myristate acetate (PMA) or the ingestion of particles including opsonized zymosan, but the amounts of O_2^- and H_2O_2 generated are quite small (Johnston *et al.*, 1978; Murray *et al.*, 1979; Nathan and Root, 1977; Nathan *et al.*, 1979). This is especially apparent when one compares the resident macrophage with cells activated *in vivo* (Table I). Macrophages elicited by proteose peptone, thioglycollate, and HIB also display relatively little H_2O_2 secretion (Murray and Cohn, 1980); therefore, in terms of both antiprotozoal and oxidative activity, these cells do not differ appreciably from normal macrophages.

Figure 2, however, clearly illustrates the enhanced rate and magnitude of the *in vivo*-activated macrophage's H_2O_2-releasing capacity induced by boosting *T. gondii*-immune mice IP with HKTs 3 days before cell harvest (Murray *et al.*, 1979). This augmented oxidative capacity is antigen specific, since boosting *T. gondii*-immune animals with heat-killed BCG, PPD, or other unrelated microbial proteins does not appreciably increase macrophage O_2^- or H_2O_2 generation.

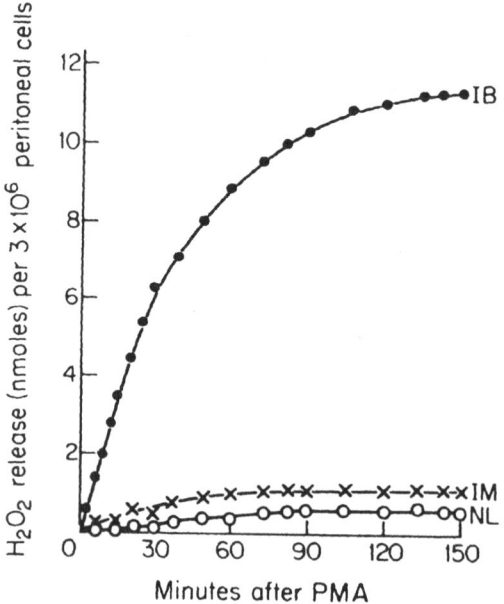

Figure 2. H_2O_2 release by normal resident (NL), *Toxoplasma gondii* immune (IM), and *T. gondii* immune-boosted (IB) macrophages after triggering with 100 ng/ml of phorbol myristate acetate (PMA). (Reproduced with permission from Murray *et al.*, 1979.)

Peritoneal macrophages from mice immunized with *T. cruzi*, BCG, or *C. parvum* also secrete increased amounts of O_2^- and H_2O_2 once effectively triggered, and this activity can also be increased by IP boosting with specific antigen (Badway and Karnovsky, 1980; Johnston *et al.*, 1978; Murray and Cohn, 1980; Nathan *et al.*, 1979). The results of this boosting technique presumably reflect the contribution of enhanced antigen-stimulated lymphokine secretion in the peritoneal cavity.

Not surprisingly, *in vitro* exposure to mitogen- and antigen-prepared lymphokines also induces normal resident macrophages to display enhanced oxidative activity (Murray and Cohn, 1980; Nathan *et al.*, 1979). In terms of the extracellular release of O_2^- and H_2O_2, however, the effects of *in vitro* lymphokine treatment on macrophage oxidative activity are not readily apparent until after 48-72 hr of stimulation, at which time O_2^- and H_2O_2 generation are increased by three- to fourfold (Table I) (Murray and Cohn, 1980). Since 24 hr (or less) of LK exposure can induce normal cells to inhibit or kill the intracellular protozoa (Murray and Cohn, 1980; Murray, 1982*a*; Nathan *et al.*, 1979), the discrepancy between the duration of stimulation required for the expression of enhanced oxidative activity versus enhanced antiprotozoal activity has raised some doubt as to the relationship between the two. However, this most likely reflects the failure of the extracellular assays for O_2^- and H_2O_2 detection to assess adequately what are primarily intracellular biochemical events. If a cell other than a peritoneal macrophage is used as the target for activation such as the oxidatively deficient J774 macrophage-like cell (Murray, 1981*b*), overnight treatment with lymphokine enhances both intracellular qualitative nitroblue tetrazolium (NBT) reduction (a O_2^--dependent reaction) and extracellular O_2^- and H_2O_2 release (Table I). In addition, using the qualitative NBT assay and toxoplasmas as the phagocytic trigger, enhanced oxidative activity induced by 18 hr of treatment with LK plus HIB can readily be demonstrated (Table I). Why 48-72 hr of LK exposure is required before enhanced extracellular O_2^- and H_2O_2 release by peritoneal macrophages is clearly detectable is unknown, but it may reflect more efficient intracellular catabolic mechanisms (Murray *et al.*, 1980), resulting in less readily detectable oxidative activity using assays that measure extracellular release.

Nevertheless, if one compares the data in Table I, it is clear that *in vivo*- and *in vitro*-activated macrophages are capable of generating enhanced amounts of oxygen intermediates and that these are the cells that display effective activity against *T. gondii* and LDA.

IV. EVIDENCE THAT OXYGEN INTERMEDIATES CONTRIBUTE TO THE MACROPHAGE'S ANTIPROTOZOAL ACTIVITY

In addition to the close correlation between the capacity of macrophages to release oxygen intermediates and their ability to display intracellular antipro-

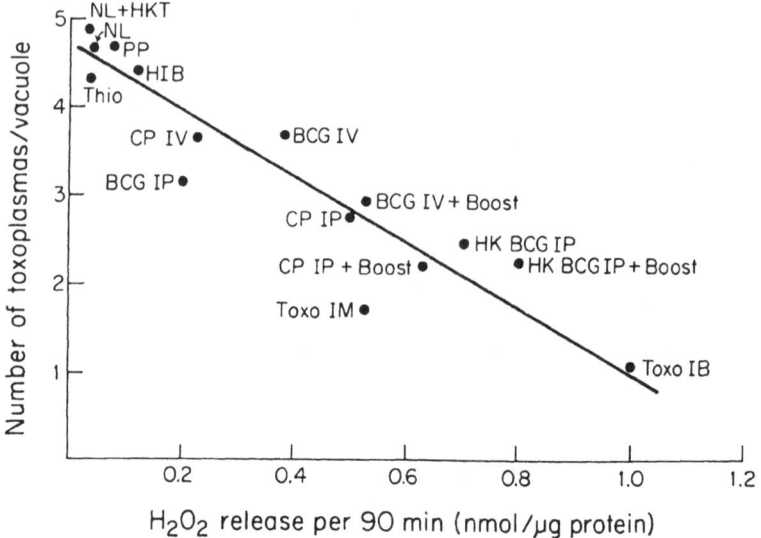

Figure 3. Correlation between the capacity of 12 macrophage populations to release H_2O_2 after phorbol myristate acetate (PMA) triggering and the ability to inhibit *Toxoplasma gondii* replication 18 hr after infection. Macrophage abbreviations: normal resident (NL), inflammatory cells elicited by intraperitoneal (IP) thioglycollate (THIO), heart infusion broth (HIB), proteose peptone (PP), and heat-killed toxoplasmas (HKT), and *in vivo* activated intravenous (IV) or IP immunization with or without IP boosting with *T. gondii* (Toxo), bacillus Calmette–Guérin (BCG), or *Corynebacterium parvum* (CP). (Reproduced with permission from Murray and Cohn, 1980.)

tozoal activity (Fig. 3), a considerable amount of more direct evidence has recently been accumulated to indicate that O_2^- and H_2O_2 (or products of their interaction) do indeed largely mediate macrophage antitoxoplasma and antileishmanial activity. A number of techniques have been employed that markedly impair the capacity of macrophages to generate O_2^- and H_2O_2, and the results of these manipulations on macrophage antiprotozoal activity are shown in Table II. Glucose deprivation (Murray *et al.*, 1979) and pretreatment with PMA (Murray, 1982*b*) or tumor cell-conditioned medium (TCM) (Szuro-Sudol and Nathan, 1982) all effectively suppress respiratory burst activity resulting in up to 90% inhibition of O_2^- and H_2O_2 release. In parallel, these techniques impair or abolish the killing of LDPs by normal macrophages, the killing of LDAs and the toxoplasmastatic activity of LK-stimulated cells, and the killing of toxoplasmas by *in vivo*-activated macrophages.

To dissect out which of the oxygen intermediates generated by macrophages are microbicidal, various cell populations have also been exposed before and during infection to high doses of scavengers of O_2^-, H_2O_2, and OH$^.$. The

Table II. Effect of Impairing the Macrophage Oxidative Burst on Antiprotozoal Activity[a, b]

Macrophage	Pretreatment of macrophages						
	Medium	No glucose	PMA	TCM	Catalase	SOD	Mannitol or benzoate
Normal resident							
Percentage of LDPs killed							
at 4–6 hr	47	33	8	11	28	51	46
at 18 hr	81	49	26	72	42	86	77
In vitro activated							
+LK, percentage of LDAs killed at 72 hr	91	0	0	—	2	84	81
+LK + HIB, Toxo/vacuole at 18–24 hr	1.4	3.5	3.5	3.1	4.4	2.9	3.0
In vivo activated							
Percentage of *Toxoplasma gondii* killed							
at 6 hr	80	18	0	0	38	41	49
at 18 hr	91	60	0	0	56	58	42

[a]From Murray *et al.*, 1979, 1980; Murray and Cohn, 1980; Murray, 1981*a, b*, and unpublished observations.
[b]*Abbreviations:* HIB, heart infusion broth; LDA, *Leishmania donovani* amastigote; LDP, *Leishmania donovani* promastigote; LK, lymphokine; PMA, phorbol myristate acetate; SOD, superoxide dismutase; TCM, tumor cell-conditioned medium.

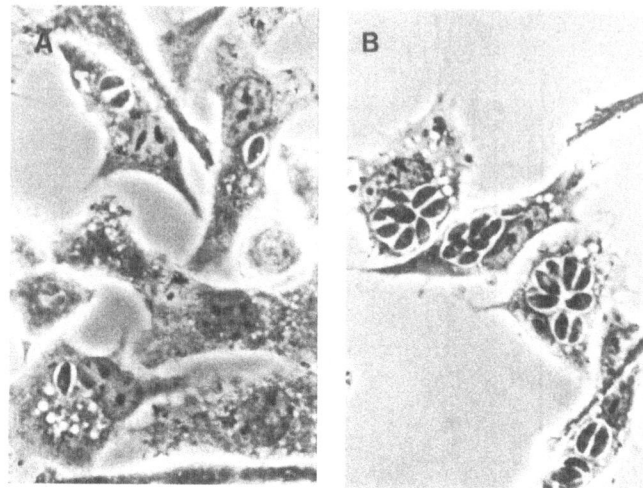

Figure 4. Macrophages activated by lymphokine plus HIB inhibit *Toxoplasma gondii* replication 18 hr after infection (A). Same cells treated with catalase (B) permit toxoplasmas to replicate freely (X600). (Reproduced with permission from Murray and Cohn, 1980.)

results of these treatments are shown in Table II and Fig. 4. As judged by the lack of effect of SOD but the inhibitory action of catalase on both the pro-mastigocidal activity of normal macrophages and the amastigocidal activity of LK-stimulated cells, it appears the H_2O_2, and not O_2^-, is the key leishman-icidal intermediate (Murray, 1981*a*, 1982*a*). OH$^{\cdot}$ scavengers (mannitol and benzoate) also had no effect in this model, suggesting that the more distal radicals in the reduction pathway of molecular oxygen were not required for antileishmanial activity. In contrast, both SOD and catalase reversed the inhibition of toxoplasma replication by LK-HIB-activated cells and the killing of toxoplasmas by *in vivo*-activated macrophages. These observations suggested that although both O_2^- and H_2O_2 were required, neither intermediate by itself was toxic to *T. gondii*, but that a product of O_2^--H_2O_2 interaction was presumably responsible (Murray *et al.*, 1979; Murray and Cohn, 1980). Indeed, OH$^{\cdot}$ scavengers also effectively inhibited the antitoxoplasma activity of the activated macrophage.

Finally, a clone of the J744 macrophage-like cell line (J774G8) was found deficient in the capacity to generate O_2^- and H_2O_2 (Murray, 1981*b*) and was thus a helpful probe to extend this analysis of the role of oxygen intermediates in macrophage antiprotozoal activity. These J774G8 cells released up to 10-fold less O_2^- and H_2O_2 than did normal resident peritoneal macrophages, and <10% of cells showed NBT reduction upon ingestion of LDPs or LDAs (Table I). Whereas normal macrophages killed >85% of LDPs by 18 hr after infection and allowed LDA to persist for 72 hr, J774G8 cells killed <5% of LDPs and permitted LDAs to replicate freely (Murray, 1981*b*, 1982*a*). Additional work with these deficient cells also strengthened the evidence that the augmentation of the macrophage oxidative capacity by LK exposure is indeed linked to enhanced antiprotozoal activity. Thus, overnight cultivation with LK resulted in a two- to threefold increase in the ability of J774G8 cells to release O_2^- and H_2O_2 after PMA triggering and enhanced J774G8 cell NBT reduction in response to LDP injection; these changes were paralleled by the acquisition of effective promastigocidal activity (Table I). As with normal peritoneal macrophages, the latter could also be inhibited by glucose deprivation and catalase, suggesting a key role for H_2O_2 (Murray, 1981*a,b*).

V. TRIGGERING OF THE NORMAL AND ACTIVATED MACROPHAGE'S RESPIRATORY BURST BY PARASITE INGESTION

In order to deliver toxic oxygen intermediates to or within a phagocytic vacuole housing an ingested protozoan, an obvious prerequisite is effective triggering of the macrophage's oxidative burst mechanism. Although normal resident macrophages and cells elicited by sterile inflammatory agents readily respond to both PMA stimulation and zymosan ingestion with qualitative NBT reduction (Table I), these cells fail to respond to toxoplasma ingestion (Murray and Cohn, 1980; Wilson *et al.*, 1980). While the molecular basis for this lack of response has not yet been determined, it is known that intracellular toxoplasmas do not inhibit the capacity of resident cells to respond normally to other stimuli and that parasite viability is not required, since formalin-killed toxoplasmas also fail to trigger HMP shunt activity or NBT reduction (Wilson *et al.*, 1980). It does seem possible, however, that it may be the organism's surface properties that permit it to evade triggering the normal macrophage's plasma membrane receptor mechanism(s) responsible for initiating the respiratory burst (Wilson *et al.*, 1980). Thus, once coated by specific antibody, toxoplasmas readily stimulate the resident cells' oxidative response, and most of the opsonized organisms are killed once ingested (Wilson *et al.*, 1980). Alternatively, the normal macrophage's membrane receptor, which mediates toxoplasma ingestion, may not be effectively linked to the cellular respiratory burst mechanism as long as the macrophage remains in the nonactivated state. Once activated *in vivo* by microbial immunization and boosting or *in vitro* by LK plus HIB, however, peritoneal macrophages clearly acquire the capacity to respond to *T. gondii* ingestion, and these are the cells able to display antitoxoplasma activity.

Normal resident cells also show differential oxidative responses to the two forms of *L. donovani* (Fig. 5). LDP ingestion readily triggers resident macro-

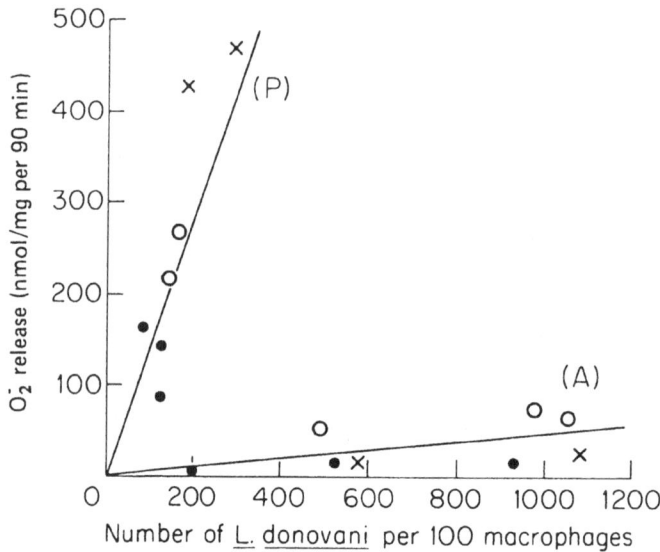

Figure 5. Differential capacity of various macrophage populations to release H_2O_2 during ingestion of *Leishmania donovani* promastigotes (P) or amastigotes (A). (Reproduced with permission from Murray, 1982*b*.)

phages to reduce NBT and secrete H_2O_2 and, as previously described, most LDPs are killed within these cells (Murray, 1981*a*). In contrast, LDAs provoke considerably less of an oxidative response from normal macrophages, and these organisms survive intracellularly unharmed. Macrophages activated by LK, however, promptly respond to LDA ingestion with respiratory burst activity and, in parallel, display potent amastigocidal effects (Murray *et al.*, 1982*a*; Murray, 1982*a*).

VI. APPARENT EXCEPTIONS

Although the relationship between (1) an intrinsically enhanced capacity to generate O_2^- and H_2O_2 and (2) the ability to respond specifically to parasite ingestion with oxidative burst activity both appear to correlate closely with the augmented antiprotozoal effects of activated macrophages, there are three specific instances in which these properties seem at first glance to be dissociated.

The first exception is the failure of *in vivo*-activated macrophages, cells capable of releasing copious amounts of O_2^- and H_2O_2 once effectively triggered, to kill intracellular LDA. It appears, however, that LDA ingestion does not adequately trip the *in vivo*-activated macrophage's plasma membrane O_2^--

and H_2O_2-generating mechanism, in contrast to agents such as PMA, zymosan, and *T. gondii*. For unknown reasons, opsonizing LDA with immune serum does little to enhance the macrophage's oxidative response (Haidaris and Bonventre, 1982). Thus, LDA displays the unique capability of being able to evade triggering a potentially lethal mechanism by cells primed to generate high levels of microbicidal oxygen intermediates. Why the *in vivo*-activated, but not the *in vitro* (LK)-activated macrophage fails to respond oxidatively to LDA ingestion is not clear, but may reflect functional or perhaps quantitative alterations in the plasma membrane mechanisms of these two particular populations of activated macrophages.

The second exception is that despite being able to generate three to four times more O_2^- and H_2O_2 after PMA triggering than untreated resident cells and despite responding to *Toxoplasma* ingestion with NBT reduction, normal macrophages stimulated for 72 hr with LK fail to display antitoxoplasma activity (Murray and Cohn, 1980; Murray *et al.*, 1980). One possible explanation for this observation may be the effect LK has on the macrophage's endogenous enzymatic scavengers of O_2^- and H_2O_2–SOD, catalase, and glutathione peroxidase (GPO). Thus, after 72 hr of exposure to LKs, resident cells exhibit considerable increases in the intracellular activities of all three enzymes (Murray *et al.*, 1980), which, although not capable of diminishing the extracellular release of O_2^- and H_2O_2 after PMA triggering, might alter the intravacuolar generation or delivery of O_2^- or H_2O_2 or products of their interaction (e.g., $OH^{\bullet}$). This speculation may be most relevant when considering the microenvironment of a phagocytic vacuole housing a recently ingested *Toxoplasma*, especially since *T. gondii* is resistant to O_2^- and H_2O_2 alone, but susceptible to $OH^{\bullet}$ (Murray and Cohn, 1979). Although the potential role of endogenous oxygen intermediate scavengers in modulating *Toxoplasma*-macrophage interaction has been considered in detail elsewhere (Murray *et al.*, 1980), it is pertinent to point out here that (1) LK-activated normal cells whose catalase has been inhibited by aminotriazole pretreatment and (2) LK-activated resident cells from acatalasemic mice are able to display levels of toxoplasmastatic activity, which untreated or normocatalasemic macrophages cannot.

The third apparent exception stems from the observation that although almost 50% of normal resident cells can respond to LDA ingestion with respiratory burst activity, as judged by the qualitative reduction of NBT (Murray, 1982*a*), these macrophages fail to kill any LDAs. It must be recalled, however, that despite a respectable response in this qualitative oxidative assay, in quantitative assays normal macrophages generate scant amounts of O_2^- and H_2O_2 irrespective of the agent used to trigger the respiratory burst. In addition, as discussed in Section VII, LDAs are considerably more resistant to H_2O_2 than are LDPs, organisms to which normal macrophages respond with vigorous oxidative activity (both qualitative and quantitative) and effective killing.

VII. SUSCEPTIBILITY OF *T. GONDII* AND *L. DONOVANI* TO OXYGEN INTERMEDIATES IN CELL-FREE SYSTEMS

Aside from the observation that normal resident peritoneal macrophages generate low levels of O_2^- and H_2O_2, another reason why these particular cells fail to kill toxoplasmas appears to be related to this parasite's intrinsic resistance to oxygen intermediates. Thus, toxoplasmas are not killed by a 1-hr exposure of up to 10^{-3} M reagent H_2O_2 or to high fluxes (15–20 nmoles/min) of H_2O_2 generated enzymatically by the glucose-glucose oxidase (GO) reaction (Table III) (Murray and Cohn, 1979; Murray, 1981a). In addition, in a cell-free xanthine–xanthine oxidase (XO) reaction that generates O_2^-, H_2O_2, $OH^·$, and probably $^1O_2$ (Rosen and Klebanoff, 1979), toxoplasmas are also resistant to O_2^- alone. These organisms are, however, killed by the XO reaction apparently because of the effects of the more distal radicals, including $OH^·$ and perhaps $^1O_2$. Toxoplasmas were found to contain high levels of SOD, catalase, and GPO, possibly explaining their intrinsic resistance to exogenous O_2^- and H_2O_2. Nevertheless, it also appears likely that this endogenous defense mechanism, which may protect *T. gondii* from oxidant-mediated injury once ingested by a normal macrophage, is not sufficient to defend the organism from the activated macrophage's capacity to generate much higher levels of O_2^- and H_2O_2.

Since normal macrophages are able to kill LDPs by what seems to be primarily at H_2O_2-dependent mechanism, it was not surprising to find that LDPs are exquisitely susceptible to low fluxes of H_2O_2 generated by GO (Table III) (Murray, 1981a). In addition, LDPs contain little catalase or GPO at levels <100 times those within *T. gondii*. LDPs do, however, possess SOD activity and are resistant to O_2^- in the XO system (Murray, 1981a). LDAs contain 3-fold more catalase and 14-fold more GPO than do LDPs and are four to five times more resistant to H_2O_2 (Murray, 1982a). These findings may also help explain why low H_2O_2-releasing normal macrophages can kill LDPs but not LDAs.

Table III. Parasite Susceptibility to H_2O_2 and Endogenous H_2O_2 Scavengers[a,b]

Protozoan	LD_{50} of H_2O_2 (nmol/min)[c]	Percentage killed by 5 nmol/min of H_2O_2[c]	Catalase (BU/mg $\times 10^{-2}$)	GPO (nmol/min per mg)
Toxoplasma gondii	<20	<5	4.8	117
LDP	1.5	90	0.05	0.5
LDA	8.1	25	0.14	7.2

[a]From Murray and Cohn, 1979; Murray, 1981a, 1982a.
[b]*Abbreviations:* BU, Bauduin units; GPO, glutathione peroxidase; LDA, *Leishmania donovani* amastigote; LDP, *Leishmania donovani* promastigote.
[c]After a 1-hr exposure at 37°C to fluxes of H_2O_2 generated by glucose oxidase.

VIII. SUMMARY OF OBSERVATIONS DERIVED FROM THE MOUSE PERITONEAL MACROPHAGE MODEL

The results of the studies presented in sections I-VII can be summarized according to the following studies:

1. From the standpoint of oxygen-dependent anti-protozoal activity, there appear to be at least three key determinants of a protozoan's intracellular fate—parasite susceptibility to H_2O_2, the magnitude of the macrophage oxidative burst capacity, and the ability of the macrophage to respond to parasite ingestion with effective oxidative burst activity.

2. Normal resident and inflammatory macrophages display little oxidative activity irrespective of the nature of the triggering agent, in particular respond poorly to *T. gondii* and LDAs, and fail to kill or inhibit either either of these pathogens. Normal cells can, however, respond to and kill ingested LDPs, an organism highly sensitive to H_2O_2.

3. Macrophages activated *in vivo* by immunization with viable or dead microbial agents display enhanced oxidative activity, and these cells respond oxidatively to and kill or inhibit *T. gondii*. LDA ingestion, however, fails to provoke a substantial respirator burst from these cells, and this protozoan persists unharmed within *in vivo*-activated macrophages.

4. Macrophages activated *in vitro* by 18 hr of exposure to prepared LKs also display enhanced oxidative activity and are able to respond to and kill LDAs. These cells, however, fail to respond to *T. gondii* ingestion and exert no antitoxoplasma activity unless a synergistic sterile inflammatory agent is included during the activation period. The resulting effects are then primarily toxoplamastatic and toxoplasmacidal. *Toxoplasma gondii*, is, however, considerably more resistant to H_2O_2 than are LDAs.

IX. STUDIES WITH ACTIVATED HUMAN MACROPHAGES

It is also appropriate to review briefly recent studies that have extended the analysis of the antiprotozoal effects of the macrophage's oxygen-dependent mechanism to models using human peripheral blood mononuclear phagocytes. Fresh human monocytes secrete copious amounts of O_2^- and H_2O_2 after triggering with PMA, readily respond to *T. gondii* ingestion with NBT reduction, and kill toxoplasmas or inhibit their replication (Murray and Cartelli, 1983; Nakagawara *et al.*, 1981; Pabst *et al.*, 1982; Wilson *et al.*, 1980). These cells also promptly generate large quantities of H_2O_2 during LDP ingestion and destroy >90% of phagocytized LDP (Murray and Cartelli, 1983). LDA ingestion, however, provokes relatively little oxidative activity on the part of human monocytes (>sixfold less H_2O_2 than stimulated by LDPs), and only 25% of LDAs are killed by the cells (Murray and Cartelli, 1983). Thus, these observations support the correlation derived from the mouse peritoneal macrophage model between antiprotozoal activity and the capacity of mononuclear phagocytes to respond to parasite ingestion with effective oxidative burst activity.

The monocyte-derived human macrophage, however, behaves quite differently than its peripheral blood precursor. Cells cultivated for 1 week or more to allow *in vitro* maturation to the macrophage stage release 90% less O_2^- and H_2O_2 and have lost most of their MPO activity (Murray and Cartelli, 1983; Nakagawara *et al.*, 1981; Pabst *et al.*, 1982). Not surprisingly, the monocyte derived macrophage fails to exert any activity against *T. gondii*, LDPs, or LDAs and supports their intracellular replication (Murray and Cartelli, 1983; Pearson *et al.*, 1981; Wilson *et al.*, 1980). The application of mitogen- or antigen-stimulated LKs, however, induces the human macrophage to display strikingly different properties. Exposure to 10% LKs for 48–72 hr enhances H_2O_2 release by > sixfold, restores oxidative responsiveness to parasite ingestion and, in parallel, activates these cells to inhibit or kill *T. gondii* and to display both promastigocidal and amastigocidal activity against *L. donovani* (Anderson *et al.*, 1976; Borges and Johnson, 1975; Murray and Cartelli, 1983; Nakagawara *et al.*, 1982). Again, these data support the close correlation between effective antiprotozoal activity and the capacity of activated macrophages to (1) secrete high levels of toxic oxygen intermediates and (2) specifically respond to parasite ingestion with O_2^- and H_2O_2 generation simultaneously.

X. MACROPHAGE OXYGEN-INDEPENDENT ANTIMICROBIAL ACTIVITY

To conclude this discussion, it is worthwhile to point out that there are also recent data that suggest that oxygen-independent mechanisms may contribute to antiprotozoal activity. For example, we have observed that oxidatively deficient mouse fibroblasts (L cells) can be activated by LK to inhibit toxoplasma replication in the absence of NBT reduction or enhanced release of H_2O_2 (Murray *et al.*, 1983). In addition, the microbistatic effects of LK-activated mouse peritoneal macrophages toward *Chlamydia psittaci*, another intracellular pathogen, also do not appear to require respiratory burst activity (Byrne and Favbion, 1982). Finally, although pretreatment with tumor cell-conditioned medium (TCM) (Szuro-Sudol and Nathan, 1982) markedly decreases the resident mouse macrophage's oxidative capacity and virtually abolishes the killing of LDPs at 6 hr, by 18 hr most LDPs are still eradicated.

Similar observations have also been made with monocytes from patients with chronic granulomatous disease (CGD), cells that generate little or no O_2^- or H_2O_2. For instance, 1-day old CGD monocytes inhibit *C. psittaci* as effectively as normal monocytes (Rothermel *et al.*, 1982). Although CGD cells are considerably less active than normal cells against LDPs during the first 24 post-phagocytic hours, CGD monocytes still kill 40% of LDPs given an additional 24 hr. This promastigocidal activity can also be strikingly enhanced by LK pretreatment (Murray and Cartelli, 1983). Fresh CGD monocytes have also been reported to kill 45% of phagocytized toxoplasmas during the first 6 hr after infection; however, at the same time, these cells also permit surviving *T. gondii* to replicate freely, while normal monocytes do not (Locksley *et al.*, 1982). We have also found that once activated by LK, monocyte-derived CGD macrophages inhibit *T. gondii* and *C. psittaci* replication and display promastigocidal and amastigostatic activity against *L. donovani* (Murray and Caretlli, 1983; Rothermel *et al.*, 1982; Murray *et al.*, 1983). In certain instances in which CGD cells have demonstrated antiprotozoal activity, however, this activity has either not been as effective (e.g., static but not cidal) or has required a longer postingestion period for its effects to become evident. Thus, while indicating that activated macrophages can also use an apparent oxygen-independent antimicrobial mechanism, these studies serve to reemphasize the key role of oxygen-dependent mechanisms.

XI. REFERENCES

Anderson, S. E., Bautista, S., and Remington, J. S., 1976, Induction of resistance to *Toxoplasma gondii* in human macrophages by soluble lymphocyte products, *J. Immunol.* 117:381–387.

Badway, J. A., and Karnovsky, M. L., 1980, Active oxygen species and the functions of phagocytic leukocytes, *Annu. Rev. Biochem.* 49:595–622.

Borges, J. S., and Johnson, W. D., 1975, Inhibition of multiplication of *Toxoplasma gondii* by human monocytes exposed to T-lymphocyte products, *J. Exp. Med.* 141:438–496.

Byrne, G. I., and Favbion, C., 1982, Lymphokine-mediated microbistatic mechanisms restrict *Chlamydia psittaci* growth in macrophages, *J. Immunol.* 128:469–474.

Diamond, R. D., Haudenschild, C. C., and Erickson, N. F., 1982, Monocyte-mediated damage to *Rhizopus oryzae* hyphae in vitro, *Infect. Immun.* 38:292–297.

Haidaris, C. G., and Bonventre, P. F., 1982, A role for oxygen-dependent mechanisms in killing of *Leishmania donovani* tissue forms by activated macrophages, *J. Immunol.* 129:350–355.

Johnston, R. B., Godzik, C. A., and Cohn, Z. A., 1978, Increased superoxide anion production by immunologically activated and chemically elicited macrophages, *J. Exp. Med.* 148:115–126.

Kazura, J. W., Fanning, M. M., Blumes, J. T., and Mahmoud, A. A., 1981, Role of cell-generated H_2O_2 in granulocyte-mediated killing of schistosomula of *Schistosoma mansoni*, *J. Clin. Invest.* 67:93–102.

Klebanoff, S. J., 1980, Oxygen metabolism and the toxic properties of phagocytes, *Ann. Intern. Med.* 93:480–489.

Locksley, R. M., Wilson, C. B., and Klebanoff, S. J., 1982, Role for endogenous and acquired peroxidase in the toxoplasmacidal activity of murine and human mononuclear phagocytes, *J. Clin. Invest.* 69:1099–1111.

Murray, H. W., 1981a, Susceptibility of *Leishmania* to oxygen intermediates and killing by normal macrophages, *J. Exp. Med.* 153:1302–1315.

Murray, H. W., 1981b, Interaction of *Leishmania* with a macrophage cell line. Correlation between intracellular killing and the generation of oxygen intermediates, *J. Exp. Med.* 153:1690–1695.

Murray, H. W., 1982a, Cell-mediated immune response in experimental visceral leishmaniasis. II. Oxygen-dependent killing of intracellular *Leishmania donovani* amastigotes, *J. Immunol.* 129:351–357.

Murray, H. W., 1982b, Pretreatment with phorbol myristate acetate inhibits macrophage activity against intracellular protozoa, *J. Reticuloendothel. Soc.* 37:479–487.

Murray, H. W., and Cartelli, D. M., 1983, Killing of intracellular *Leishmania donovani* by human mononuclear phagocytes: Evidence for oxygen-dependent and -independent leishmanicidal activity, *J. Clin. Invest.* 72:32–41.

Murray, H. W., and Cohn, Z. A., 1979, Macrophage oxygen-dependent antimicrobial activity. I. Susceptibility of *Toxoplasma gondii* to oxygen intermediates, *J. Exp. Med.* 150:983–979.

Murray, H. W., and Cohn, Z. A., 1980, Macrophage oxygen-dependent antimicrobial activity. III. Enhanced oxidative metabolism as an expression of macrophage activation, *J. Exp. Med.* 152:1596–1609.

Murray, H. W., Juangbhanich, C. W., Nathan, C. F., and Cohn, Z. A., 1979, Macrophage oxygen-dependent antimicrobial activity. II. The role of oxygen intermediates, *J. Exp. Med.* 150:950–864.

Murray, H. W., Nathan, C. F., and Cohn, Z. A., 1980, Macrophage oxygen-dependent antimicrobial activity. IV. The role of endogenous scavengers of oxygen intermediates, *J. Exp. Med.* 152:1610–1624.

Murray, H. W., Masur, H., and Keithly, J. S., 1982, Cell-mediated immune response in experimental visceral leishmaniasis. I. Correlation between resistance to *Leishmania donovani* and lymphokine-generating capacity, *J. Immunol.* 129:344–350.

Murray, H. W., Byrne, G. I., Rothermel, C. D., and Cartelli, D. M., 1983, Lymphokine enhances oxygen-dependent activity against intracellular pathogens, *J. Exp. Med.* 158:234–235.

Nacy, C. A., Meltzer, M. S., Leonard, E. J., and Wyler, D. J., 1981, Intracellular replication and lymphokine-induced destruction of *Leishmania tropica* in C3H/Hen mouse macrophages, *J. Immunol.* 127:2381–2386.

Nakagawara, A., Nathan, C. F., and Cohen, Z. A., 1981, Hydrogen peroxide metabolism in human monocytes during differentiation in vitro, *J. Clin. Invest.* 68:1243–1252.

Nakagawara, A., DeSantis, N. M., Nogueira, N., and Nathan, C. F., 1982, Lymphokines enhance the capacity of human monocytes to secrete reactive oxygen intermediates, *J. Clin. Invest.* 70:1042–1048.

Nathan, C. F., and Root, R. K., 1977, Hydrogen peroxide release from mouse peritoneal macrophages, *J. Exp. Med.* 146:1648–1662.

Nathan, C. F., Nogueira, N., Juangbhanich, C., Ellis, J., and Cohn, Z. A., 1979, Activation of macrophages *in vivo* and *in vitro*. Correlation between hydrogen peroxide release and killing of *Trypanosoma cruzi*, *J. Exp. Med.* 149:1056–1068.

Nathan, C. F., Murray, H. W., and Cohn, Z. A., 1980, The macrophage as an effector cell, *N. Engl. J. Med.* 303:622–626.

North, R. J., 1978, The concept of the activated macrophage, *J. Immunol.* 121:806–808.

Pabst, M. J., Hedegaard, H. B., and Johnston, R. B., 1982, Cultured human monocytes require exposure to bacterial products to maintain an optimal oxygen radical response, *J. Immunol.* 128:123–128.

Pearson, R. D., Romito, R, Symes, P. H., and Harcus, J. L., 1981, Interaction of *Leishmania donovani* promastigotes with human monocyte-derived macrophages: Parasite entry, intracellular survival, and multiplication, *Infect. Immun.* **32**:1249-1253.

Rosen, H., and Klebanoff, S. J., 1979, Bactericidal activity of a superoxide anion-generating system. A model for the polymorphonuclear leukocyte, *J. Exp. Med.* **149**:27-40.

Rothermel, C. D., Byrne, G. I., and Murray, H. W., 1982, Growth of *Chlamydia psittaci* in human monocytes (abstract), in: *22nd Interscience Conference on Antimicrobial Agents and Chemotherapy, Miami, 1982*, No. 36.

Sasada, M., and Johnston, R. B., 1980, Macrophage microbicidal activity. Correlation between phagocytosis-associated oxidative metabolism and the killing of *Candida* by macrophages, *J. Exp. Med.* **152**:85-99.

Szuro-Sudol, A., and Nathan, C. F., 1982, Suppression of macrophage oxidative metabolism by products of malignant and non-malignant cells, *J. Exp. Med.* **156**:945-961.

Szuro-Sudol, A., Murray, H. W., and Nathan, C. F., 1983, Supression of macrophage antimicrobial activity by a tumor cell product, *J. Immunol.* **131**:384-385.

Walker, L., and Lowrie, D. B., 1981, Killing of *Mycobacterium microti* by immunologically activated macrophages, *Nature* **293**:69-70.

Wilson, C. B., Tsai, V., and Remington, J. S., 1980, Failure to trigger the oxidative burst by normal macrophages. Possible mechanism for survival of intracellular pathogens, *J. Exp. Med.* **151**:328-346.

Chapter 6

Activation of Mononuclear Phagocytes for the Destruction of Intracellular Parasites: Studies with Trypanosoma cruzi

Nadia Nogueira and Zanvil A. Cohn

Department of Cellular Physiology and Immunology
The Rockefeller University
New York, New York 10021

I. INTRODUCTION

Trypanosoma cruzi is an obligate intracellular parasite of man and other mammals. In man, it produces a chronic infection which, in its more severe form, leads to cardiac damage and failure (Chagas disease).

The initial motivation for our work was the desire to understand the role of cell-mediated immunity in resistance to this infection. In addition, we wanted to investigate the basic cell biology of the interaction of this parasite with macrophages. Such studies led to the development of an experimental model extremely suitable for the study of the mechanisms involved in the process of macrophage activation.

A number of factors made this organism quite useful in the study of the mechanisms of macrophage microbicidal activity:

1. Availability of defined forms of the organism
2. Large size and prompt interiorization by macrophages, permitting an easy evaluation of its uptake and fate within mononuclear phagocytes by direct microscopic examination
3. Obligate intracellular nature and resistance to antibiotics, which facilitated tissue culture conditions, without interfering with the microbe's fate
4. Susceptibility of mice to infection by this parasite permitting us to study the process *in vitro* as well as *in vivo* in a well-characterized animal model.

This chapter summarizes some of the information provided by this model on the mechanisms of macrophage activation to a microbicidal state by lymphocyte products.

II. INTERIORIZATION AND FATE WITHIN NORMAL MONONUCLEAR PHAGOCYTES

A. Insect Stages

The insect stages of *Trypanosoma cruzi* are interiorized by mononuclear phagocytes within a parasitophorous vacuole. This uptake is mediated by receptor-like peptides on the macrophage surface. Treatment of the macrophage surface with proteolytic enzymes (trypsin and chymotrypsin) has been shown to result in abolition of the uptake of *T. cruzi* by mouse peritoneal macrophages (Nogueira and Cohn, 1976). The uptake is mediated by nonspecific receptors and does not involve C_3 or Fc receptors. Interiorization of the organisms is inhibited by cytochalasin B (Fig. 1).

Metacyclic trypomastigotes, the infective stages of *T. cruzi,* however, reside

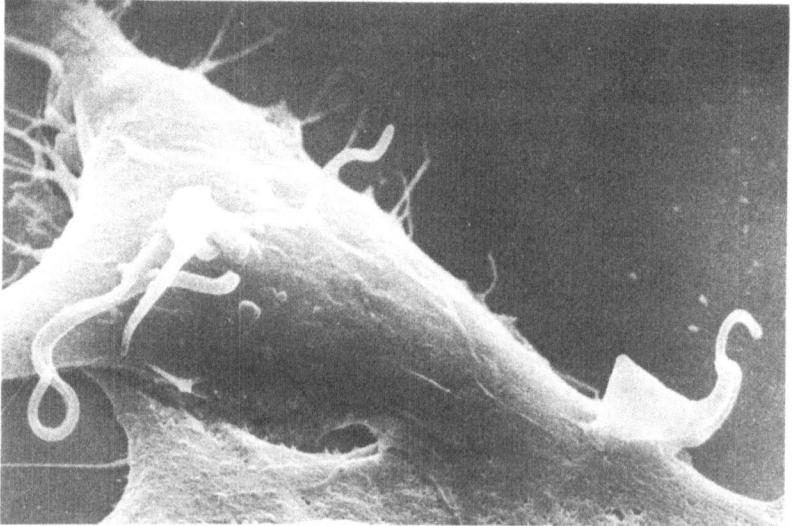

Figure 1. Uptake of metacyclic trypomastigotes of T. *cruzi* by a resident mouse peritoneal macrophage. (Scanning electron micrograph kindly taken by Dr. Gilla Kaplan.)

for only a limited time within the confines of these vacuoles. Fusion with lyso-somes seems to occur, followed by escape of the organisms into the cytosol, where they replicate. The mechanism underlying the escape from phagocytic vacuole is still unknown.

In contrast to trypomastigotes, epimastigotes, the replicative forms in the in-vertebrate host, are unable to leave the parasitophorous vacuole; there they will be killed and rapidly digested. This fact, their inability to enter other cell types, and their susceptibility to lysis mediated by complement (Nogueira *et al.*, 1975) explain their lack of infectivity for mammalian hosts.

B. Mammalian Stages

Trypomastigotes of *T. cruzi* are able to replicate in many other cell types in addition to mononuclear phagocytes. Avoiding interiorization by the macrophage therefore seems advantageous for the parasite, since this cell type is one of the few with microbicidal potential. Indeed, blood-form trypomastigotes (BFTs) pos-sess on their surface an antiphagocytic molecule that reduces its phagocytosis by macrophages (Nogueira *et al.*, 1980; M. Reesink and N. Nogueira, unpublished re-sults). Removal of this surface glycoprotein by proteolytic treatment or opsoniza-tion with IgG class antibodies neutralizes the antiphagocytic effect and results in interiorization of the organisms. Interiorization occurs without modifying its intra-cellular fate within normal macrophages.

III. *IN VIVO* ACTIVATION: ROLE OF INFECTION AND SECONDARY CHALLENGE

In contrast to the quantitative survival of trypomastigotes of *T. cruzi* within normal or inflammatory phagocytes (Nogueira and Cohn, 1976), macrophages obtained from infected mice were able to destroy the majority of the ingested trypomastigotes (Nogueira *et al.*, 1977*a*).

We tested the ability of macrophages from *T. cruzi*-infected mice to deal with trypomastigotes *in vitro*, comparing the resident cells of infected animals (im-mune) with those evoked by a secondary intraperitoneal challenge with heat-killed *T. cruzi* or other nonspecific antigens (immune boosted). In addition, animals were also infected with bacillus Calmette-Guérin (BCG) to examine the specificity of the acquired immunity. The results obtained are outlined in Table I. Resident macrophages from *T. cruzi*- or BCG-infected animals (immune) showed no try-panocidal activity, as compared with macrophages from noninfected controls. However, when these animals are challenged with the specific antigen, i.e., heat-killed trypanosome (HKT) or H37Ra, intraperitoneally (immune boosted), their

Table I. Fate of Trypomastigotes of *Trypanosoma cruzi* in Different
Populations of Cultivated Macrophages[a]

Macrophage type	Treatment of mouse	Parasites/100 MACs at 24 hr % initial inoculum[b]	No. infected cells at 24 hr % initial value[b]
Resident	None	118	104
Inflammatory	Proteose peptone i.p.	102	100
Inflammatory	HKT i.p.	97	81
Immune	3-week *T. cruzi* infection	97	95
Immune	3-week BCG infection	92	82
Immune-boosted	3-week *T. cruzi* infection + HKT challenge i.p.	35	19
Immune-boosted	3-week BCG infection + H37Ra challenge i.p.	53	42

[a]Swiss mice infected intraperitoneally with 5×10^6 live culture forms of *T. cruzi* y strain, or intravenously with $2-6 \times 10^7$ viable BCG (strain 1011, Trudeau Institute, Saranac Lake, N.Y.) and, when indicated, challenged intraperitoneally 3 weeks later with 5×10^6 heat-killed trypanosomes (HKT) or 400 μg of *Mycobacterium tuberculosis* strain H37Ra. Cells were harvested 2 days later, and macrophages cultivated for 24 hr before infection. Proteose peptone, 1 ml of 1% solution. (From Nogueira and Cohn, 1976; Nogueira *et al.*, 1977a).
[b]Cells were exposed to purified metacyclic trypomastigotes at a 1:1 or 1:2 parasite:cell multiplicity for 3 hr. Under these conditions, 30–35% of the original macrophage population is infected. Values represent numbers at (24 hr/3 hr) $\times$ 100.

Abbreviations: BCG, bacillus Calmette–Guérin; HKT, heat-killed trypanosomes; i.p., intraperitoneal; MAC, macrophage.

macrophages were able to destroy 50–70% of the interiorized organisms. The injection of HKT, proteose peptone, or H37Ra alone was without significant influence on their microbicidal activity.

When followed up for several days, immune resident peritoneal macrophages from *T. cruzi*- or BCG-infected mice displayed a trypanostatic rather than trypanocidal activity. In addition, it was evident that the trypanocidal activity of the immune-boosted cells and the trypanostatic activity of the immune cells waned upon *in vitro* cultivation. After 72 hr in culture, the surviving trypomastigotes in these cells began to multiply at normal rates and were able to survive when entering uninfected cells in the monolayer (Fig. 2).

Other parameters of macrophage function were also modified in this activated population, as summarized in Table II. As shown, the conditions leading to microbicidal activity also resulted in enhanced secretion of plasminogen activator. These two properties are not related, however, since thioglycollate-activated macrophages have been shown to secrete large amounts of plasminogen activator

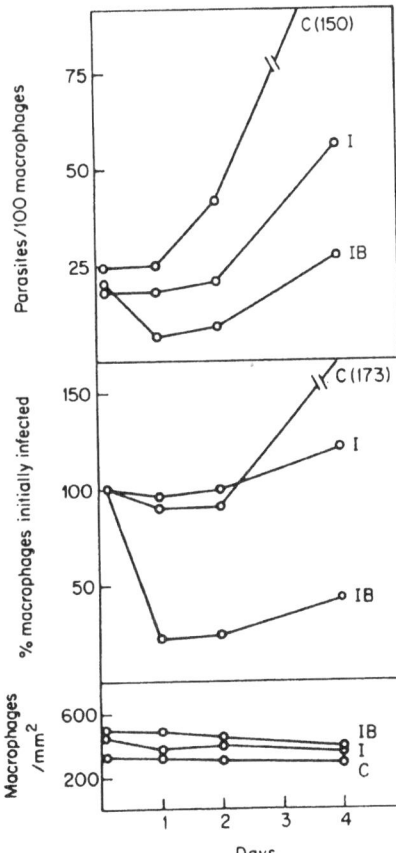

Figure 2. Fate of trypomastigotes of *T. cruzi* in three different macrophage populations. Peritoneal macrophages from *T. cruzi*-infected Swiss mice were harvested 3 weeks after infection without (I) or with (IB) an intraperitoneal challenge with *T. cruzi* antigen. Control macrophages (C) were resident peritoneal macrophages from noninfected mice; 24-hr explanted macrophages were exposed to trypomastigotes at a 1:1 parasite:cell ratio for 3 hr.; I, immune; IB, immune boosted. (From Nogueira *et al.*, 1977*a*.)

(Unkeless *et al.*, 1974), but do not display trypanocidal activity (Nogueira and Cohn, 1976).

IV. *IN VITRO* ACTIVATION: MODULATION OF MACROPHAGE FUNCTION

Resident mouse peritoneal macrophages as well as inflammatory macrophages can be activated completely under test tube conditions to display microbicidal activity against *T. cruzi*. This can be accomplished by the daily exposure of macrophages to products generated by antigen-stimulated sensitized spleen cells from mice infected with *T. cruzi* or BCG (Nogueira and Cohn, 1978). The gener-

Table II. Properties of Activated Macrophages[a]

Macrophage type	Microbicidal activity against Trypanosoma cruzi	Secretion plasminogen activator (% radioactivator released/4 hr per 10^6 cells)[b]	Complement-mediated ingestion (% ingestion of E(IgM)C)[c]	Rate of H_2O_2 release (nmol/μg protein per 20 min)
Resident	None	1.8	5	0.005
Inflammatory (PP)	None	1.9	28	0.01
Immune (T. cruzi infection)	No (microbistatic)	2.1	24	ND
Immune (BCG infection)	No (microbistatic)	1.3	51	ND
Immune-boosted (T. cruzi-HKT)	Yes	23.9	76	0.09
Immune-boosted (BCG-PPD)	Yes	18.3	54	0.06
Activated by lymphokine[d]	Yes	12.0	ND	0.05

[a] Activities measured in parallel cultures at 48 hr after harvesting. (From Nathan et al., 1979; Nogueira and Cohn, 1976, 1978; Nogueira et al., 1977a,b.)

[b] Plasminogen activator measured by fibrinolysis on [^{125}I]fibrinogen coated plates as described by Unkeless et al. (1974).

[c] Cells kept for 24 hr in medium before incubation for 60 min with 0.1% E(IgM)C. E(IgM)C ingestion expressed as percentage of macrophages ingesting more than three sheep erythrocytes.

[d] Resident macrophages activated by exposure to 25% antigen- or mitogen-induced lymphokine.

Abbreviations: BCG, bacillus Calmette–Guérin; HKT, heat-killed trypanosomes; PP, propeose peptone.

ation of these lympokines from T cells required immunologic specificity between antigen and sensitized spleen cells. Once generated, the lymphocyte product leads to the global activation of macrophages and their ability to kill a variety of organisms. This is in agreement with what was known for cell-mediated immunity as expressed in the whole animal (Mackaness, 1964). Optimal induction of trypanocidal activity was obtained by preincubation of the cells for 24 hr before infection with the active spleen cell supernatants followed by daily addition of fresh supernatants. Fresh medium alone or control supernatants were without significant effect on the antitrypanosomal activity of cultured macrophages.

Induction of microbicidal activity could be induced in both inflammatory and resident mouse peritoneal macrophages. Inflammatory macrophages responded more promptly, whereas an additional 24-hr exposure was needed to activate resident cells. Microbicidal activity could also be induced in the same populations with supernates from concanavalin A- (Con A) and lipopolysaccharide- (LPS) -stimulated normal spleen cells.

Exposure of macrophages to immune lymphokine (LK) leads to a variety of structural and metabolic sequellae: (1) increased spreading on glass and plastic substrates with symmetric spreading, ruffling of plasma membrane, and the presence of increased pinocytic vesicles and lysosomes; and (2) induction of plasminogen activator synthesis and secretion. This property could also be induced with a dose–response titration quite similar to that observed for the induction of microbicidal activity (Nogueira *et al.*, 1977*b*). They reach a plateau at concentrations of 25%, but at 2.5% they already produce significant responses.

The responses can be abolished by pretreatment of the total spleen cells with anti-Thy serum and complement, a finding that implicates T lymphocytes in the generation of these cytokines.

Similar induction of trypanocidal activity could be obtained when using cells derived from human peripheral blood. The exposure of monocyte-derived macrophages to *T. cruzi*-stimulated peripheral blood lymphocytes (PBL) from 14 patients with chronic Chagas disease led to the activation of the microbicidal state. This activity was also nonspecific, since it could be induced by lymphokines generated with unrelated antigens, i.e., PPD-positive subjects exposed to heat-killed BCG or Con A-stimulated normal lymphoid cells. In contrast, supernatants generated by the stimulation of lymphoid cells from normal donors by *T. cruzi* antigen did not result in macrophage microbicidal activity (Nogueira *et al.*, 1982; Williams and Remington, 1977). The induction process in human cells presented a few differences from those seen in the mouse model:

1. Lack of requirement for the daily addition of soluble factors
2. Shift to lower concentrations of lymphokines to obtain similar responses, i.e., concentrations as low as 2.5% already resulted in induction of maximal microbicidal activity
3. Direct correlation between the lymphokine activity generated by antigen-

stimulated PBLs and their proliferative responses to antigen (One-half the patients studied showed low proliferative responses and correspondingly lower lymphokine activity. In contrast, mitogen responses were normal in all patients. In mice infected with *T. cruzi*, lymphokines were generated from antigen-stimulated spleen cells despite a complete suppression of their proliferative responses to antigen and mitogens.)

V. MECHANISM OF INTRACELLULAR KILLING: ROLE OF REACTIVE OXYGEN INTERMEDIATES

The mechanism of *Trypanosoma cruzi* killing within activated macrophages was examined next. Our attention was focused on oxygen intermediates in view of the formation of both superoxide anion and hydrogen peroxide by activated macrophage populations (Nathan and Root, 1977; Johnston *et al.*, 1978). We first performed correlative studies in which the microbicidal activity and H_2O_2 production was examined in cell populations. The results of such studies indicated that macrophages activated *in vivo* by infection and challenge with *T. cruzi* (immune boosted) or *in vitro*, by exposure to active lymphokines, both kill *T. cruzi* and secreted high levels of H_2O_2 when suitably triggered (see Table II)(Nathan *et al.*, 1979).

The correlation between H_2O_2 production and *Trypanosoma cruzi* killing was observed under a wide range of experimental conditions. The microbicidal activity of *in vivo*-activated cells wanes upon *in vitro* cultivation, but can be either maintained or restored by exposure to active lymphokine. These responses are paralleled by similar changes in H_2O_2 release.

Similarly, *in vitro* activation of macrophages by LK also results in the enhanced secretion of H_2O_2; this activity also returns to control levels upon removal of the lymphokines. Removal of LK before all intracellular parasites have been destroyed results in subsequent replication of the remaining trypanosomes.

Similarly, addition of lymphocyte mediators to human monocyte-derived macrophages in which the monocyte-type oxidative capacity has already waned restores part of the capacity to release reactive oxygen intermediates (Nakagawara *et al.*, 1982). This activity could explain the induction of microbicidal activity of these cells to *T. cruzi* (Nogueira *et al.*, 1982).

Interesting correlation also exists between the sensitivity of the life stages of *T. cruzi* to H_2O_2, revealing that trypomastigotes, the infective forms of the organisms, are relatively more resistant to H_2O_2 than are epimastigotes. We have already shown that epimastigotes are promptly killed by even normal resident macrophages, suggesting that resident cells may generate small fluxes of H_2O_2 sufficient to kill this highly susceptible organism.

Additional evidence for the role of H_2O_2 in macrophage antimicrobial action has been provided by studies with eosinophil peroxidase (EPO) (Nogueira *et al.*, 1982). This peroxidase, derived from eosinophil granules, is a basic protein that can adsorb to the surface of the parasite without affecting viability.

Metacyclic or blood-form trypomastigotes coated with EPO were killed intracellularly by normal macrophages. Both mouse and human mononuclear phagocytes are unable to kill the uncoated organisms. The inhibition of the killing by catalase and azide implicates the production of H_2O_2 by the resident cells in the killing process.

EPO-coated organisms could also be killed extracellularly when exposed to normal macrophages at high parasite:cell ratios or when a high phagocytic load of another particle was given simultaneously. Again, cytotoxicity was inhibitable by azide and catalase, but not by superoxide dismutase.

These studies provide evidence for, and another interesting demonstration of, H_2O_2 release after binding and interiorization of *T. cruzi* by mononuclear phagocytes. The concentration of H_2O_2 generated by normal macrophages upon phagocytosis of the organisms is not toxic to the untreated parasites, but is sufficient to kill them if they are coated with EPO. The susceptibility of EPO-coated trypanosomes to H_2O_2 toxicity provided a highly sensitive means of detecting H_2O_2 release during the phagocytic event.

VI. CONCLUSIONS

The process of macrophage activation is complex. Our understanding of it is still very limited, in both functional and biochemical terms. We have described an experimental system that permitted us to ask some of the questions about this process. As usual, however, more questions were raised by these experiments than were answered by them.

The nature of the lymphokines and the molecular effects they could have on mononuclear phagocytes are still poorly understood. The possible role of suppressor mechanisms has not yet been studied in detail. The role of oxygen intermediates has been investigated as a possible biochemical basis for the microbicidal activity of macrophages in this system, but we know little about other mechanisms that are oxidative and nonoxidative in nature. The molecular nature of the cellular components responsible for their generation and the mechanism by which lymphocyte mediators trigger their action are still unknown.

Answers to these questions are being actively pursued in our laboratory. It is an exciting period, since these answers will have a considerable impact on our ability to interfere with many chronic infections and inflammatory diseases.

VII. REFERENCES

Johnston, R. B, Godzik, C. A., and Cohn, Z. A., 1978, Increased superoxide anion production by immunologically activated and chemically elicited macrophages, *J. Exp. Med.* 148:115-127.

Mackaness, G., 1964, The immunological basis of acquired cellular resistance, *J. Exp. Med.* 120:105-120.

Nagakawara, A., De Santis, N., Nogueira, N., and Nathan, C., 1982, Lymphokines enhanced the capacity of human mononuclear phagocytes to secrete reactive oxygen intermediates, *J. Clin. Invest.* 70:1042-1048.

Nathan, C. F., and Root, R. K., 1977, Hydrogen peroxide release from mouse peritoneal macrophages. Dependence on sequential activation and triggering, *J. Exp. Med.* 146: 1648-1662.

Nathan, C., Nogueira, N., Juangbhanich, C., Ellis, J., and Cohn, Z., 1979, Activation of macrophages *in vivo* and *in vitro*. Correlation between H_2O_2 release and killing of *T. cruzi*, *J. Exp. Med.* 149:1056-1068.

Nogueira, N., and Cohn, Z. A., 1976, *Trypanosoma cruzi:* Mechanism of entry and intracellular fate in mammalian cells, *J. Exp. Med.* 143:1402-1420.

Nogueira, N., and Cohn, Z. A., 1978, *Trypanosoma cruzi: In vitro* induction of macrophage microbicidal activity, *J. Exp. Med.* 148:288-300.

Nogueira, N., Bianco, C., and Cohn, Z. A., 1975, Studies on the selected lysis and purification of *Trypanosoma cruzi*, *J. Exp. Med.* 142:224-229.

Nogueira, N., Gordon, S., and Cohn, Z. A., 1977a, *Trypanosoma cruzi*: Modification of macrophage function during infection, *J. Exp. Med.* 146:157.

Nogueira, N., Gordon, S., and Cohn, Z., 1977b, The immunological induction of macrophage plasminogen activator requires thymus-derived lymphocytes, *J. Exp. Med.* 146:172-183.

Nogueira, N., Chaplan, S., and Cohn, Z. A., 1980, *Trypanosoma cruzi*: Factors modifying ingestion and fate of blood form trypomastigotes, *J. Exp. Med.* 152:447-451.

Nogueira, N., Klebanoff, S., and Cohn, Z. A., 1982, *T. cruzi*: Sensitization to macrophage killing by eosinophil peroxidase, *J. Immunol.* 128:1705.

Nogueira, N., Chaplan, S., Reesink, M., Tydings, J., and Cohn, Z. A., 1982, *Trypanosoma cruzi*: Induction of microbicidal activity in human mononuclear phagocytes, *J. Immunol.* 128:2142-2146.

Unkeless, J. C., 1979, Characterization of a monoclonal antibody directed against mouse macrophage and lymphocyte Fc receptors, *J. Exp. Med.* 150:580-596.

Unkeless, J. C., Gordon, S., and Reich, E., 1974, Secretion of plasminogen activator by stimulated macrophages, *J. Exp. Med.* 139:834-850.

Williams, D. M., and Remington, J. S., 1977, Effects of human monocytes and macrophages on *T. cruzi*, *Immunology* 32:19-23.

Chapter 7

Expression and Development of Macrophage Activation for Tumor Cytotoxicity

William J. Johnson and Scott D. Somers

Department of Pathology
Duke University Medical Center
Durham, North Carolina 27710

and

Dolph O. Adams

Departments of Pathology and Microbiology-Immunology
Duke University Medical Center
Durham, North Carolina 27710

I. INTRODUCTION

Mononuclear phagocytes of the tissues, when exposed to the numerous humoral mediators found at sites of inflammation, and particularly when exposed to lymphokines, undergo profound alterations in their cellular physiology (for review, see Adams and Marino, 1984). These intricate, tightly coordinated changes in metabolism and structure, broadly termed activation even though they are often defined in more restrictive ways, are essential to the contributions made by mononuclear phagocytes in maintenance of homeostasis, in metabolism of endogenous and exogenous toxins, in regulation of inflammation, and in the immune response and defense against tumors and microbes. Since the metabolic consequences of inductive signals leading to activation are extremely complex, examination of macrophage activation for a single function can provide a useful model for studying the cell biology of activation (Adams and Marino, 1984).

Macrophages taken from sites of inflammation containing numbers of committed T lymphocytes, when cocultivated for several days with neoplastic cells, dramatically lyse almost all the tumor targets. Since this lytic phenomenon was initially described (Alexander and Evans, 1971; Hibbs *et al.*, 1972), macrophage-mediated tumor cytotoxicity (MTC) has been of great interest because of its relevance to host defense against neoplasia *in vivo* and because of its striking efficiency and target selectivity *in vitro* (see Adams and Marino, 1984). Specifically, one activated macrophage can lyse multiple targets; these destructive effects are directed almost solely at neoplastic rather than nonneoplastic cells, even when both types of targets are cocultivated with the activated macrophages (see Adams and Marino, 1984). Numerous laboratories have clearly documented that MTC requires intimate contact between macrophages and tumor cells and does not depend on either phagocytosis of the tumor cells or the demonstrable presence of lytic mediators within the culture fluids (Adams and Marino, 1984).

Over the past several years, our laboratory has found that macrophages activated for tumor cytotoxicity have two properties that clearly distinguish them from their noncytolytic counterparts: activated macrophages vigorously bind tumor cells to their surface and copiously secrete a novel proteinase, cytolytic proteinase (CP), that selectively and potently lyses tumor cells. The capacity to mediate selective binding and the capacity to secrete CP are both necessary in order for activated macrophages to complete tumor cytotoxicity and indeed interact with one another in a defined sequence to mediate lysis. Despite this intimate interrelationship, the capacity for selective binding and that for secreting CP are clearly distinct and independently regulated properties of activated macrophages. It has recently become clear that the signals that induce these two capacities are closely related to the signals that induce the activation of macrophages for cytolysis. This chapter reviews studies supporting these conclusions and provides evidence suggesting that induction of activation for tumor cytolysis is equivalent to induction of those capacities necessary for that function.

II. TARGET BINDING BY ACTIVATED MACROPHAGES

Observation of neoplasms *in vivo* often reveals substantial numbers of intratumoral macrophages in close proximity to or in contact with the tumor cells (e.g., Alexander, 1976). A correlate of this relationship has been observed *in vitro* in several laboratories: Activated macrophages cluster around neoplastic cells before they induce lysis of the targets (see Adams and Marino, 1984; Evans and Alexander, 1976). These microscopic observations were made quantitative by Piessens (1978), who devised a method for measuring the extent of macrophage–tumor cell interaction. When radiolabeled neoplastic cells were added to cultures of macrophages, guinea pig macrophages treated *in vitro* with lymphokines bound more tumor cells than did control macrophages (Piessens, 1978).

Macrophage-tumor cell interactions have also been studied in our laboratory by use of murine macrophages activated *in vivo* and *in vitro* (Marino and Adams, 1980a; Somers *et al.*, 1983). The assay consists of establishing monolayers of macrophages to which are added target cells previously labeled with radioisotope. The cultures are then incubated for varying periods of time, and unbound targets are removed by vigorous washes. The radioactivity remaining bound to the monolayer is determined, and the number of bound targets is then quantified.

A. Two Types of Binding

For several years, we have observed two general types of binding between macrophages and target cells *in vitro* (Marino and Adams, 1980a; Somers *et al.*, 1983). First, nonselective binding of low extent (i.e., 4-16% of added targets bound or 2-8 $\times$ 10^4 targets bound/10^6 macrophages) is observed between either resident or inflammatory macrophages and either neoplastic cells, lymphocytes, or lymphoblasts (Marino and Adams, 1980a). No matter how many targets are added to such a monolayer of macrophages, the binding does not become saturated (Marino and Adams, 1980a). This binding does not lead to target lysis, and no functional consequences of such binding have yet been observed in the macrophages.

Second, selective binding of an appreciable extent (i.e., 25-60% of added targets bound or 18-30 $\times$ 10^4 targets/10^6 macrophages) is observed between activated macrophages and a wide range of nonadherent, neoplastic targets (Marino and Adams, 1980a). More than 20 murine neoplastic cell lines of disparate origins have been demonstrated to bind significantly to macrophages activated for cytotoxicity *in vivo* or *in vitro* (Marino and Adams, 1980a, 1982; Adams and Dean, 1982). The amount of target cell binding, which is directly proportional to the number of interacting macrophages and of target cells, can be readily saturated by addition of excess targets (Marino and Adams, 1980a).

The low-level and high-level binding are also found to have different morphologies when viewed by either phase-contrast or scanning electron microscopy (Somers *et al.*, 1983; Marino and Adams, 1980b). The low-level binding is characterized by a random distribution of a few targets to all types of macrophages in the monolayer; rarely is more than one target cell adherent per macrophage (Marino and Adams, 1980a); the few targets that are bound are located around the periphery of the macrophages (Fig. 1A). By contrast, the high-level binding to activated macrophages is characterized by clustering of four to six targets over the main body of the most highly spread macrophages (six shown in Fig. 1B). High-level binding results in a close apposition of the bacillus Calmette-Guérin (BCG)-activated macrophages to a portion of the plasma membranes of the tumor targets (Marino and Adams, 1980a). Transmission electron

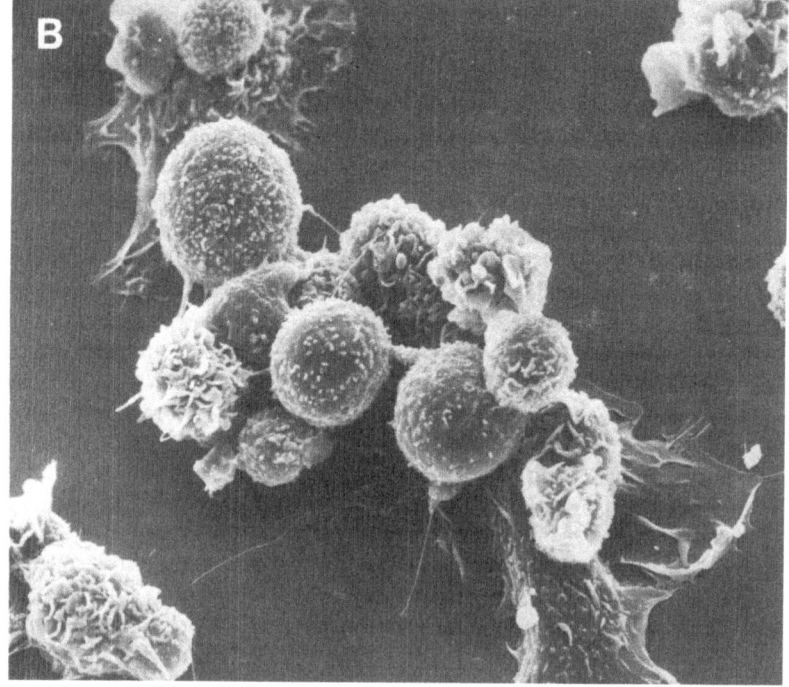

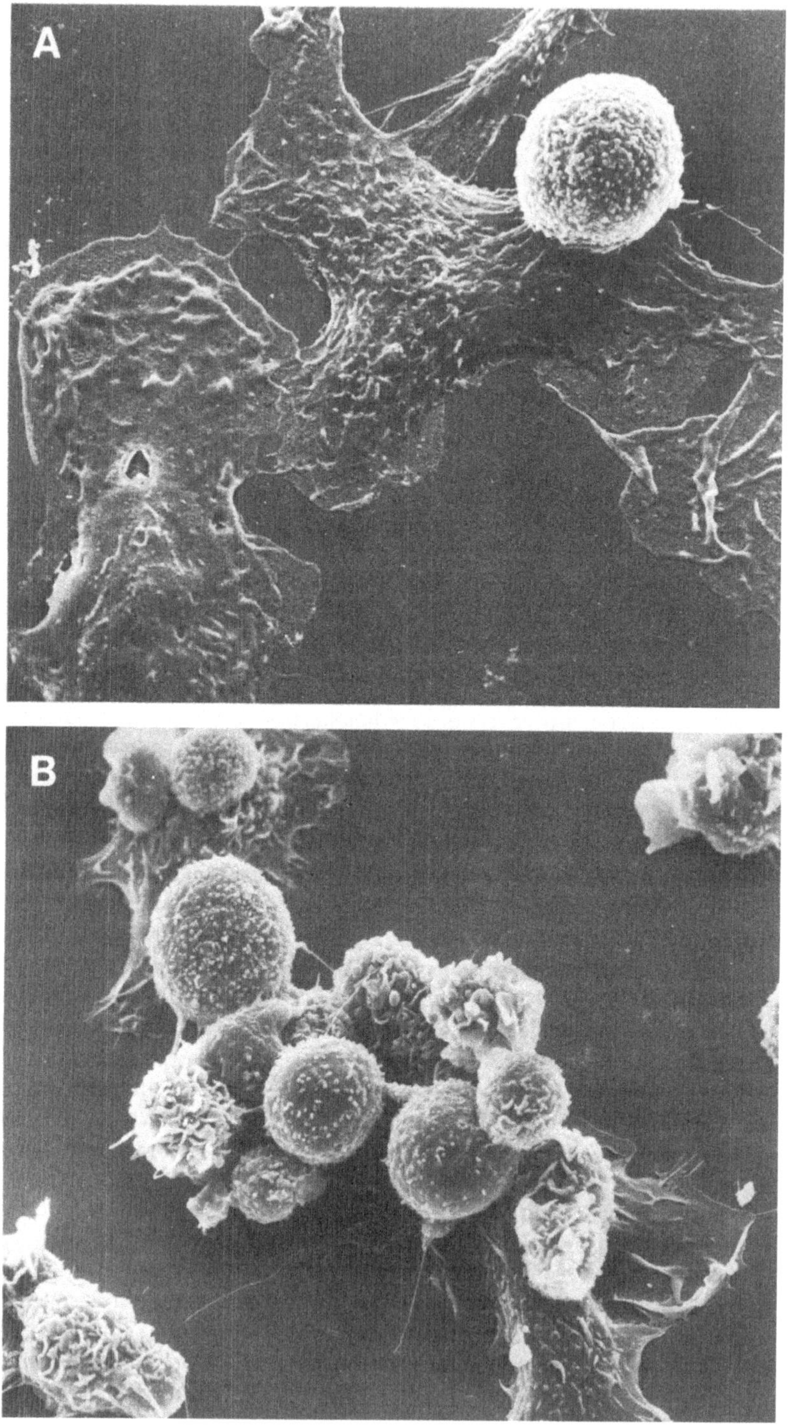

microscopy of the activated macrophage–tumor cell interface reveals a zone of very close contact between the two cells. At this interface, the membranes of the two cells are closely apposed to one another to create a space ~100 Å in width. At the periphery are areas of even closer contact (~20-30 Å wide) (Marino and Adams, 1980a).

The augmented binding of tumor cells to activated macrophages requires the presence of divalent cations, while the low-level binding to inflammatory macrophages does not (Marino and Adams, 1980a; Somers et al., 1983). The binding site on activated macrophages is trypsin sensitive. Pretreatment of activated macrophages with trypsin reduces tumor cell binding capacity, but the low-level binding of targets to inflammatory macrophages is resistant to trypsin pretreatment (Somers et al., 1983). Of particular interest, either trypsinization of the activated macrophage or removal of divalent cations reduces the number of targets bound by activated macrophages to levels comparable with those for inflammatory macrophages. Both types of binding also differ in requirements for energy, microtubules, and microfilaments and in the strength of binding once it has fully developed (see Section II.D).

Another profound difference between low-level and high-level binding is that the latter can be competitively inhibited (Marino et al., 1981; Somers et al., 1983). During binding, addition of either unlabeled targets or of membranes from tumor targets inhibits high-level but not low-level binding (Marino and Adams, 1980b; Marino et al., 1981). When excess unlabeled targets are added to macrophages with targets already bound, targets are competitively removed from BCG-activated macrophages but not from inflammatory macrophages (Somers et al., 1983).

The latter two observations suggest that in some regards, macrophage–tumor cell binding resembles the specific binding between ligands and cellular receptors. The usual method for quantifying specific binding between a ligand and its receptor is to determine binding of the radiolabeled ligand in the presence of a large excess of unlabeled ligand (Hollenberg and Cuatrecassas, 1979). In this circumstance, almost all the specific binding sites are occupied by unlabeled ligand, and any labeled ligand attached to the cells is nonspecifically bound (Hollenberg and Cuatracasas, 1979). We have used a similar method to estimate the specific binding of tumor targets to macrophages (Somers et al., 1983). Use of ^{111}In-oxine to label targets allows a much higher specific activ-

---

Figure 1. Morphology of the interaction of P815 tumor targets with macrophages. After 1-hr incubation at 37°C between macrophages and P815 tumor targets, unbound targets were removed by vigorous washings. The cultures were fixed with glutaraldehyde and then dehydrated, critical-point-dried, coated with platinum, and viewed with a scanning electron microscope. (A) Only one tumor cell at the periphery of the inflammatory macrophage is seen (×2290). (B) Six P815 tumor targets are clustered on the central portion of the BCG-activated macrophages in the middle of the field (×2240). (Reprinted by permission from Somers et al., 1983).

ity in the targets and thus quantifying the binding of low numbers of probe cells. Employing 400 unlabeled targets to one labeled target, binding of tumor targets to inflammatory macrophages is almost entirely nonspecific, while binding to activated macrophages is mostly (i.e., ~75%) specific (Somers *et al.*, 1983). The binding of lymphocytes to either type of macrophage is almost entirely nonspecific.

The differences between selective and nonselective binding are summarized in Table I. Of note, the binding of any probe cell tested thus far to inflammatory macrophages is almost entirely nonselective. By contrast, binding of tumor cells to activated macrophages comprises selective and nonselective elements though the former predominates by far.

Of further interest, selective binding has many characteristics generally associated with the specific binding of ligand to a cell-surface receptor (see Table I) (Hollenberg and Cuatrecasas, 1979). Specifically, ligand selectivity, saturability, high affinity, and susceptibility to competitive inhibition are characteristic of

Table I. Characteristics of Tumor Cell Binding by Activated and Inflammatory Mϕ[a]

Property	Selective binding to activated Mϕ	Nonselective binding inflammatory Mϕ	Reference
Total capacity for binding	Low	High	Somers *et al.* (1983)
Affinity for binding	Relatively higher	Relatively lower	Somers *et al.* (1983)
Saturable	Yes	No	Marino and Adams (1980a), Somers *et al.* (1983)
Competitively inhibitable			
During binding of targets	Yes	No	Marino and Adams (1980a)
After binding complete	Yes	No	Somers *et al.* (1983)
Cation dependent	Yes	No	Marino and Adams (1980a), Somers *et al.* (1983)
Dependent on trypsin-sensitive surface components on the Mϕ	Yes	No	Marino and Adams (1980a), Somers *et al.* (1983)
Dependent on metabolic energy	Yes	No	Marino and Adams (1980a), Somers *et al.* (1983)
Dependent on cytoskeletal components	Yes	No	Somers *et al.* (1983)
Target selective	Yes	No	Marino and Adams (1980a), Marino *et al.* (1981)
Functional consequences	Yes	No	Marino and Adams (1980b), Johnson *et al.* (1981)
Type	Mostly specific	Mostly nonspecific	Somers *et al.* (1983)

[a]Mϕ, macrophage.

ligand–receptor interactions. Although this interesting possibility awaits critical verification, it is of particular interest that binding of either tumor cells or membranes thereof to activated macrophages induces two functional alterations in the macrophages: initiation of cytolysis and secretion of CP (see Section III.A).

B. Relationship of Target Binding to Cytolysis

Several lines of evidence indicate that selective target binding is an essential part of macrophage-mediated cytolysis (Adams and Marino, 1981):

1. The selectivity of augmented binding mimics that of cytolysis, i.e., only neoplastic cells are bound and are bound only by activated macrophages (Marino and Adams, 1980b).
2. Addition of unlabeled tumor cells or of tumor cell membranes inhibits selective binding and proportionately decreases the extent of cytolysis (Marino and Adams, 1980b; Adams et al., 1981b). Other experimental manipulations, which increase or decrease the extent of binding, either increase or decrease the extent of cytolysis, respectively (Adams and Marino, 1981).
3. BCG-elicited macrophages from A/J mice are genetically deficient in their capacity to bind neoplastic cells and are incapable of effecting macrophage-mediated cytolysis (Adams et al., 1981b). This lack of cytolysis is in spite of the fact that the macrophages produce large amounts of cytolytic protease (see below).
4. Placement of a porous filter between activated macrophages and neoplastic targets abrogates both binding and cytolysis (Adams and Marino, 1981).

The binding of targets to macrophages, however, does not necessarily lead to target destruction. For example, macrophages primed in vivo with pyran copolymer or BCG or in vitro by lymphokines do bind targets, but do not lyse them (Adams and Marino, 1981; Adams and Dean, 1982). In summary, the binding of targets to activated macrophages is necessary but not sufficient for completion of macrophage-mediated cytolysis (Table II).

Macrophages elicited by BCG injection in A/J mice can be used to test critically the hypothesis that selective binding is necessary for cytolysis. A/J mice are genetically deficient in macrophage-mediated cytolysis because of an inability to bind tumor cells (Adams et al., 1981b). The deficit in lytic capacity, however, is almost total, whereas the deficit in total binding capacity is but partial (i.e., macrophages from A/J mice bind at 25-33% of control levels). When examined in detail, macrophages from A/J mice bind tumor cells nonselectively (Somers et al., 1983).

Table II. Evidence That Selective Binding Is Necessary for MTC[a]

The selectivity of binding, in regard to type of Mϕ and type of target, resembles precisely that of lysis.

Experimental manipulations that either increase or decrease selective binding, respectively, increase or decrease lysis and do so proportionately.

Mϕ deficient in capacity for selective binding are deficient in capacity for lysis.

Abrogation of selective binding abrogates lysis.

Development of binding capacity during activation correlates with development of lytic capacity.

[a]*Abbreviations:* Mϕ, macrophage; MTC, macrophage-mediated tumor toxicity.

C. Induction of Binding Capacity

The ability to complete selective or augmented binding of tumor cells is an acquired property of macrophages. Specifically, only macrophages in the last two stages of activation (i.e., primed macrophages and activated macrophages) can complete binding. Resident peritoneal macrophages and inflammatory macrophages are unable to mediate selective binding (Marino and Adams, 1980*a*; Johnson *et al.*, 1983).

The capacity for augmented binding is induced principally, if not entirely, by lymphokine(s) (Marino and Adams, 1982). Exposure of young mononuclear phagocytes from sites of inflammation (i.e., responsive macrophages), although not of resident peritoneal macrophages, to lymphokine(s) induces this capacity (Marino and Adams, 1982). This effect is dependent on the concentration of lymphokine(s); optimum induction of binding capacity requires 8-12 hr of interaction between macrophages and lymphokine(s). Only lymphokine(s) appear to be required; a requirement for any additional signal, including endotoxin, has yet been established. Of interest, the requirements and kinetics of induction closely resemble those for induction of priming for cytolysis (Marino and Adams, 1982). The precise molecular signals that induce binding capacity and the precise alterations in macrophage physiology required for completion of selective binding remain to be established.

D. Cell Biology of Binding

To understand the cell biology of selective binding, it is instructive to compare binding by activated macrophages with that by inflammatory macrophages. Binding of tumor targets to BCG-activated macrophages is time dependent, and maximum binding occurs after 1-3 hr of interaction (S. D. Somers, C. Whisnant, and D. O. Adams, unpublished observations, 1983). These workers have found that the binding of targets to inflammatory macrophages, however, reaches its maximum within 15 min.

We have recently begun to study the forces required to dislodge targets bound to macrophages (S. D. Somers, C. Whisnant, and D. O. Adams, unpublished observations, 1983). Our data already suggest the interaction of targets with macrophages occurs in at least two steps. Initially, a weak interaction occurs very rapidly after probe cells are brought into contact with macrophage monolayers. This interaction occurs between all types of macrophages and either tumor targets or lymphocytes. If the macrophages are activated and if the probe cells are neoplastic, the weak interactions are then converted over 1–3 hr to a much stronger interaction. This conversion is inhibited at 4°C.

Augmented binding requires metabolically active macrophages (Marino and Adams, 1980a; Somers et al., 1983). Pretreatment of BCG macrophages with 2-deoxyglucose, iodoacetic acid, or sodium azide, at concentrations sufficient to reduce phagocytosis of IgG-coated sheep red blood cells, significantly inhibits tumor cell binding. These inhibitors do not reduce the low-level binding of target cells to inflammatory macrophages, although the phagocytic activity of such macrophages is significantly decreased (Somers et al., 1983). Cycloheximide does not inhibit augmented binding, but intact microtubules and microfilaments are required for binding of tumor cells to BCG-activated macrophages. Specifically, concentrations of colchicine or cytochalasin B, which significantly inhibit phagocytosis of opsonized sheep red blood cells, inhibit tumor cell binding by activated macrophages (Somers et al., 1983). By contrast, the low-level binding of tumor cells to inflammatory macrophages is not inhibited by either colchicine or cytochalasin B (Somers et al., 1983).

Transmethylation reactions are also required for tumor cell binding by activated macrophages (Adams et al., 1981c). Of note, methylation reactions, which are necessary for phagocytosis of targets bound via the Fc receptor, are not necessary for binding of targets to Fc receptors. Although transmethylation reactions are involved in a variety of cellular functions, we have questioned whether their effects on membrane fluidity might play a role in regulatory binding. We have subsequently observed that treatments that decrease the membrane microviscosity of activated macrophage membranes enhance binding capacity (S. D. Somers and D. O. Adams, unpublished observations). If a membrane receptor on macrophages is involved in binding (see Section II.A), altered exposure, number, affinity, or mobility of the putative tumor recognition site within the macrophage plasma membrane could result. All these possibilities have resulted from the alteration of membrane fluidity in other biologic systems (for review, see Shinitsky and Henkart, 1980).

The molecular basis of recognition between activated macrophages and neoplastic cells is not yet established. The possibility that a receptor on macrophages is involved in the binding (Section II.A) remains to be critically tested. The structure(s) on tumor cells responsible for the recognition of these cells by activated macrophages also remain(s) to be identified. Recent evidence indicates that structures contained within the plasma membranes of three murine tumor cell lines may be responsible for binding of these cells to activated macrophages (Marino *et al.*, 1981). Intact and unlabeled tumor targets readily prevent binding of labeled targets to activated macrophages (Marino and Adams, 1980*a*). Membrane preparations from a leukemia, lymphoma, and a mastocytoma bind to BCG-activated macrophages and competitively inhibit binding of homologous and heterologous targets (Marino *et al.*, 1981). Membrane preparations from normal lymphocytes, at equal concentrations of membrane proteins, do not inhibit such binding. The low-level binding of tumor cells and lymphocytes to inflammatory macrophages, however, is not inhibited by the membrane preparations. The inhibitory activity of membrane preparations may be adsorbed by passage over BCG-activated but not over inflammatory macrophages. Taken together, the data suggest that a structure or structures responsible for binding of these three neoplastic targets are contained within plasma membranes of the tumor cells.

These initial studies on the cell biology of binding do not as yet permit formulation in precise detail of how stable attachment of tumor cells to activated macrophges is achieved. Our working hypothesis is as follows: Contact between diverse cell pairs leads to rapid but unsubstantial attachment attributable to weak forces such as elecrostatic interactions or van der Waals forces. Activated macrophages possess a trypsin-sensitive receptor(s) in the plasma membrane that recognizes components within the plasma membrane of tumor cells. Interaction between these structures and the receptor(s) on macrophages, in cooperation with microtubules and microfilaments of the macrophages and with alterations in fluidity of macrophage membranes (alterations in the function/mobility of macrophage membrane proteins or receptors?) leads to the gradual development of firm attachment of tumor cell to an activated macrophage. It is worth noting that the binding of antibody-coated tumor cells to activated macrophages via Fc receptors does not have such complex requirements for the development of stable binding (S. D. Somers and D. O. Adams, unpublished observations, 1982). We are now critically examining this hypothesis and attempting to define those capacities possessed by activated macrophages that permit the development of firm binding. It will be of particular interest to define the relationship between these capacities and the physiologic alterations in macrophages induced by lymphokines.

III. SECRETION OF A CYTOLYTIC PROTEASE BY ACTIVATED MACROPHAGES

Supernatants from BCG-activated macrophages, when collected in the absence of serum, have significant lytic activity against neoplastic cells (Adams *et al.*, 1980). The lytic principle contained in these supernatants, termed CP, is quantified by culturing radioactively labeled tumor targets with various dilutions of the macrophage supernatants and then determining the release of radiolabel from the killed targets (Adams *et al.*, 1980). Lytic activity is readily detected in supernatants collected from cultures of BCG-activated macrophages, but not in supernatants from resident peritoneal macrophages, inflammatory macrophages, or plastic culture vessels (Adams *et al.*, 1980).

CP is potent—dilutions of up to 1:200 of supernatants from a monolayer of BCG-activated macrophages will readily lyse more then 50,000 tumor cells (Adams *et al.*, 1980). Specifically, the concentration of CP that will lyse one-half of a population of MCA-I sarcoma target cells is estimated to be $\sim 10^{-9}$ M; CP is $\sim 10^5$ times more cytolytic than are equally proteolytic amounts of trypsin (Adams *et al.*, 1982). This lytic activity is selective, however, in that neoplastic cells are killed but nonmalignant cells are spared under normal conditions. More then 20 different neoplastic targets have been tested for susceptibility to CP, and all are susceptible, although the degree of sensitivity can vary from target to target (Adams *et al.*, 1980). The neoplastic targets tested have included sarcomas, carcinomas, lymphomas, leukemias, and melanomas. In contrast, six nonneoplastic targets have been tested and none was found susceptible to CP (Adams *et al.*, 1980).

When tested against a susceptible cell line, lytic activity of CP as monitored by release of ^{51}Cr from labeled tumor cells becomes apparent after 5-6 hr of interaction with the targets (Adams *et al.*, 1980; Johnson *et al.*, 1982*b*). Target lysis then proceeds progressively in a sigmoidal manner until maximum lysis is observed at 16-20 hr. The final degree of lysis observed can reach 100% and is dependent on the concentration of CP used (Adams *et al.*, 1980; Johnson *et al.*, 1982*b*). The actual amount of lysis produced by a given dilution of CP varies, depending on the potential of the macrophage for tumor cytolysis (Johnson *et al.*, 1981). In addition, maximum lysis is often observed at dilutions not containing the maximum concentration of conditioned medium. For these reasons, data are usually reported as the dilution of CP at which maximum lysis is observed. The content of CP in a given supernatant can also be readily estimated by probit analysis of lytic curves (Johnson *et al.*, 1981).

The lytic effects of CP depends on the direct interaction of CP with the targets and not on interaction with medium components such as arginine (Adams *et al.*, 1980; Johnson *et al.*, 1982*b*). CP does not appear to induce programming of tumor targets for lysis, as incubation of tumor targets with CP for less than 6 hr followed by removal of CP does not result in appreciable lysis of targets (Johnson *et al.*, 1982*b*). After 6 hr, lysis begins but is dependent on the continued presence of CP. CP is not adsorbed by either normal or neoplastic cells; studies with trypan blue indicate further that target cell injury and death occur well before adherent targets are released from the culture vessels (Johnson *et al.*, 1982*b*). In addition, the action of CP on tumor targets does not appear to be dependent on cofactor(s) secreted by the tumor cells themselves (Johnson

Table III. Some Characteristics of Cytolytic Protease[a]

Inhibited by three inhibitors of serine proteases as well as by fetal calf serum

$M_r \sim 40,000$

Cochromatographs with a novel protease secreted only by activated macrophages

Kills tumor cell targets selectively as opposed to nonneoplastic targets

CD_{50}: $\sim 10^{-9} M$

Requires 16 hr to induce lysis

Secretion regulated in two steps: preparation and release

[a]*Abbreviations:* CD_{50}, cytolytic dose at 50%; CP, cytolytic protease; M_r, molecular weight.

et al., 1982*b*). Taken together, these data suggest that the interaction of tumor cells with CP has several characteristics of an enzyme–substrate interaction.

Biochemical characterization of CP is as yet incomplete, but several lines of evidence suggest that CP is a neutral serine protease:

1. The lytic effects of secreted CP are readily inhibited by fetal calf serum, by heating the CP to $56°C$ for 30 min, and by the serine protease inhibitors bovine pancreatic trypsin inhibitor (BPTI), diisopropylflurophosphate (DFP), and α_2-macroglobulin (Adams, 1980; Johnson *et al.*, 1981). CP is not, however, inhibited by catalase, arginine, glucose, protein hydrolysate, supernatants of inflammatory macrophages, or extensive (>99%) dialysis against fresh medium (Adams *et al.*, 1980).

2. Upon separation of supernatants of activated macrophages by gel filtration, the lytic activity is contained in a fraction with a M_r of $\sim 40,000$ (Adams *et al.*, 1980). The lytic activity cochromatographs with a unique neutral protease secreted by activated, but not resident or inflammatory macrophages (Adams *et al.*, 1980). The lytic activity of the separated protease is inhibited by BPTI and DFP (Adams *et al.*, 1980).

Thus, CP appears to be a neutral serine protease of $M_r \sim 40,000$ (Table III).

A. Relationship of Cytolytic Protease Secretion to Tumor Cytolysis

The secretion of CP correlates strongly with macrophage competence for tumor cytolysis (Table IV). Macrophages from unmanipulated mice neither lyse tumor targets nor secrete CP. Nor do inflammatory macrophages, such as those elicited by thioglycollate broth, kill tumor cells or secrete CP, except in the presence of large (microgram) quantities of bacterial lipopolysaccharide (LPS). Macrophages, primed for cytolysis *in vivo* by BCG or pyran copolymer or *in vitro* by lymphokines, do not kill tumor cells or secrete CP unless small (nanogram) quantities of LPS are present (Adams *et al.*, 1980, 1983; Johnson

Table IV. Evidence That Secretion of CP is Necessary for MTC[a]

Spontaneous and stimulated secretion of CP correlate closely with spontaneous and stimulated MTC.

The amount of CP secreted is consistent with the amount required for completion of MTC (in terms of number of targets killed).

Low-molecular-weight inhibitors of CP inhibit MTC and do so by acting at the target injury stage.

Mϕ deficient in release of CP are deficient in MTC. Restoration of CP secretion to such Mϕ by addition of alternative or appropriate triggers restores capacity for MTC.

Binding of tumor cells to activated Mϕ stimulates secretion of CP.

The time course, morphology, and target susceptibility to lysis with CP correlate closely with these characteristics of MTC.

---

[a]*Abbreviations:* CP, cytolytic protease; Mϕ, macrophage; MTC, macrophage-mediated tumor toxicity.

et al., 1983). The amounts of LPS required to induce secretion of CP and tumor cytolysis by primed macrophages are virtually identical. Macrophages activated for spontaneous tumor cytolysis by BCG or by *Propionibacterium acnes* (*Corynebacterium parvum*) secrete CP spontaneously (Adams *et al.*, 1980; Johnson *et al.*, 1983b). In addition, BCG-activated macrophages lose the capacity to lyse tumor cells and to secrete CP when they are held in culture for 24 hr. Pulsing these macrophages with nonogram amounts of LPS restores both functions (Adams and Marino, 1981). Finally, BCG-elicited macrophages from C3H/HeJ mice have a defect in their ability to respond to LPS, do not lyse tumor cells, and do not secrete CP (Adams *et al.*, 1981b).

Further evidence that secretion of CP plays a role in macrophage-mediated tumor cytolysis comes from experiments with protease inhibitors (Adams, 1980; Adams and Marino, 1981). BPTI and DFP inhibit macrophage-mediated cytotoxicity at concentrations that inhibit CP. Both inhibitors are effective only if added to the macrophages during cocultivation with tumor cells; pretreatment of macrophages with the inhibitors is not effective (Adams, 1980). The inhibitors are now known to act at the target injury phase of tumor cytolysis (Adams and Marino, 1981). Importantly, BPTI does not inhibit the release of hydrogen peroxide from macrophages (Goldstein *et al.*, 1979) and does not inhibit the cytolysis of antibody-coated tumor targets (Adams *et al.*, 1984) a process in which lysis has been attributed to secreted hydrogen peroxide (Nathan and Cohn, 1980). In addition, the kinetics of tumor cell lysis by CP are quite similar to those by intact macrophages (Adams *et al.*, 1980).

A final line of evidence that CP plays an important role in macrophage-mediated tumor cytolysis comes from examining the relationship between target binding by macrophages and secretion of CP. When BCG-activated macrophages are pulsed with tumor targets or with plasma membrane preparations from such tumors, the macrophages are triggered to secrete augmented amounts of CP

(Johnson *et al.*, 1981). The augmented secretion is selective in that only tumor targets added to activated macrophages result in augmented secretion. When tumor targets are added to inflammatory macrophages or when normal targets are added to activated macrophages, augmented secretion of CP is not seen. The augmented secretion of CP occurs quite rapidly, beginning in fewer than 30 min after binding of tumor cells to the activated macrophages. Augmented secretion is selective itself in that secretion of plasminogen activator, another serine protease, is not increased (Johnson *et al.*, 1981).

B. Relationship between Selective Binding and Secretion of Cytolytic Protease

Although target binding and CP secretion are intimately related, several lines of evidence establish that these two functions are separate and independently regulated:

1. Certain inbred strains of mice are genetically deficient in their ability to mediate macrophage–tumor cytolysis (Meltzer *et al.*, 1979). Macrophages elicited by BCG from C3H/HeJ mice are fully competent to bind neoplastic targets but are markedly deficient in their ability to secrete CP (Adams *et al.*, 1981*b*). Conversely, BCG-elicited macrophages from A/J mice have normal capacity for secreting CP, but do not bind tumor cells appreciably (Adams *et al.*, 1981*b*).
2. Activated macrophages held in culture for 1 day lose their ability to secrete CP and to lyse tumor cells; binding capacity is not, however, appreciably diminished (Adams and Marino, 1981). Restoration of secretion of CP is synonymous with restoration of lytic capacity.
3. Development of the ability to bind tumor cells and to secrete CP occurs at distinct stages of activation (Johnson *et al.*, 1983*b*). Macrophages acquire the capacity for tumor cytolysis in a defined progression (Hibbs *et al.*, 1978; Meltzer *et al.*, 1979). Resident and responsive macrophages cannot mediate selective binding of tumor cells, whereas primed and activated macrophages can (see Section V). Resident macrophages do not secrete CP, even when large (microgram) amounts of LPS are needed; responsive macrophages secrete CP only when microgram amounts of LPS are added. Primed macrophages can secrete CP when exposed to nanogram amounts of LPS, while activated macrophages secrete CP spontaneously (Johnson *et al.*, 1983).

Thus, the paths for acquisition of capacity for selective binding and capacity for secretion of CP are distinct.

C. Regulation of Secretion of Cytolytic Protease

Secretion of CP, like many other secretory products of macrophages, is regulated in two steps: (1) an initial signal primes or prepares the macrophages for release of CP, and (2) a second signal triggers actual release of CP (Johnson *et al.*, 1983).

Preparation for secretion of CP appears to be regulated principally by lymphokine(s) (Johnson *et al.*, 1983). Macrophages elicited by fetal calf serum are unable to secrete CP spontaneously or after application of triggering signals (Johnson *et al.*, 1983). When cultured with lymphokine(s) for 7-10 hr, the macrophages gain competence to secrete CP if challenged with a signal such as nanogram amounts of LPS plus tumor cells (Johnson *et al.*, 1983). Macrophages can thus be primed for CP secretion *in vitro*. The dose curve and kinetics of lymphokine priming for CP secretion are similar to those for induction of binding and of priming for cytolysis of tumor cells (Johnson *et al.*, 1983). Macrophages can be primed for CP secretion *in vivo*, if taken from sites of inflammation where committed T cells are active. Studies with fractions of partially purified lymphokines indicate that activity for inducing binding and priming for cytolysis elute in the same fractions (D. O. Adams and M. S. Meltzer, unpublished observations). The lymphokines that induce these capacities need to be defined precisely.

The diverse signals that trigger release of CP from primed macrophages include LPS, binding of tumor cells, and binding of membranes of tumor cells. It is highly probable that other substances, such as *Listeria monocytogenes*, the lipid A portion of LPS, and high concentrations of lymphokines that trigger cytolysis by primed macrophages, also trigger release of CP although this remains to be documented formally.

Recent studies have demonstrated yet another system that regulates CP secretion from primed macrophages: engagement of scavenger receptors on the macrophages (Johnson *et al.*, 1982*a*). Receptors on macrophages that bind and internalize maleylated or acetylated proteins are important in homeostasis because they can rid the extracellular compartment of lipoproteins (Brown *et al.*, 1980). When receptors for maleylated or acetylated proteins are triggered by maleylated bovine serum albumin (Mal-BSA) or by fucoidin, an algal polysaccharide recognized by the Mal-BSA receptor, it has been found that a dose-dependent increase in CP secretion is induced (Johnson *et al.*, 1982*a*). This increase occurs when either primed macrophages with a low basal secretory level or activated macrophages with a high basal secretory level are examined (Johnson *et al.*, 1982*a*). In contrast to secretion triggered by bound tumor cells (Johnson *et al.*, 1981), engagement of the Mal-BSA receptor does not result in selective secretion of CP as total neutral caseinases and plasminogen activator secretion are also triggered. Selectivity is evident, however, in that secretion of the lysosomal hydrolase acid phosphatase is not triggered (Johnson *et al.*, 1982a). Tumor cytolysis is also induced by engagement of these receptors (Johnson *et al.*, 1982a). In fact, the lytic deficit in macrophages elicited by BCG from endotoxin-resistant C3H/HeJ mice can be overcome. BCG-elicited macrophages from these mice secrete CP and become tumoricidal when the Mal-BSA receptor is triggered (Johnson *et al.*, 1982a).

Of interest, macrophage receptors for α_2-macroglobulin-protease complexes regulate protease secretion and tumoricidal activity; both functions are decreased in a dose-dependent manner after the addition of α_2-macroglobulin-trypsin complexes (Johnson *et al.*, 1982a).

IV. MACROPHAGE-MEDIATED CYTOLYSIS AS A MULTISTEP PROCESS

Evidence presented to this point suggests that in order for macrophage-mediated tumor cytolysis to occur, at least two events must take place: (1) selective binding of tumor cells to macrophages, and (2) secretion of cytolytic effector molecules, including CP, by the macrophages. Several lines of evidence indicate that these two steps occur in cytolysis, that binding precedes secretion of CP, that both are necessary but not sufficient for completion of cytolysis, and that the second step (or the step of target injury) is induced by agents that induce release of CP and is blocked by agents that inhibit CP (see Adams and Marino, 1984; Adams *et al.*, 1982; Johnson and Adams, 1984; Johnson *et al.*, 1982).

These observations are clarified by experiments that employ porous filters interposed between activated macrophages and tumor target cells (Adams and Marino, 1981). When serum is present in the medium and such filters separate the two cell types, binding and ctyolysis are both abrogated completely (Adams and Marino, 1981). If the same experiment is conducted in the absence of serum, binding remains abrogated, but cytolysis is restored (Adams and Marino, 1981). If the protease inhibitor BPTI is added to the serumless medium in a similar experiment, binding is still abolished and cytolysis is again inhibited (Adams and Marino, 1981). These observations suggest that secretory products from macrophages do not cause target cytolysis in the presence of serum and that the predominant lytic product is a protease. It is not surprising that serum inhibits CP, since serum contains large amounts of protease inhibitors, including α_2-macroglobulin, which is known to inhibit CP (Johnson *et al.*, 1981). The data further suggest that binding of tumor cells by macrophages provides a mechanism to abrogate serum-mediated inhibition of secretory components of macrophages. When macrophages selectively bind neoplastic targets, a space is created between the contact points of the two cell types (Marino and Adams, 1980a). The restoration of cytolysis by binding would be explained if this space were diffusion limited to macromolecules (for details, see Adams *et al.*, 1982).

V. MODELS OF MACROPHAGE-MEDIATED CYTOLYSIS AND OF ACTIVATION FOR CYTOLYSIS

A model of cytolysis in MTC can be constructed as follows: First, activated macrophages bind tumor cells vigorously. After firm binding between macrophages and tumor cells is completed, a contact zone between the two cells is created, and metabolic events leading to secretion of lytic substances such as CP are initiated in the macrophages. The contact zone is probably diffusion limited to exclude large molecules such as the antiproteases of serum. Binding thus selects targets for lysis and creates a microenvironment from which macromolecular inhibitors are excluded and in which favorable concentrations of lytic effectors can be achieved.

The next step in MTC would be secretion of lytic effector molecules, including CP, into the contact zone between the macrophages and tumor cells; these molecules produce target injury and eventual target death. A point worth emphasizing is that a number of effectors probably contribute to target cell injury and death. Numerous toxic substances, such as thymidine, arginase, complement components, and hydrogen peroxide, are secreted by macrophages (see Adams and Marino, 1984; Adams and Nathan, 1983). Different effector mechanisms may exist for different tumor targets and several effector molecules may cooperate synergistically with one another. Recent evidence demonstrates that synergy exists between CP and hydrogen peroxide (Adams *et al.*, 1981*a*).

Current models of the activation of murine mononuclear phagocytes for tumor cytotoxicity encompass several stages, which are defined by the sequence of inductive signals that must be applied to the macrophages to induce lytic competence (Hibbs *et al.*, 1978; Meltzer *et al.*, 1979; Russell *et al.*, 1977). Responsive macrophages—generally young macrophages taken from sites of inflammation—are not directly cytolytic and do not respond to triggering signals such as traces of endotoxin. After several hours of incubation with lymphokines, responsive macrophages gain competence in responding to second signals. These latter macrophages, termed primed macrophages, are not cytolytic but become so upon application of a second signal, such as traces of endotoxin. Activated macrophages are fully competent to undergo lysis and require no further signaling.

We have now defined how the capacity for selective binding and the capacity for secreting CP are acquired (Table V). Lymphokine(s) appear to have at

Table V. Capacity for Binding Tumor Cells and Secreting CP by
Various Populations of Mφ[a]

Type of Mφ	Selective tumor binding	Prepared to secrete CP	Secrete CP	Prepared for tumor cell kill	Kill tumor cells
Resident Mφ	−	−	−	−	−
Responsive Mφ	−	−	−	−	−
Primed Mφ	+	+	−	+	−
Activated Mφ	+	+	+	+	+

[a]*Abbreviations:* CP, cytolytic protease; Mφ, macrophage.

least two effects: They induce capacity for selective binding and they prepare macrophages for release of CP. Application of a second signal to lymphokine-treated macrophages induces lysis of tumor cells, since these macrophages can bind targets and are prepared to release CP. These observations indicate why activating signals must be applied in the defined sequence of lymphokine followed by endotoxin: Application of endotoxin to responsive macrophages does not induce release of CP and responsive cells cannot complete selective binding. The general hypothesis that activation for a given function is equivalent to inducing the various capacities necessary for effecting that function was recently reviewed by Adams and Marino (1984).

The biologic basis of macrophage tumor cytolysis appears to require, at minimum, full expression of at least two capacities: selective binding and secretion of CP. The development of these capacities is closely related to the development of competence for tumor cytolysis. Since exposure of responsive macrophages to crude preparations of lymphokines induces capacity for selective binding and preparation for secretion of CP, the precise signals that induce these capacities are currently under investigation. The relevance of other biologic changes in macrophages accompanying activation (see Adams and Marino, 1984) to lytic function now need to be established. Of particular interest will be the determination of how macrophages complete other tumor-destructive functions, such as cytostasis or tumor cytolysis in the presence of anti-tumor antibodies, i.e., antibody-dependent cell-mediated cytotoxicity (ADCC); the cellular physiology of lymphokine-induced alterations; and whether this model of activation is pertinent to activation for other macrophage functions as well.

ACKNOWLEDGMENTS

The work described in this chapter was supported in part by USPHS grants CA29584, CA16784, CA14236, and ES02922 and by a grant from the Kroc Foundation. W.J.J. is a Fellow of the Leukemia Society of America.

VI. REFERENCES

Adams, D. O., 1980, Effector mechanisms of cytolytically activated macrophages. I. Secretion of neutral proteases and effect of protease inhibitors, *J. Immunol.* **124**:286–292.

Adams, D. O., and Dean, J., 1982, Analysis of macrophage activation and of biologic response modifiers by use of objective markers to characterize the stages of activation, in: *NK Cells and Other Natural Effector Cells* (R. B. Herberman, ed.), pp. 511–518, Academic Press, New York.

Adams, D. O., and Marino, P. A., 1981, Evidence for a multistep mechanism of cytolysis by BCG-activated macrophages: The interrelationship between the capacity for cytolysis, target binding, and secretion of cytolytic factor, *J. Immunol.* **126**:981–987.

Adams, D. O., and Marino, P.A., 1984, The activation of mononuclear phagocytes for destruction of tumor cells as a model for the study of macrophage development, in: *Contemporary Hematology/Oncology,* Vol. 3 (A. S. Gordon, R. Silber, and J. LoBue, eds.), pp. 69–136, Plenum Medical Book Company, New York.

Adams, D. O., Kao, K. J., Farb, R., and Pizzo, S. V., 1980, Effector mechanisms of cytolytically activated macrophages. II. Secretion of a cytolytic factor by activated macrophages and its relationship to secreted neutral proteases, *J. Immunol.* **124**:293–300.

Adams, D. O., Johnson, W. J., Fiorito, E., and Nathan, C. F., 1981*a*, Hydrogen peroxide and CF can interact synergistically in effecting cytolysis of neoplastic targets, *J. Immunol.* **127**:1973–1977.

Adams, D. O., Marino, P. A., and Meltzer, M. S., 1981*b*, Characterization of genetic defects in macrophage tumoricidal capacity: Identification of murine strains with abnormalities in secretion of cytolytic factor and ability to bind neoplastic targets, *J. Immunol.* **126**:1843–1847.

Adams, D. O., Pike, M. C., and Snyderman, R., 1981*c*, The role of transmethylation reactions in regulating the binding of BCG-activated murine macrophages to neoplastic targets, *J. Immunol.* **127**:225–230.

Adams, D. O., Johnson, W. J., and Marino, P. A., 1982, Mechanisms of target recognition and destruction in macrophage-mediated tumor cytotoxicity, *Fed. Proc.* **41**:2212–2221.

Adams, D. O., Johnson, W. J., Marino, P. A., and Dean, J. H., 1983, The effect of pyran copolymer on activation of murine macrophages: Evidence for incomplete activation by use of functional markers, *Cancer Res.* **43**:3633–3637.

Adams, D. O., Cohen, M. S., and Koren, H. S., 1984, Activation of mononuclear phagocytes for cytotoxicity: Parallels and contrasts between tumor cytotoxicity and antibody-dependent cytotoxicity, in: *Macrophage Activation and ADCC* (H. S. Koren, ed.), pp. 155–168, Marcel Dekker, New York.

Alexander, P., 1976, The functions of the macrophage in malignant disease, *Annu. Rev. Med.* **27**:207–224.

Alexander, P., and Evans, R., 1971, Endotoxin and double stranded RNA render macrophages cytotoxic, *Nature (New Biol.)* **232**:76–78.

Brown, M. S., Basu, S. K., Falck, J. R., Ho, Y. K., and Goldstein, J. L., 1980, The scavenger cell pathway for lipoprotein degradation: Specificity of the binding site that mediates the uptake of negatively-charged LDL by macrophages, *J. Supramol. Struct.* **13**:67–81.

Evans, R., and Alexander, P., 1976, Mechanisms of extracellular killing of nucleated mammalian cells by macrophages, in: *Immunobiology of the Macrophage* (D. S. Nelson, ed.), pp. 536–576, Academic Press, New York.

Goldstein, B. D., Witz, G., Amoruso, M., and Troll, W., 1979, Protease inhibitors antagonize the activation of polymorphonuclear leukocyte oxygen consumption, *Biochem. Biophys. Res. Commun.* **88**:854–858.

Hibbs, J. B., Lambert, L. H., and Remington, J. S., 1972, Possible role of macrophage mediated nonspecific cytotoxicity in tumor resistance, *Nature (New Biol.)* **235**:48–50.

Hibbs, J. B., Chapman, H. A., and Weinberg, J. B., 1978, The macrophage as an antineoplastic surveillance cell: Biological perspectives, *J. Reticuloendothel. Soc.* 24:549-569.

Hollenberg, M. D., and Cuatrecasas, P., 1979, Distinction of receptor from nonreceptor interaction in binding studies, in: *The Receptors*, Vol. 1, *General Principles and Procedures* (R. D. O'Brien, ed.), pp. 193-214, Plenum Press, New York.

Johnson, W. J. and Adams, D. O., 1984, Activation of mononuclear phagocytes for tumor cytolysis: Analysis of inductive and regulatory signals, in: *Mononuclear Phagocyte Biology* (A. Volkman, ed.), Marcel Dekker, New York (in press).

Johnson, W. J., Whisnant, C. C., and Adams, D. O., 1981, The binding of BCG-activated macrophages to tumor targets stimulates secretion of cytolytic factor, *J. Immunol.* 127:1787-1792.

Johnson, W. J., Pizzo, S. V., Imber, M. J., and Adams, D. O., 1982*a*, Receptors for maleylated proteins regulate neutral protease secretion by murine macrophages, *Science* 218:574-576.

Johnson, W. J., Weiel, J. E., and Adams, D. O., 1982*b*, The relationship between secretion of a novel cytolytic protease and macrophage mediated tumor cytotoxicity, in: *NK Cells and Other Natural Effector Cells* (R. B. Herberman, ed.), pp. 949-954, Academic Press, New York.

Johnson, W. J., Marino, P. A., Schrieber, R. D., and Adams, D. O., 1983, The sequential activation of murine mononuclear phagocytes for tumor cytolysis: Expression of objective markers by macrophages in the several stages of activation, *J. Immunol.* 131:1038-1043.

Marino, P. A., and Adams, D. O., 1980*a*, Interaction of Bacillus Calmette-Guerin activated macrophages and neoplastic cells *in vitro*. I. Conditions of binding and its selectivity, *Cel. Immunol.* 54:11-25.

Marino, P. A., and Adams, D. O., 1980*b*, Interaction of Bacillus Calmette-Guerin activated macrophages and neoplastic cells *in vitro*. II. The relationship of selective binding to cytolysis, *Cell. Immunol.* 54:26-35.

Marino, P. A. and Adams, D. O., 1982, The capacity of activated murine macrophages for augmented binding of neoplastic cells: Analysis of induction by lymphokine containing MAF and kinetics of the reaction, *J. Immunol.* 128:2816-2823.

Marino, P. A., Whisnant, C., and Adams, D. O., 1981, The binding of BCG-activated macrophages to tumor targets: Selective inhibition by membrane preparations from homologous and heterologous cells, *J. Exp. Med.* 154:77-87.

Meltzer, M. S., Ruco, L. P., Boraschi, D., and Nacy, C. A., 1979, Macrophage activation for tumor cytotoxicity: Analysis of intermediary reactions, *J. Reticuloendothel. Soc.* 26:403-415.

Nathan, C. F., and Cohn, Z. A., 1980, Role of oxygen dependent mechanisms in antibody-induced lysis of tumor cells by activated macrophages, *J. Exp. Med.* 152:198-208.

Nathan, C. F., Murray, H. W., and Cohn, Z. A., 1980, Current concepts: The macrophage as an effector cell, *N. Engl. J. Med.* 303:622-626.

Piessens, W. F., 1978, Increased binding of tumor cells by macrophages activated *in vitro* with lymphocyte mediators, *Cell. Immunol.* 35:303-317.

Russell, S. W., Doe, W. F., and McIntosh, A. J., 1977, Functional characterization of a stable, non-cytolytic stage of macrophage activation in tumors, *J. Exp. Med.* 146:5111-1520.

Shinitzky, M., and Henkart, P., 1980, Fluidity of cell membranes—current concepts and trends, *Int. Rev. Cytol.* 60:121-147.

Somers, S. D., Mastin, P., and Adams, D. O., 1983, The binding of tumor cells by murine mononuclear phagocytes can be divided into qualitatively distinct types, *J. Immunol.* 131:2086-2093.

Chapter 8

Activation of Macrophages to Kill Rickettsiae and Leishmania: Dissociation of Intracellular Microbicidal Activities and Extracellular Destruction of Neoplastic and Helminth Targets

Carol A. Nacy and Charles N. Oster,

Department of Immunology
Walter Reed Army Institute of Research
Washington, D.C. 20307

Stephanie L. James

Laboratory of Parasitic Diseases
National Institute of Allergy and Infectious Diseases
National Institutes of Health
Bethesda, Maryland 20205

and

Monte S. Meltzer

Department of Immunology   and   *Department of Dermatology*
Walter Reed Army Institute of   *Walter Reed Army Medical Center*
 Research   *Washington, D.C. 20307*
Washingtion, D.C. 20307

I. INTRODUCTION

Analysis of *Listeria* infection in mice (Mackaness, 1962) clearly defined host-cell interactions that occur during resolution of this infectious disease: T lympho-cytes develop specific immune responses to the invading microorganism, but

The views of the authors do not purport to reflect the position of the Department of the Army or the Department of Defense.

macrophages are the proximate effector cells that eliminate the pathogen. Parenthetically, these early studies also showed that macrophages recovered from infected animals can destroy microorganisms antigenically unrelated to *Listeria*. In fact, these "activated" macrophages develop the capacity to kill a variety of intracellular and extracellular targets. The observation that mediators released by specifically sensitized T lymphocytes confer enhanced but entirely nonspecific killing properties on macrophages focused attention on a new group of important immunoregulatory mechanisms. Initial attempts to demonstrate activation of macrophages for enhanced microbicidal activities *in vitro*, however, were frustrated by a variety of technical difficulties inherent in these facultative intracellular bacteria: adherence to macrophage surfaces, extracellular replication, and short generation times obscured important details of intracellular events. Only recently have we and several other groups developed techniques to analyze macrophage antimicrobial activities *in vitro* (Anderson and Remington, 1974; Hinrichs and Jerrels, 1976; Nogueira and Cohn, 1978; Mauel *et al.*, 1978; Nacy and Meltzer, 1979; Nacy *et al.*, 1981*b*). The key to success of these later studies is the target cell: obligate intracellular microorganisms. The use of obligate intracellular parasites circumvents the major problem associated with facultative microorganisms—extracellular replication—and permits quantitative analysis of events that lead to macrophage activation for microbicidal activities.

Obligate intracellular parasites, particularly those that infect and replicate in macrophages, create special problems for the immune system of infected individuals (Nacy and Groves, 1981; Nacy *et al.*, 1982, 1983*a,c*). Sequestered in the intracellular environment of the major participant of both humoral and cellular immune reactions, these macrophage parasites are protected from many host-defense mechanisms that develop during infectious diseases. Parasite survival is ensured by the evolutionary acquisition of evasive mechanisms that subvert the potent microbicidal armamentarium of the resting macrophage (Nacy and Osterman, 1979; Nacy and Diggs, 1981). Resolution of disease, then, is dependent on extraordinary alterations in the intracellular environment of the macrophage, or elimination of the infected cell itself (Meltzer and Nacy, 1980; Nacy and Meltzer, 1982; Nacy *et al.*, 1983*c*).

Mediators released by antigen-stimulated leukocytes induce profound changes in the biochemical and physiologic repertoire of resting macrophages (North, 1978; Cohn, 1978; Karnovsky and Lazdins, 1978). These changes, in turn, dramatically alter the capacity of obligate intracellular parasites to infect and replicate within macrophages. This chapter describes the events that induce activated macrophages with microbicidal activities against two very different obligate intracellular parasites, *Rickettsiae* and *Leishmania*, and compares these events with those that regulate extracellular cytolytic reactions. Intracellular and extracellular effector activities of activated macrophages share many common characteristics, but can ultimately be dissociated on the basis of responsive cell populations and inductive signals.

II. ACTIVATION OF MACROPHAGES FOR TUMORICIDAL AND MICROBICIDAL EFFECTOR ACTIVITIES

Macrophages activated *in vivo* during immune reactions in neoplastic or infectious diseases, or *in vitro* after treatment with lymphokine-rich culture fluids of antigen-stimulated leukocytes (LK) develop capacity to kill tumor cells and exert potent microbicidal activity against the intracytoplasmic bacteria, *Rickettsia tsutsugamushi* (Nacy and Meltzer, 1979; Nacy *et al.*, 1982), intraphagolysosomal protozoa, *Leishmania tropica* (Mauel *et al.*, 1978; Nacy *et al.*, 1981*b*; Pappas *et al.*, 1983*a,b*), and skin-stage schistosomulum of the extracellular helminth, *Schistosoma mansoni* (James *et al.*, 1982*a,c*). These cytotoxic reactions define a particular subset of activated macrophage effector functions that share several basic regulatory mechanisms. This subset can be clearly distinguished from several other effector reactions mediated by these activated cells, e.g., antibody-dependent cellular cytotoxicity.

A. Common Regulatory Events in Macrophage Activation for Microbicidal and Tumoricidal Activities

1. Time Course for Induction and Expression of Effector Activities

The cytotoxic activity of activated macrophages and the capacity of cells to be activated for these effector mechanisms are short-lived cell functions. In fact, the time course for acquisition and loss of cytotoxicity is the single most distinctive characteristic of this subset of reactions (Ruco and Meltzer, 1977; Nacy and Meltzer, 1979; Bout *et al.*, 1981; Nacy *et al.*, 1981*b*; Oster *et al.*, 1984). Macrophage cytotoxic activity is evident within 4 hr of LK treatment, reaches maximal levels by 10–12 hr, and progressively diminishes to control levels by 24 hr (Fig. 1). A similar time course has been reported for killing of *Schistosoma mansoni* larvae (Bout *et al.*, 1981). Although mechanisms underlying the decay of macrophage cytotoxic activity remain unclear, the loss of cytotoxicity is not caused by cell death: changes in vital dye uptake, phagocytic capacity, or production of O_2^- after phorbol myristate acetate (PMA) treatment are not significantly different in 1–48 hr cells. Nor can loss of activity be attributed to depletion of active lymphokine (LK): replacement of LK at any time does not affect the rate of decay. Decrease in macrophage killing capacity with time in culture could reflect either the loss of a labile effector mechanism or, alternatively, loss of macrophage responsiveness to LK signals. If killing mechanisms of activated macrophages were short lived, then cytotoxicity of all macrophages treated with LK immediately after addition of targets should be identical. Figure 1 demonstrates, however, that both tumoricidal and microbicidal activities decline rapidly when macrophage cultures are treated with medium for increasing time before addition of LK and target cells. Thus, the decay of nonspecific microbicidal and tumoricidal

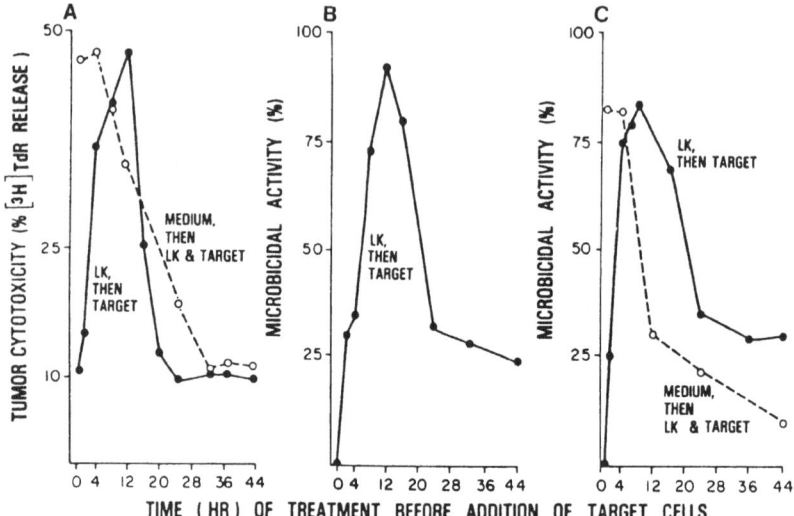

Figure 1. Induction of tumoricidal and microbicidal activities by treatment of macrophage cultures with lymphokine (LK) or medium for various times before addition of target cells. (——) Macrophages cultures treated with 1 : 20 dilution of LK for various times (1–44 hr) before addition of (A) TU-5 tumor, (B) *Rickettsia tsutsugamushi*, or (C) *Leishmania tropica* target cells. (- - -) Cultures treated with medium for 1–44 hr before addition of target and 1 : 20 LK.

activities by LK-activated cells, as well as loss of macrophage responsiveness to activation signals, are irreversible events that may be secondary to macrophage maturation or to differentiation *in vitro*, or both. It is interesting to note that macrophages that have become unresponsive for these short-lived effector activities retain cytotoxic activities by other unrelated mechanisms. Antibody-mediated cellular cytotoxicity (ADCC) reactions with the same target and cytotoxicity against certain other targets in the presence of PMA are intact (Koren *et al.*, 1981).

2. Genetic Control of Macrophage Activation

A second common feature of this subset of effector reactions is that cytotoxicity against tumor or microbial targets fails to develop with macrophages from certain strains of mice[*] (Table I) (Boraschi and Meltzer, 1979*a*; Nacy *et al.*, 1982; Nacy and Meltzer, 1982; Pappas *et al.*, 1983*a*; James *et al.*, 1983*b*).

[*]In conducting research described in this report, the investigators adhered to the Guide for Laboratory Animal Facilities and Care as promulgated by the Committee for Laboratory Animals and Care, of the Institute of Laboratory Animal Resources, National Academy of Sciences, National Research Council.

Table I. Tumoricidal and Microbicidal Activities of Macrophages from
BCG-Treated Mice[a]

Source of macrophages[b]	Target of macrophage cytotoxicity			
	TU-5 tumor cells (%)	*Schistosoma mansoni* schistosomula (%)	*Rickettsia tsutsugamushi* (%)	*Leishmania tropica* (%)
C3H/HeN	63[c]	97[c]	90[c]	94[c]
C3H/HeJ	8	3	21	28
A/J	10	2	14	37
P/J	12	3	ND	13

[a]*Abbreviations:* BCG, bacillus Calmette–Guérin; ND, not done.
[b]Macrophages were obtained from mice inoculated intraperitoneally with 1×10^6 viable BCG 8 days before harvest.
[c]Cytotoxicity is estimated by (1) percentage release of total incorporated tritiated thymidine at 48 hr from prelabeled tumor TU-5 cells, (2) percentage schistosomula that did not exclude vital dye or that were not motile at 48 hr, (3) percentage decrease in macrophages infected with *Rickettsia* at 24 hr, and (4) percentage decrease in macrophages infected with *Leishmania* amastigotes at 72 hr.

1. Macrophages from mice with the Lps^d gene for unresponsiveness to the lipid A region of bacterial endotoxic LPS (C3H/HeJ) do not develop cytotoxic activities after *in vivo* treatment with bacillus Calmette-Guérin (BCG) or *Corynebacterium parvum* or after *in vitro* treatment with LK (Ruco and Meltzer, 1978a). The cytotoxic defect in this mouse strain is controlled by a single gene closely linked with or identical to the *Lps* gene on chromosome 4 (Ruco *et al.*, 1978).

2. Macrophages from A-strain (A/J)-derived mice are also inefficient at nonspecific killing activities. The phenotypic expression of the A/J defect is similar to that of C3H/HeJ mice but is not controlled by the *Lps* gene (Boraschi and Meltzer, 1979b,c, 1980a) and may be the result of a different mechanism (Adams *et al.*, 1981).

3. Finally, cells from the P/J mice are among the least reactive for this subset of effector reactions (Boraschi and Meltzer, 1980b,c; Haverly *et al.*, 1983; Nacy *et al.*, 1983c; James *et al.*, 1983). The genetic basis of the P/J defect is different from either the C3H/HeJ or the A/J strains.

These cytotoxic defects of macrophages from nonresponsive mouse strains are selective, however. Cell responses to phagocytic and chemotactic stimuli, or production of O_2^- after PMA stimulation, are normal. In fact, macrophages from strains with these defects can often kill the same target through PMA or ADCC mechanisms.

Genetic defects that regulate macrophage activation provide unique tools for detailed analysis of events that lead to microbicidal and tumoricidal activities. That these defects influence response of macrophages to LK activation signals, and not the cytolytic mechanism itself, is best illustrated in the analysis of the reaction sequence for macrophage activation shown in the following section.

3. Regulation of Cytotoxic Activities by a Two-Stage Reaction Sequence

Within the characteristic time course shown in Fig. 1, development of the transient, nonspecific cytotoxic responses of activated macrophages can be separated into at least two reaction stages (Meltzer, 1981). Cells that are exposed to one stimulus enter a primed, noncytotoxic state; these cells can now respond to other trigger stimuli to develop full functional activity. The activation of defective C3H/HeJ macrophages for tumoricidal and microbicidal activities illustrates this priming and triggering sequence (Table II) (Ruco and Meltzer, 1978c,d; Pappas *et al.*, 1983a). The priming signal in these studies is provided by an immune response to BCG infection *in vivo*. Peritoneal macrophages from C3H/HeJ mice express little or no cytotoxic reactivity after this treatment (Table I). When the cells from BCG-treated mice are further exposed to LK *in vitro*, however, they develop

Table II. Activation of C3H/HeJ Macrophages for Tumoricidal and Microbicidal Activities by a Two-Stage Reaction Sequence[a]

Mouse strain	BCG treatment *in vivo*[d]	LK treatment *in vitro*[e]	TU-5 tumor cells (%)	*Leishmania tropica* amastigotes (%)
			\multicolumn{2}{Target of macrophage cytotoxicity[b,c]}	
C3H/HeN	−	−	10	0
	−	+	45*	85*
	+	−	50*	90*
	+	+	55*	95*
C3H/HeJ	−	−	10	0
	−	+	15	30
	+	−	10	20
	+	+	55*	90*

[a] *Abbreviation:* BCG, bacillus Calmette–Guérin.
[b] Tumor cytotoxicity is measured by estimation of percentage total release of tritiated thymidine from prelabeled TU-5 tumor cells at 48 hr; intracellular destruction of amastigotes is estimated by the decrease in percentage of infected macrophages in treated cultures compared with control untreated macrophages at 72 hr.
[c] Asterisk (*) denotes significant microbicidal and tumoricidal activity ($p \leqslant 0.05$).
[d] Mice were infected with 1×10^6 viable BCG intraperitoneally 8 days before macrophage harvest.
[e] Macrophage cultures were exposed to 1:20 dilution of LK supernatants for the duration of the assay.

strong microbicidal and tumoricidal activity after only a brief (10 min) exposure to the second signal. Table III lists the factors that can serve as second signals in this reaction sequence. Effective signals include bacterial cell wall products, plant lectins with unrelated sugar specificities, and certain LK factors.

The priming and triggering activation sequence can also be demonstrated *in vitro* by LK treatment of C3H/HeN (LK responsive) macrophages (Fig. 2) (Meltzer, 1981). In these experiments, the priming signal is supplied by LK at a concentration that does not by itself induce cytotoxicity (1:500). Only those cells that receive an appropriate trigger signal after LK priming develop the capacity to kill tumor cells and intracellular parasites. Macrophages exposed to trigger signals themselves, alone or in combination, or cells treated with trigger signals before LK priming are not cytotoxic.

Although several different molecules can be used to trigger cytotoxic activities (Table III), factor(s) that prime macrophages for intracellular and extracellular

Table III. Trigger Signals for Induction of Nonspecific Tumoricidal and Microbicidal Activities by *in Vivo* BCG-Primed or *in Vitro* LK-primed Macrophages[a]

Effective trigger signals	Ineffective trigger signals
LPS (1 ng/ml)	Freeman polysaccharide of LPS
Phenol-extracted LPS	Latex beads (1000:1)
Butanol-extracted LPS	EAIgG (100:1)
Lipid A	Ascorbic acid (25 μg/ml)
Lipid A-associated proteins	$BeSO_4$ (0.1 μg/ml)
	$BaSO_4$ (20 μg/ml)
LK (1:20 dilution)	Dimethyl sulfoxide (1.0%)
	Ethanol (0.5%)
Various plant lectins	β-IFN (1000 IU/ml)
Arachis hypogaea	Cholera toxin B (10 μg/ml)
Glycine max	Epidermal growth factor (50 ng/ml)
PHA[b]	Nystatin (50 U/ml)
Concanavalin A[b]	Amphotericin B (5 μg/ml)
	PMA (0.2–2 μg/ml)
	N-acetylmuramyl-L-analyl-D-isoglutamine (25 μg/ml)
	Tumor cells (P815, P388, 1023, L929)
	Granulocyte–macrophage colony stimulating factor from L929 or WEHI-3

[a]*Abbreviations:* BCG, bacillus Calmette–Guérin; IFN, interferon; LK, lymphokine; LPS, lipopolysaccharide; PHA, phytohemagglutinin; PMA, phorbol myristate acetate.
[b]PHA and concanavalin A are active only for induction of microbicidal activities; activity is likely a result of LK production over 72-hr assay by the mixed cell cultures.

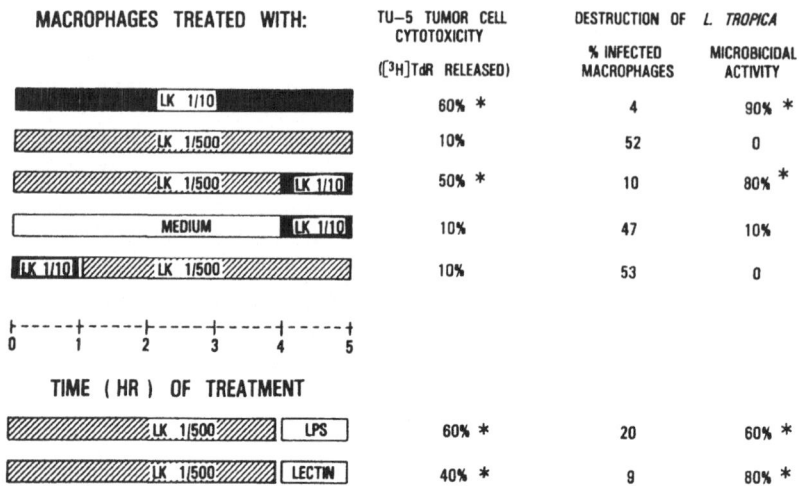

Figure 2. Analysis of priming and triggering *in vitro* with lymphokine (LK)-responsive C3H/ HeN macrophages. Cells were exposed to dilutions of LK, medium, and trigger signals for a total of 5 hr before addition of TU-5 fibrosarcoma cells or *Leishmania tropica* amastigotes. Tumor cytotoxicity was estimated by release of radiolabel from tritiated thymidine prelabeled TU-5 cells in triplicate cultures at 48 hr, and expressed as percentage of total SDS release. Microbicidal activity was expressed as the percentage decrease in infected macrophages in treated cultures compared with untreated control cells at 72 hr. Asterisk (*) denotes significant microbicidal and tumoricidal activity ($p \leqslant 0.05$).

killing mechanisms have only been found in LK-rich supernatants (Table IV). Reagents known to induce an array of effector activities in macrophages all fail to initiate macrophage activation for the nonspecific transient cytotoxic responses (Meltzer, 1981; James *et al.*, 1983a; Ralph *et al.*, 1983; Nacy *et al.*, 1983d). The single exception is the activation of macrophages to kill schistosomula of *S. mansoni* by LPS; this reaction may, however, occur through LPS- induced LK production during the 48-hr assay for larvacidal activity (James *et al.*, 1983a).

LK supernatants contain both priming (effective at low concentrations) and triggering (effective at only high concentrations) signals; these factors, however, are physicochemically indistinguishable thus far (Meltzer, 1982). In fact, it is possible that the entire sequence of activation events is initiated and amplified by a single molecular species—in much the same manner as that described for insulin and interferon reactions. Binding of the priming molecule may induce reactive changes in the macrophage (number or affinity of receptors, enhanced transduction of receptor signals) such that a subsequent interaction with a high concentration of the same molecule initiates the effector reaction. Regardless of the mechanism, the two-stage (at least) reaction initiated by LK priming and triggering signals forms the basis of a regulatory system that controls induction and expression of the transient, nonspecific macrophage effector activities.

Table IV. Development of Nonspecific Tumoricidal and Microbicidal
Activities After *in Vitro* Treatment with a Variety of Bioactive Agents[a]

	Target of macrophage cytotoxicity[c]			
Agents used to treat macrophage[b]	TU-5 tumor cells (%)	*Schistosoma mansoni* schistosomula (%)	*Rickettsia tsutsugamushi* (%)	*Leishmania tropica* (%)
LK				
1:20	45*	100*	85*	85*
1:40	40*	100*	75*	60*
1:80	20	40*	30*	10
1:160	10	10	0	0
Medium	10	5	0	0
β-IFN				
10^4 IU/ml	10	5	10	0
10^3 IU/ml	10	5	0	0
Colony-stimulating factor	10	5	ND[d]	0
Interleukin (2×10^6 U/ml)	10	10	ND	ND
PMA, 1 μg/ml	10	10	ND	10
LPS				
20 ng/ml	10	100*	0	0
10 ng/ml	10	10	0	0

[a]*Abbreviations:* IFN, interferon; LPS, lipopolysaccharide; PMA, phorbol myristate acetate.
[b]Agents were present continuously in treated cultures. Casein-elicited macrophages were used for 48 hr extracellular killing assays; unfractionated resident peritoneal cells were used for 24–72 hr intracellular killing assays.
[c]Asterisk (*) denotes significant microbicidal and tumoricidal activities ($p \leqslant 0.05$).
[d]Not done.

B. Dissociation of Macrophage Microbicidal and Tumoricidal Activities

Despite broad commonality for regulation of transient nonspecific macrophage effector activities, the details of these cytotoxic reactions are unique for each target. This subset of activated macrophage functions can be dissociated by (1) responsiveness of macrophage populations to different LK treatment sequences, (2) characterization of responsive macrophage populations, or (3) identification of LK activation signals.

1. Macrophage Activation by LK Pulse or by Continuous LK Stimulation

Lymphokine treatment of macrophages before addition of targets was used to define events that regulate macrophage nonspecific tumoricidal and microbicidal activities (Figs. 1 and 2). Alteration of this treatment sequence to include continuous LK stimulation provides interesting new infromation that dissociates responsiveness of macrophage populations for intracellular and extracellular cytolytic processes.

 a. Intracellular Killing by Activated Macrophages. At least two different activities are responsible for the increase in parasite-free macrophages observed in in cultures pulsed with LK before exposure to *Rickettsia* or *Leishmania* (Fig. 3) (Nacy and Meltzer, 1979; Nacy *et al.*, 1981*b*): an early initial alteration in macrophage–parasite interaction that results in fewer infected macrophages immediately (1 hr) after infection, followed by the second response, destruction of intracellular organisms over 24-72 hr. The effector function that can be measured at 1 hr represents an apparently stable change in a subpopulation of peritoneal macrophages: activity persists for as long as 72 hr after addition of LK, and the capacity of cells to respond to signals that induce this activity is not substantially altered with time in culture (Oster *et al.*, 1984). Although this effector activity has been observed with a number of microorganisms (Nacy and Meltzer, 1979, 1982; Nacy *et al.*, 1981; Horwitz and Silverstein, 1981), the precise mechanism that affects parasite entry into macrophages is completely unknown. LK-induced decrease in infected cells is not associated with changes in nonspecific (Latex bead) or receptor-mediated (Fc or C3) phagocytosis (Nacy *et al.*, 1981*b*; Ralph *et al.*, 1983), nor is it a result of residual toxic LK (Nacy *et al.*, 1981*b*). For want of a

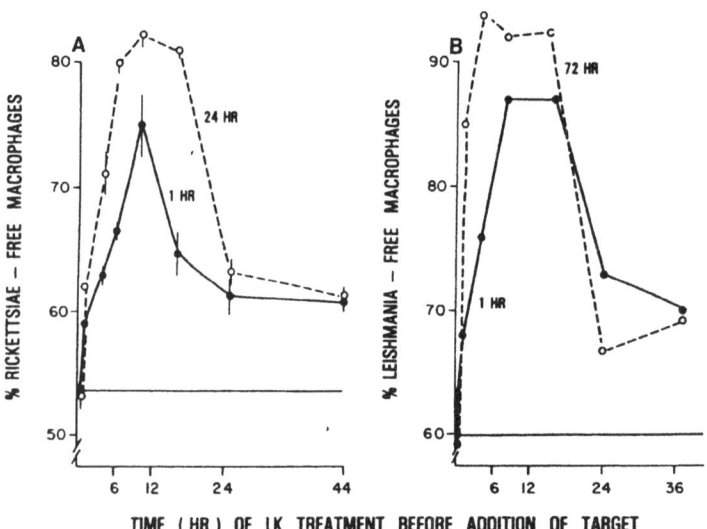

TIME (HR) OF LK TREATMENT BEFORE ADDITION OF TARGET

Figure 3. Kinetics of microbicidal activities of macrophages treated with lymphokine (LK) before addition of *Rickettsia* or *Leishmania*. Resident peritoneal macrophages were treated with 1:20 dilution of LK for ≤44 hr before addition of *R. tsutsugamushi* (A) or *L. tropica* (B). Samples were removed 1 hr after addition of targets (——) and again at 24 hr (*Rickettsia*) or 72 hr (*Leishmania*) (- - -). Microbicidal activity is expressed as the percentage decrease in infected macrophages of treated cultures compared with untreated control cells at each time.

better descriptor, we refer to this activity of activated macrophages as resistance to infection.

Superimposed on resistance to infection in macrophages pretreated with LK is a significant increase in parasite-free macrophages by 24–72 hr. This reflects destruction of microorganisms intracellular at 1 hr (Fig. 3). That these organisms in 1-hr infected cells are in fact viable, and not simply phagocytized dead parasites, is demonstrated in Table V. With a modified plaque assay for *Rickettsiae* (Nacy and Oaks, 1981), we can demonstrate decreases in infectious *R. tsutsugamushi* recovered from lysed macrophages that exactly correspond to decreases in percent infected cells. Thus, any viable intracellular rickettsiae present in macrophages at 1 hr are killed by these LK activated cells within 24 hr.

Although this intracellular killing by LK-pretreated macrophages followed the general kinetics of nonspecific tumor cytotoxicity (i.e., a return to 1-hr levels by 24 hr), the long-lived resistance to infection made analysis of the decay in activity rather complex. We took advantage of the macrophage defect of C3H/HeJ mice to confirm the kinetic data obtained with LK-responsive C3H/HeN macrophages (Fig. 4). Macrophages from C3H/HeJ mice are completely unresponsive to LK signals that induce resistance to infection; they are also defective for intracellular killing of *Rickettsia* and *Leishmania*. The defect for killing of *L. tropica*, however, can be overcome with very high concentrations of LK (1:10) (Nacy *et al.*, 1983c). The kinetics of C3H/HeJ macrophage microbicidal activity (Fig. 4) suggest that intracellular killing is, in fact, a short-lived effector function. Parenthetically, the stable nature of resistance to infection in a subpopulation of peritoneal macrophages explains prior observations that decay in microbicidal activity of LK responsive macrophages reached a plateau before returning to control levels (Fig. 1, 3, and 4).

In an effort to analyze induction of intracellular killing independent of effects

Table V. Titration of Viable *Rickettsiae* in Lysed Peritoneal Cell Cultures by
Plaque Assay in Irradiated L929 Cell Monolayers[a]

Sample time (hr)	Peritoneal cell culture treatment	Plaque titer ($\times 10^4$)	Reduction in viable *Rickettsia* titer (%)	Microbicidal activity, decrease in infected cells (%)
1	Medium	10.4	–	–
	LK before infection	6.0	39	35
18	Medium	9.2	–	–
	LK after infection	2.2	76	80
	LK before and after infection	0.9	91	90

[a]*Abbreviation:* LK, lymphokine.

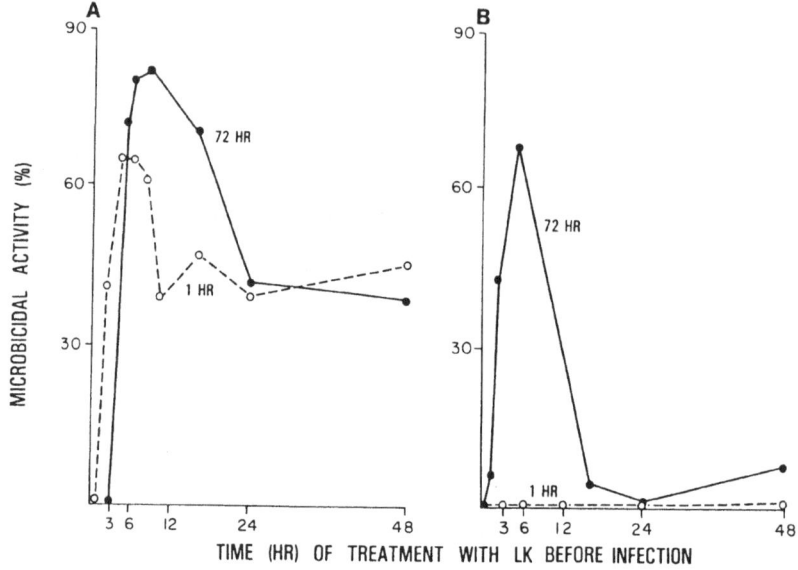

Figure 4. Microbicidal activities of C3H/HeN and C3H/HeJ macrophages treated with lymphokine (LK) before addition of *Leishmania tropica*. Resident peritoneal macrophages of C3H/HeN (A) and C3H/HeJ (B) mice were treated with 1:6 dilution of LK for ≤48 hr before addition of amastigotes. Samples were removed at 1 hr (---) and again at 72 hr (——). Microbicidal activity is expressed as percentage decrease in infected macrophages of treated cultures compared with untreated control cultures at each time.

on macrophage ingestion of parasites, we exposed cells to *Rickettsia* or *Leishmania* first, then add LK after the 1-hr infection period (Fig. 5) (Nacy and Meltzer, 1979; Nacy *et al.*, 1981*b*). In this way, we demonstrated that maximal intracellular killing activity is induced with as little as 4-hr exposure to LK after infection; levels of this intracellular destruction of parasites are equivalent to those induced by optimal LK treatment before infection (Fig. 1). Thus, cells that can respond to LK signals for antimicrobial effects against *Rickettsia* and *Leishmania* will do so with 4- to 8-hr treatment before or after infection, either as a pulse (LK removed) or with continuous LK stimulation (Figs. 4 and 5).

   b. Extracellular Cytotoxicity. In contrast to the induction of intracellular killing activities, extracellular destruction of neoplastic or larval targets is markedly potentiated when macrophages are exposed to LK throughout the 48-hr cytotoxicity assays (Fig. 6) (Nakamura and Meltzer, 1981; James *et al.*, 1983*a*). The difference in killing activity is quantitative: When cytotoxicity of cultures with continuous LK stimulation is compared with cytotoxicity of cultures optimally treated by LK pulse, the number of cytotoxic events is increased approx-

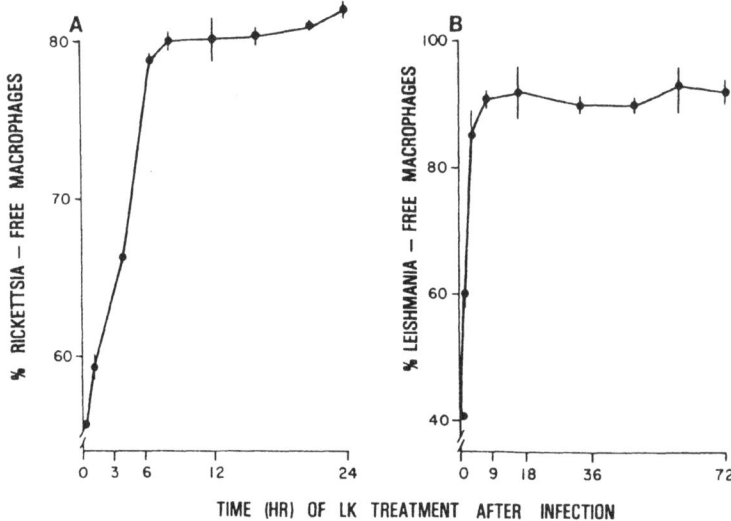

Figure 5. Induction of intracellular killing activities in macrophages exposed to lymphokine (LK) after infection with *Rickettsia tsutsugamushi* or *Leishmania tropica*. Cells were exposed to *Rickettsia* (A) or *Leishmania* (B) for 1 hr, washed, then treated with 1:20 dilution of LK for various times. Cells were washed free of LK after 1–72 hr of treatment and incubated in medium for the remaining time of the assay (24 hr for *Rickettsia*, 72 hr for *Leishmania*). Microbicidal activity is expressed as the percentage decrease in infected macrophages in treated cultures compared with untreated control cultures at 24 or 72 hr.

imately threefold. The precise nature of enhanced extracellular killing events that occur when cells, targets, and LK are present at the same time is not yet clear. It is fascinating, however, that the increase in cytotoxic events with continuous LK stimulation follows the same kinetics of induction and decay and is regulated by the same signals as are events induced by LK pulse before addition of targets (Nakamura and Meltzer, 1981). The 3-fold increase in cytotoxic events might reflect the activation of greater numbers of macrophages, or might be the result of increased activity of each activated cytolytic macrophage.

The LK-induced expression of transient nonspecific effector activities by macrophages differs with the nature of the target: Maximal antimicrobial activities against obligate intracellular parasites can be induced with 4- to 8-hr LK treatment, regardless of the treatment sequence; maximal extracellular killing of neoplastic cells and schistosomula of *S. mansoni* requires the simultaneous presence of macrophage, target cell, and LK activation signals. This dissociation of macrophage intracellular and extracellular effector activities is complicated by the fact that cell populations effectively performing these functions are quite different.

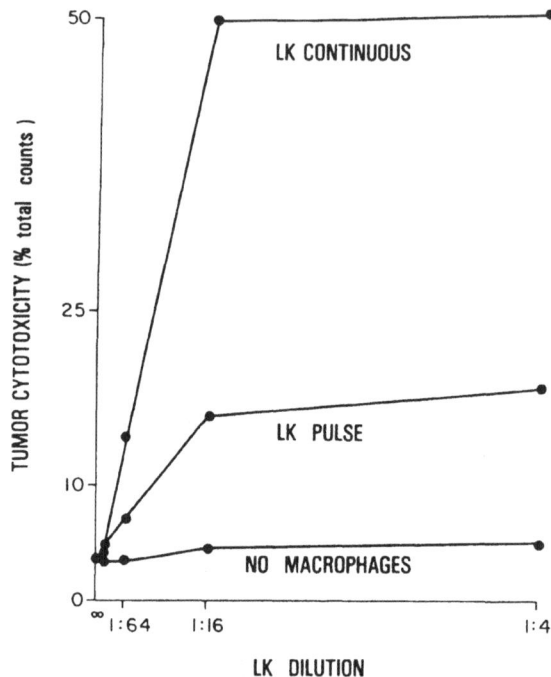

Figure 6. Tumor cytotoxicity by macrophages continuously exposed or pulsed with lymphokine (LK). Adherent peritoneal cells (PEC) from casein-injected mice were cultured in LK dilutions for 5 hr. Certain cultures were washed with medium. Tumor cells prelabeled with tritiated thymidine were added to washed (LK pulse) and unwashed (continuous) monolayers. Cytotoxicity was estimated by measurement of radiolabel release in triplicate cultures at 48 hr and expressed as a percentage of SDS total counts. Release of radiolabel from tumor cells incubated with LK alone through 48 hr is designated by the curve for no macrophages.

2. *Macrophage Populations That Respond to LK for Expression of Intracellular and Extracellular Cytolytic Activities*

Mononuclear phagocytes undergo a series of morphologic and biochemical changes during differentiation from precursor cells to resident tissue macrophages. The capacity of these cells to respond to a particular stimulus may change with cell differentiation.

a. Nonspecific Tumoricidal Activity of Activated Macrophages. Early studies on activated macrophage tumor cytotoxicity showed that inflammatory macrophages, cells that migrate to the peritoneal cavity in response to a sterile irritant (starch, casein), are not by themselves cytotoxic. They are, however, much more responsive than resident peritoneal macrophages to the LK signals that induce tumor cell killing (Ruco and Meltzer, 1978*b*). The number of macrophages within an irritant-induced inflammatory exudate able to respond to LK and develop tumoricidal activity is 10- to 20-fold more than that within the more differentiated resident macrophage population. These large differences in tumor cytotoxicity between differentiated and immature cell populations are observed with macrophage activation by both LK pulse (Ruco and Meltzer, 1978*b*) and continuous LK stimulation (Nakamura and Meltzer, 1981) (Fig. 7A).

    b. Extracellular Destruction of Schistosomula. Studies on LK activation to kill *S. mansoni* schistosomula, another extracellular target, use primarily inflammatory exudate macrophages. We do, however, observe larvacidal activity in resident peritoneal cells equivalent to that of inflammatory cells exposed to the same high concentration of LK (Fig. 7B). In these experiments, LK is added to treated cultures for the entire 48 hr (James *et al.*, 1983*a*). Since the resident cell experiments were performed with LK concentrations well into the LK dose-response plateau, it is not yet clear whether the differentiated and immature cell populations contain the same number of reactive macrophages for this cytolytic effector activity (Bout *et al.*, 1981).

    There is another instance, however, in which macrophage activation for tumoricidal and schistosomulicidal activities, both extracellular cytolytic effector functions, appear to dissociate. Macrophages from *S. mansoni*-infected mice are strongly tumoricidal *in vitro*. The levels of cytotoxicity by these cells are identical, over a broad macrophage dose response, to those of cells from BCG or *C. parvum*-treated mice (James *et al.*, 1982*a*). Activated tumoricidal macrophages recovered from the peritoneal cavity during murine schistosomiasis, however, differ from activated macrophages of BCG-infected mice in several important respects:

1. *S. mansoni*-activated macrophages are indistinguishable from resident peritioneal macrophages in ectoenzyme profile, granular peroxidase content, glucose oxidation during phagocytosis, and PMA-induced O_2^- production (James *et al.*, 1982*a*).
2. BCG-activated macrophages consistently kill more than 75% of schistosomula *in vitro*; three times as many macrophages from *S. mansoni*-infected mice kill only 20% of larvae (James *et al.*, 1982*b*).
3. This low level of schistosomulicidal activity cannot be increased by further LK treatment of cells *in vitro*.

Thus, chronic murine schistosomiasis generates a peculiar mononuclear phagocyte population that is strongly activated for one extracellular cytolytic function (tumor cytotoxicity), yet shows only weak cytotoxicity against schistosomula. These observations could reflect different killing mechanisms effective against tumor cells and larvae. Alternatively, they could reflect a difference in the ability of the two targets (one a single cell and the other a complex multicellular organism) to resist the same effector mechanism. In the latter case, the observed dissocia-

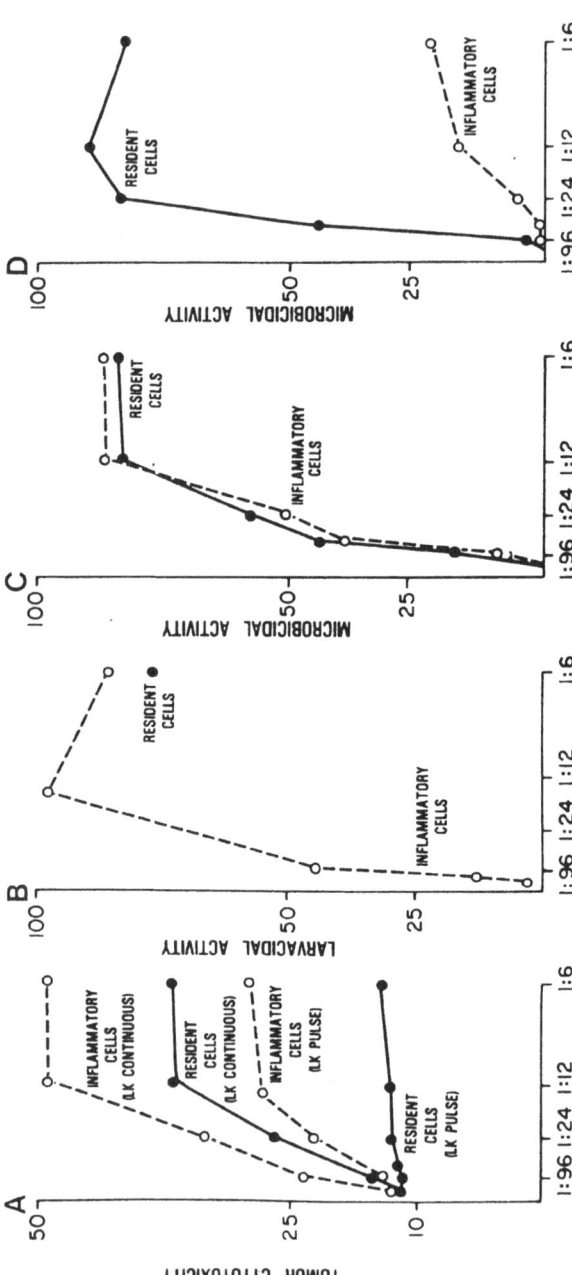

LYMPHOKINE CONCENTRATION

Figure 7. Tumoricidal and macrobicidal activities of inflammatory exudate cells and resident peritoneal macrophages treated with various concentrations of lymphokine (LK). (A, B) Extracellular targets; (C, D) intracellular targets. Casein-elicited cells (- -) and resident peritoneal macrophages (——) were treated with different concentrations of LK as a 4-hr pulse (A, tumor cytotoxicity) or continuously throughout the microbicidal and tumoricidal assays. (A) Tumor cytotoxicity is expressed as percentage total SDS release of radiolabel from prelabeled TU-5 tumor cells. (B) Larvacidal activity is expressed as percentage nonviable schistosomula, as assessed by uptake of vital dyes and motility. Microbicidal activities for (C) *Rickettsia* and (D) *Leishmania* are expressed as percentage decrease in infected macrophages in treated cultures compared with untreated control macrophages.

tion might be attributable to differences in the number of activated macrophages in inflammatory-like cell populations recovered from BCG-infected mice and the resident-like populations recovered from *S. mansoni*-infected mice.

Like the *S. mansoni* larvae, elimination of intracellular rickettsiae by activated macrophages can be demonstrated in both resident and inflammatory cell populations.

c. Intracellular Destruction of Rickettsiae. The number of macrophages that become infected with *R. tsutsugamushi* in resident and inflammatory cell populations is quite similar (Nacy *et al.*, 1982). Intracellular destruction of the rickettsiae is also indistinguishable in these two cell populations over a broad LK dose response (Fig. 7C). Thus, the mechanism(s) responsible for killing this intracytoplasmic bacterium can be demonstrated in both immature and differentiated cells. This is not the case, however, for another obligate intracellular parasite, *L. tropica*.

d. Intracellular Destruction of Leishmania. *L. tropica* amastigotes do infect macrophages in inflammatory exudates. In fact, the parasite enters younger, peroxidase granule-containing macrophages in numbers disproportionate to that expected for random entry into cells (Fortier *et al.*, 1982). Replication of the parasite once inside the cell, however, is not different from that observed with resident peritoneal macrophages (Nacy and Diggs, 1981). Although inflammatory macrophages are more sensitive than resident cells to LK that induce nonspecific tumor cytotoxicity and can kill a different obligate intracellular parasite, *Rickettsia*, only the resident cell can be induced by LK to destroy intracellular *L. tropica* (Nacy *et al.*, 1981*b*). There is a negative correlation between intracellular killing activity and peroxidase positive (young) cells in inflammatory exudates, i.e., activity of inflammatory macrophages for intracellular destruction of amastigotes is one-third that of resident peritoneal cells (Fig. 7D). That this inefficient killing of *L. tropica* represents the lack of an effective killing mechanism, rather than an unresponsiveness to LK activation signals, is suggested by studies of both resistance to infection in inflammatory macrophages (essentially equivalent to that of resident cells) and activation of mouse monocytes (poorly unresponsive, if at all). It is likely that macrophage killing mechanisms effective against the *Leishmania* amastigote develop only after cell differentiation. The inability of inflammatory macrophages to kill intracellular *L. tropica*, coupled with the increased infection rate of younger macrophages, provides a survival advantage for the parasite during evolution of the cutaneous lesion, and brings up an intriguing possibility, i.e., inflammation may contribute to the pathogenesis of leishmanial disease by supplying lymphokine-unresponsive host cells that sequester the parasite from effective immune responses.

Superimposed on this diversity in responsive macrophage effector cells for nonspecific intracellular and extracellular cytotoxic activities is the heterogeneity in lymphokine mediators that regulate these functions.

3. Lymphokine Signals for the Induction of Intracellular and Extracellular Cytolytic Activities

Analysis of LK activities in supernatant fluids from antigen-stimulated spleen cells suggests that the factors for regulation of transient cytotoxic reactions against various targets may be different. With the same but quite heterogeneous cell population (resident peritoneal macrophages), we can induce both tumoricidal and microbicidal activities with spleen cell-derived LK. Further physicochemical characterization of the LK activities, however, showed that several factors for induction of microbicidal activity against intracellular targets (*R. tsutsugamushi*, *L. tropica*) are completely ineffective in inducing tumoricidal and larvacidal activities against extracellular targets (TU-5, *S. mansoni*) (Fig. 8) [(Leonard *et al.*, 1978; Nakamura and Meltzer, 1981; Nacy *et al.*, 1981*a*; Meltzer *et al.*, 1982*b*; James *et al.*, 1983*a*)] . Activities in LK for intracellular destruction of *Rickettsia* and *Leishmania* elute from Sephadex G-200 in three distinct regions of approximately 130,000, 50,000, and 10,000 MW. In contrast, LK that activate macro-

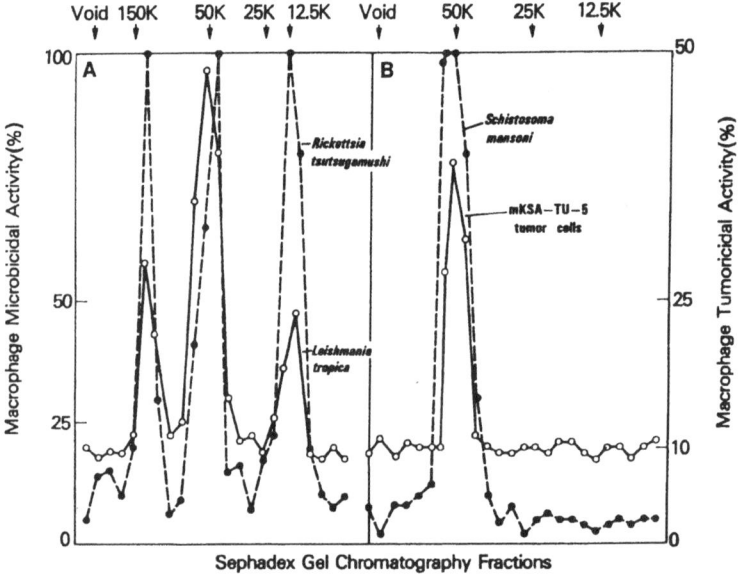

Figure 8. Induction of macrophage microbicidal and tumoricidal activities by lymphokine (LK) in antigen-stimulated spleen cell cultures. Resident peritoneal macrophages were treated with 1:20 dilution of Sephadex G-200 fractions of LK from tuberculin-stimulated bacillus Calmette–Guérin (BCG)-immune spleen cell cultures. (A) Cells were assayed for microbicidal activity against *Rickettsia tsutsugamushi* (- - -) and *Leishmania tropica* (——). (B) Cells were assayed for lavacidal activity against *Schistosoma mansoni* (- - -) and for tumoricidal activity against TU-5 fibrosarcoma cells (——). Molecular-weight markers are shown above.

phages for extracellular destruction of tumor and larval targets elute as a single peak of activity with ~45,000–50,000 MW.

The recent report of LK from a continuous T-cell line gave us the unique opportunity to examine the question of diverse regulatory signals with another source of activation stimuli (Meltzer *et al.*, 1982a). Active supernatant fluids are prepared by treating a variant of the EL-4 thymoma cell line, EL-4 Farrar, with PMA for 48 hr; EL-4 factors are separated by chromatography on Sephadex G-100 columns. Macrophages treated with column fractions of PMA-stimulated EL-4 supernatants develop the capacity to kill both intracellular and extracellular targets (Fig. 9) (Nacy *et al.*, 1983b,d). Activities for destruction of *S. mansoni* schistosomula cofractionate with activities for nonspecific tumor cytotoxicity: both extracellular killing activities are induced with factors in fractions of 45,000- and 23,000-MW regions of the column. In contrast, EL-4-derived factors that activate macrophages for microbicidal activity against the intracellular amastigote

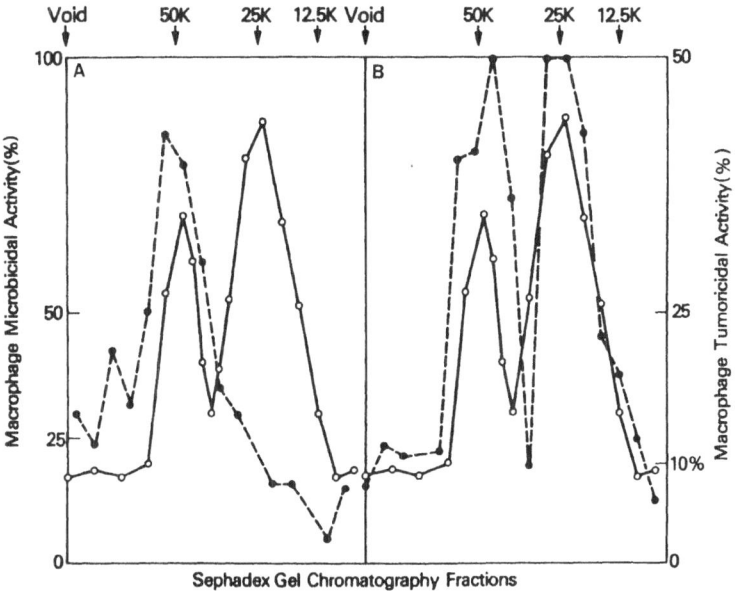

Figure 9. Induction of macrophage tumoricidal and microbicidal activities by lymphokine (LK) in phorbol myristate acetate (PMA)-stimulated EL-4 Farrar cell cultures. Resident peritoneal macrophages or PEC were treated with 1:20 dilution of Sephadex G-200 fractions of LK from PMA-stimulated EL-4 Farrar cell cultures. (A) cells were assayed for microbicidal activity aginst *Leishmania tropica* (- - -) and for tumoricidal activity against TU-5 fibrosarcoma cells (——). (B) Cells were assayed for microbicidal activity against *S. mansoni* (- - -) and for tumoricidal activity against TU-5 fibrosarcoma cells (——). Molecular weights are shown above.

Table VI. Induction of Macrophage Microbicidal and Tumoricidal Activity
by Culture Fluids from Various Cell Sources[a]

Source of culture fluids	IFN titer (IU/ml)	Target of macrophage-mediated cytotoxicity		
		Tumor cells (%)	*Leishmania tropica* (%)	*Schistosoma mansoni* (%)
PMA-stimulated				
EL-4 Farrar				
45,000 MW	15	60	90	100
23,000 MW	0	80	0	100
L929 (β-IFN)	1000	10	0	5
Antigen-stimulated spleen cells, LK				
1:20	100	50	85	100
1:40		25	40	70
1:80		10	15	25
γ-IFN (monkey kidney cells)	200	40	70	ND
	100	10	10	ND
	50	10	0	ND

[a]*Abbreviations:* IFN, interferon; LK, lymphokine; ND, not done.

form of *L. tropica* are found only in fractions of the 45,000-MW region. The 23,000-MW activity remains inactive for induction of microbicidal activity against *Leishmania* at concentrations 50-fold higher than that found in the unfractionated culture fluids. Interferon (IFN) activity is also detected in the 45,000-MW fractions that contain factor(s) for induction of both extracellular and intracellular killing activities. In fact, activities in this MW region share several biologic, physicochemical, and antigenic properties with γ-IFN: antiviral activity, induction of macrophage Ia antigen, equivalent molecular weight and isoelectric point (pI 4.2–5.6), pH 2 lability, and neutralization by anti-γ-IFN serum (Meltzer *et al.*, 1982*a,b*; Nacy *et al.*, 1983*d*). Further confirmation of a possible role for γ-IFN as a mediator of macrophage activation by EL-4 supernatants awaits availability of monoclonal antibodies and further purification. Cloned partially purified γ-IFN does, however, induce macrophage killing of tumor cells and *Leishmania* (Table VI).

Thus, with LK signals from two different sources, we can dissociate extracellular microbicidal and tumoricidal activities from intracellular killing activities of activated macrophages. Destruction of two different extracellular targets—a fibrosarcoma cell and a schistosomulum—is induced by the same LK factors, regardless of the source of these factors. Moreover, destruction of two intracellular microorganisms that reside in different cellular compartments (an intracytoplasmic bacterium, *R. tsutsugamushi*, and an intraphagolysosomal protozoan parasite, *L.*

tropica) is also induced by similar LK factors. These factors are different, however, from those that regulate extracellular killing. LK regulation of cytotoxicity may be dictated not by the target per se, but by the site, intracellular or extracellular, of the cytotoxic reaction.

III. SUMMARY

Mononuclear phagocytes undergo dramatic changes during differentiation from bone marrow stem cells to resident tissue macrophages. Throughout differentiation, cells lose or acquire numerous morphologic, metabolic and functional capacities such that mature, resident macrophages of one tissue often bear little resemblance to resident cells of another. Superimposed on the intrinsic continuum of mononuclear phagocyte differentiation are the reactive changes in macrophages induced by endogenous and exogenous stimuli: the ability of mononuclear phagocytes to respond to a particular stimulus may also change with cell differentiation. This dynamic interaction of cell differentiation and response to a microenvironment, and the resulting heterogeneity among mononuclear phagocytes for many functional characteristics, is clearly illustrated by the effector activities of activated macrophages that we describe in this report. Despite the common regulatory events for induction and expression of transient nonspecific cytotoxic reactions effective against such diverse targets as rickettsiae, leishmania, schistosomula, and neoplastic cells, these effector functions can be dissociated by the cells that perform the effector activity, and the signals that regulate these activities. The differential susceptibility of the various targets to particular killing mechanisms induced by LK in responsive populations only adds to the complexity of these *in vitro* analyses. The details of effector functions of activated macrophages are unique for each target.

IV. REFERENCES

Adams, D. O., Marino, P. A., and Meltzer, M. S., 1981, Characterization of genetic defects in macrophage tumoricidal capacity: Identification of murine strains with abnormalities in secretion of cytolytic factors and ability to bind neoplastic targets, *J. Immunol.* **126**: 1843–1847.

Anderson, S. E., and Remington, J. S., 1974, Effect of normal and activated human macrophages on *Toxoplasma gondii*, *J. Exp. Med.* **139**:1154–1174.

Boraschi, D., and Meltzer, M. S., 1979a, Macrophage activation for tumor cytotoxicity: Genetic variation in macrophage tumoricidal capacity among mouse strains, *Cell. Immunol.* **45**:188–194.

Boraschi, D., and Meltzer, M. S., 1979b, Defective tumoricidal capacity of mice from A/J mice I. Characterization of the macrophage cytotoxic defect after *in vivo* and *in vitro* stimuli, *J. Immunol.* 122:1587-1591.

Boraschi, D., and Meltzer, M. S., 1979c, Defective tumoricidal capacity of mice from A/J mice II. Comparison of the macrophage cytotoxic defect of A/J mice with that of lipid A-unresponsive C3H/HeJ mice, *J. Immunol.* 122:1592-1597.

Boraschi, D., and Meltzer, M. S., 1980a, Defective tumoricidal capacity of mice from A/J mice III. Genetic analysis of the cytotoxic defect, *J. Immunol.* 124:1050-1053.

Boraschi, D., and Meltzer, M. S., 1980b, Defective tumoricidal capacity of mice from P/J mice I. Characterization of the macrophage cytotoxic defect after *in vivo* and *in vitro* activation stimuli, *J. Immunol.* 125:771-776.

Boraschi, D., and Meltzer, M. S., 1980c, Defective tumoricidal capacity of macrophages from P/J mice. II. Tumoricidal defect involved abnormalities in lymphocyte-derived stimuli and in mononuclear phagocyte responsiveness, *J. Immunol.* 125:777-782.

Bout, D. T., Joseph, M., David, J. R., and Capron, A. R., 1981, *In vitro* killing of *S. mansoni* schistosomula by lymphokine-activated mouse macrophages, *J. Immunol.* 127:1-5.

Cohn, Z. A., 1978, The activation of mononuclear phagocytes: Fact, fancy, and future, *J. Immunol.* 121:813-816.

Fortier, A. H., Hoover, D. L., and Nacy, C. A., 1982, Intracellular replication of *Leishmania tropica* in mouse peritoneal macrophages: Amastigote infection of resident cells and inflammatory exudate macrophages, *Infect. Immun.* 38:1304-1308.

Haverly, A. H., Pappas, M. G., Henry, R. R., and Nacy, C. A., 1983, *In vitro* macrophage antimicrobial activities and *in vivo* susceptibility to *Leishmania tropica* infection, in: *Host Defenses to Intracellular Pathogens* (T. K. Eisenstein, H. Friedman, P. Actor, eds.), pp. 433-440, Plenum Press, New York.

Hinrichs, D. J., and Jerrells, T. R., 1976, *In vitro* evaluation of immunity to *Coxiella burnettii, J. Immunol.* 117:996-1003.

Horwitz, M. A., and Silverstein, S. C., 1981, Activated human monocytes inhibit the intracellular multiplication of Legionnaire's disease bacteria, *J. Exp. Med.* 154:1618.

James, S. L., Lazdins, J. K., Meltzer, M. S., and Sher, A., 1982a, Macrophages as effector cells of protective immunity in murine schistosomiasis I. Activation of peritoneal macrophages during natural infection, *Cell. Immunol.* 67:255-266.

James, S. L., Sher, A., Lazdins, J. K., and Meltzer, M. S., 1982b, Macrophages as effector cells of protective immunity in murine schistosomiasis. II. Killing of newly transformed schistosomula *in vitro* by macrophages activated as a consequence of *S. mansoni* infection, *J. Immunol.* 128:1535-1540.

James, S. L., Leonard, E. J., and Meltzer, M. S., 1983a, Macrophages as effector cells of protective immunity in murine schistosomiasis. IV. Coincident induction of macrophage activation for extracellular killing of schistosomula and tumor cells, *Cell. Immunol.* 74: 86-96.

James, S. L., Skamene, E., and Meltzer, M. S., 1983b, Macrophages as effector cells of protective immunity in murine schistosomiasis. V. Vaiation in macrophage schistosomulicidal and tumoricidal activities among mouse strains and correlation with resistance to reinfection, *J. Immunol.* 131:948-953.

Karnovsky, M. L., and Lazdins, J. K., 1978, Biochemical criteria for activated macrophages, *J. Immunol.* 121:809-812.

Koren, H. S., Anderson, S. J., and Adams, D. O., 1981, Studies on the antibody-dependent cell-mediated cytotoxicity (ADCC) of thioglycollate-stimulated and BCG-activated peritoneal macrophages, *Cell Immunol.* 57:51-61.

Leonard, E. J., Ruco, L. P., and Meltzer, M. S., 1978, Characterization of macrophage activation factor, a lymphokine that causes macrophages to become cytotoxic for tumor cells, *Cell. Immunol.* 41:347-357.

Mackaness, G. B., 1962, Cellular immunity to infection, *J. Exp. Med.* 116:381-406.

Mauel, J., Buchmuller, Y., and Behin, R., 1978, Studies on the mechanisms of macrophage activation. I. Destruction of *Leishmania enriettii* in macrophages activated by cocultivation with stimulated lymphocytes, *J. Exp. Med.* 148:393-407.

Meltzer, M. S., 1981, Macrophage activation for tumor cytotoxicity: Characterization of priming and triggering signals during lymphokine activation, *J. Immunol.* 127:179-183.

Meltzer, M. S., and Nacy, C. A., 1980, Macrophage in resistance to rickettsial infection: Susceptibility to the lethal effects of *Rickettsia akari* infection in mouse strains with defective macrophage function, *Cell. Immunol.* 54:487-490.

Meltzer, M. S., Benjamin, W. R., and Farrar, J. J., 1982a, Macrophage activation for tumor cytotoxicity: Induction of macrophage tumoricidal activity by lymphokines from EL-4, a continuous T cell line, *J. Immunol.* 129:2802-2807.

Meltzer, M. S., Nacy, C. A., James, S. L., Benjamin, W. R., and Farrar, J. J., 1982b, Transient cytotoxic responses of activated macrophages: Characterization of signals that regulate cytotoxic activity, in: *Advances in Immunopharmacology 2* (J. W. Hadden, ed.), pp. 229-234, Pergamon Press, Oxford, England.

Nacy, C. A., and Diggs, C. L., 1981, Intracellular replication of *Leishmania tropica* in mouse peritoneal macrophages: Comparison of amastigote replicaton in adherent and nonadherent macrophages, *Infect. Immun.* 34:310-313.

Nacy, C. A., and Groves, M. G., 1981, Macrophages in resistance to rickettsial infections: Early host defense mechanisms in experimental scrub typhus, *Infect. Immun.* 31:1239-1250.

Nacy, C. A., and Meltzer, M. S., 1979, Macrophages in resistance to rickettsial infections: macrophage activation *in vitro* for killing of *Rickettsia tsutsugamushi*, *J. Immunol.* 123:2544-2549.

Nacy, C. A., and Meltzer, M. S., 1982, Macrophages in resistance to rickettsial infections: Strains of mice susceptible to the lethal effects of *Rickettsia akari* show defective macrophage rickettsiacidal activity *in vitro*, *Infect. Immun.* 36:1096-1101.

Nacy, C. A., and Oaks, S. C., 1981, Destruction of rickettsiae, in: *Methods for Studying Mononuclear Phagocytes* (D. O. Adams, P. J. Edelson, and H. Koren, eds.), pp. 725-743, Academic Press, New York.

Nacy, C. A., and Osterman, J. V., 1979, Host defenses in experimental scrub typhus: role of normal and activated macrophages, *Infect. Immun.* 26:744-750.

Nacy, C. A., Leonard, E. J., and Meltzer, M. S., 1981a, Macrophages in resistance to rickettsial infections: Characterization of lymphokines that induce rickettsiacidal activity in macrophages, *J. Immunol.* 126:204-207.

Nacy, C. A., Meltzer, M. S., Leonard, E. J., and Wyler, D. J., 1981b, Intracellular replication and lymphokine-induced destruction of *Leishmania tropica* in C3H/HeN mouse macrophages, *J. Immunol.* 127:2381-2386.

Nacy, C. A., Leonard, E. J., and Meltzer, M. S., 1982, Role of activated macrophages in resistance to rickettsial infections, in: *Phagocytosis—Past and Future* (M. Karnovksky and and L. Bolis, eds.), pp. 475-504, Academic Press, New York.

Nacy, C. A., Meltzer, M. S., Leonard, E. J., Stevenson, M. M., and Skamene, E., 1983a, Activation of macrophages for killing of rickettsiae: Analysis of macrophage effector function after rickettsial inoculation of inbred mouse strains, in: *Host Defenses to Intracellular Pathogens* (T. K. Eisenstein, H. Friedman, and P. Actor, eds.), pp. 441-459, Plenum Press, New York.

Nacy, C. A., Hockmeyer, W. T., Benjamin, W. R., Farrar, J. J., James, S. L., and Meltzer, M. S., 1983b, Lymphokines from the EL-4 T cell line induce macrophage microbicidal and tumoricidal activities, in: *Interleukins, Lymphokines, and Cytokines* (E. Pick, J. J. Oppenheim, and M. Landy, eds.), pp. 617-624, Academic Press, New York.

Nacy, C. A., Fortier, A. H., Pappas, M. G., and Henry, R. R., 1983*c*, Susceptibility of inbred mice to *Leishmania tropica* infection: Correlation of susceptibility with *in vitro* defective macrophage microbicidal activities, *Cell. Immunol.* 77:298–307.

Nacy, C. A., Hockmeyer, W. T., Benjamin, W. R., Farrar, J. J., James, S. L., and Meltzer, M. S., 1983*d*, Activation of macrophages for microbicidal and tumoricidal effector functions by soluble factors from EL-4, a continuous T cell line, *Infect. Immun.* 40:820–824.

Nakamura, R. M., and Meltzer, M. S., 1981, Macrophage activation for tumor cytotoxicity: Control of the cytotoxic activity by the time interval effector and target cells are exposed to lymphokines, *Cell. Immunol.* 65:52–65.

Nogueira, N., and Cohn, Z. A., 1978, *Trpanosoma cruz: In vitro* induction of macrophage microbicidal activity, *J. Exp. Med.* 148:288–300.

North, R. J., 1978, The concept of the activated macrophage, *J. Immunol.* 121:806–808.

Oster, C. N., and Nacy, C. A., 1984, Macrophage activation to kill *Leishmania tropica:* Kinetics of macrophage response to lymphokines that induce antimicrobial activities against *Leishmania tropica* amastigotes, *J. Immunol.* (in press).

Oster, C. N., Hockmeyer, W. T., and Nacy, C. A., 1984, Macrophage activation to kill *Leishmania tropica*: Analysis of activation steps in development of macrophage microbicidal activities against *Leishmania tropica*, *J. Immunol.* (submitted).

Pappas, M. G., Oster, C. N., and Nacy, C. A., 1983*a*, Intracellular destruction of *Leishmania tropica* by macrophages activated *in vivo* with *Mycobacterium bovis* strain BCG, in: *Host Defenses to Intracellular Pathogens* (T. K. Eisenstien, H. Friedman, and P. Actor, eds.), pp. 425–432, Plenum Press, New York.

Pappas, M. G., and Nacy, C. A., 1983*b*, Intracellular destruction of *Leishmania tropica* by C3H/HeN and C3H/HeJ mouse macrophages activated with *Mycobacterium bovis* strain BCG, *Cell. Immunol.* 80:217–222.

Ralph, P., Nacy, C. A., Meltzer, M. S., Williams, N., Nakoinz, I., and Leonard, E. J., 1983, Colony stimulating factors and regulation of macrophage tumoricidal and microbicidal activities, *Cell. Immunol.* 76:10–21.

Ruco, L. P., and Meltzer, M. S., 1977, Macrophage activation for tumor cytotoxicity: Induction of tumoricidal macrophages by supernatants of PPD-stimulated Bacillus Calmette-Guérin-immune spleen cell cultures, *J. Immunol.* 119:889–896.

Ruco, L. P., and Meltzer, M. S., 1978*a*, Defective tumoricidal capacity of macrophages from C3H/HeJ mice, *J. Immunol.* 120:329–334.

Ruco, L. P., and Meltzer, M. S., 1978*b*, Macrophage activation for tumor cytotoxicity: Increased lymphokine responsiveness of peritoneal macrophages during acute inflammation, *J. Immunol.* 120:1054–1062.

Ruco, L. P., and Meltzer, M. S., 1978*c*, Macrophage activation for tumor cytotoxicity: Tumoricidal activity by macrophages from C3H/HeJ mice requires at least two activation stimuli, *Cell. Immunol.* 41:35–51.

Ruco, L. P., and Meltzer, M. S., 1978*d*, Macrophage activation for tumor cytotoxicity: Development of macrophage cytotoxic activity requires completion of a sequence of short-lived intermediary interaction, *J. Immunol.* 121:2035–2042.

Ruco, L. P., Meltzer, M. S., and Rosenstreich, D. L., 1978, Macrophage activation for tumor cytotoxicity: control of macrophage tumoricidal capacity by the *Lps* gene, *J. Immunol.* 121:543–548.

Chapter 9

Identification of Gamma-Interferon as a Murine Macrophage-Activating Factor for Tumor Cytotoxicity

Robert D. Schreiber

Research Institute of Scripps Clinic
Scripps Clinic
La Jolla, California 92037

I. INTRODUCTION

Over the past decade, work from a number of laboratories has indicated that, under the proper conditions, the macrophage can express nonspecific effector cell function toward a variety of neoplastic cells (reviewed in Den Otter, 1981; Fidler and Raz, 1981; Keller, 1981; Lohmann-Matthes *et al.*, 1981; Meltzer, 1981*b*; Piessens *et al.*, 1981). While this type of macrophage tumoricidal activity does not require antibody or antibodylike specificity, it is nevertheless able to discriminate between normal cells and tumor cells (Hibbs, 1974; Piessens *et al.*, 1975; Fidler *et al.*, 1976). The currently accepted concept is that two signals are required to activate macrophages for tumor cell killing (Russell *et al.*, 1977; Ruco and Meltzer, 1978*a,b*; Weinberg *et al.*, 1978; Weinberg and Hibbs, 1979; Meltzer, 1981*a,b*; Pace and Russell, 1981). The first signal is a lymphokine that primes the macrophage and makes it receptive to a diverse group of substances that can act as second signals to trigger the development of full cytocidal activity. The lymphokine, denoted macrophage-activating factor (MAF), has been reported to alter a number of functional and biochemical properties in macrophage populations (reviewed in David and Remold, 1979; Rocklin *et al.*, 1980). These alterations include increases in endocytic, biosynthetic, secretory, and effector cell functions as well as changes in membrane physiology and composition. However, since it is unclear whether all these alterations are effected by the same molecular species, it has become critical to define MAF in the context

171

of the activity it induces. This chapter defines MAF exclusively as the factor that primes macrophages for nonspecific tumoricidal activity.

The molecular identity of MAF has remained obscure. In particular, a great deal of uncertainty exists as to the relationship between MAF and other lymphokines such as γ-interferon (γ-IFN), which can regulate macrophage function. This uncertainty has gone unresolved primarily because of the unavailability of purified lymphokine preparations. During the past several years, a number of advances in the fields of cell biology, biochemistry, and molecular biology have taken place that have circumvented these problems and have for the first time, allowed definitive experiments to be performed.

This chapter presents the evidence accumulated from a number of laboratories indicating that γ-IFN can function as an MAF to prime macrophages for nonspecific tumoricidal activity. Special emphasis is placed on the characterization of a MAF produced by a murine T-cell hybridoma and the identification of this activity as γ-IFN.

II. γ-IFN AS A MACROPHAGE-ACTIVATING FACTOR: HISTORICAL PERSPECTIVES

The history of the elucidation of a role for γ-IFN in the induction of macrophage tumoricidal activity stretches from 1970 to the present, when its participation was clearly established. During the early 1970s a number of laboratories reported that macrophages isolated from animals undergoing allograft or tumor rejection (Evans and Alexander, 1970; Lohmann-Matthes et al., 1973) or recovering from infection with facultative intracellular pathogens (Hibbs et al., 1972; Cleveland et al., 1974; Ruco and Meltzer, 1977) were able to express tumoricidal activity toward neoplastic cells. Subsequent experiments revealed that induction of cytotoxicity required the participation of T lymphocytes (Evans and Alexander, 1972; Lohmann-Matthes et al., 1973; Ruco and Meltzer, 1977). In this process the T cells were found to regulate macrophage function through the elaboration of a soluble mediator. This lymphokine has since become known as MAF.

In 1973, Nathan et al. published an initial characterization of a lymphokine that activated macrophages to display increased adherence, phagocytosis, and glucose metabolism. The macrophage activating activity was found to be similar to another activity termed migration inhibition factor (MIF) on the basis of molecular weight (25,000-65,000), buoyant density, and neuraminidase sensitivity. That same year, Younger and Salvin (1973) reported that MIF activity displayed similar properties to a lymphocyte-derived form of IFN (Wheelock, 1965). These properties included trypsin sensitivity, a molecular weight of 45,000-80,000, partial stability to incubation at 56°C, and complete lability at pH 2.0. Type II IFN (or γ-IFN) was thus physiochemically distinct from the type I interferons, which were produced by fibroblasts (β-IFN) or leukocytes (α-IFN), an observation that was also supported by immunochemical analysis.

Four years later, Schultz and co-workers demonstrated that macrophages treated with partially purified preparations of murine β-IFN could induce cytostasis of a murine leukemia cell line (Schultz et al., 1977; Schultz and Chirigos, 1978). However, β-IFN did not appear to be the active species in lymphokine-rich supernatants. This fact became apparent from the work of Leonard et al. (1978), who established a number of physicochemical characteristics of a MAF produced by antigen-stimulated immune splenic cells. MAF activity was found to display a molecular weight of 55,000 ± 1600 on gel-filtration columns but eluted in a heterogeneous manner when subjected to anion-exchange chromatography. Activity was protease sensitive, labile at pH values below 4.0 or above 10.0, and partially destroyed at 56°C. MAF was not bound by concanavalin A (Con A) columns but was adsorbed to columns bearing certain immobilized dyes. However, at that time no attempts were made to simultaneously compare the behaviors of the MAF and the antiviral activities present in the lymphokine-containing supernatants. In 1981 Kniep et al. reported the separation of MAF and antiviral activities from one another by chromatography of culture supernatants on polynucleotide columns. Many of the MAF characteristics obtained in this study were substantially different from those reported by Leonard et al. (1978), however. The possibility must be considered that the different assay systems employed in the two studies detected different activating molecules.

In 1982 several studies confirmed the work of Leonard et al. (1978). In addition, two studies found that the MAF produced by mitogen-stimulated splenic cell cultures was indistinguishable from γ-IFN by a variety of criteria including (1) kinetics of temperature and pH inactivation, (2) immunochemical reactivity with polyvalent rabbit antiserum to partially purified murine γ-IFN and (3) chromatography on a variety of separatory media (Roberts and Vasil, 1982; Kleinschmidt and Schultz, 1982).

One obstacle that had impeded the aforementioned studies was the lack of reproducible, large-scale sources of MAF and γ-IFN. These materials became available in 1982 when a number of MAF-producing T-cell lines, T-cell hybridomas, and murine tumor cell lines were identified. Using normal T-cell lines, Kelso and colleagues (Kelso et al., 1982; Kelso and MacDonald, 1982) and Prystowsky et al. (1982) were unable to distinguish between MAF and γ-IFN on a biosynthetic basis. However, these same studies were able to differentiate clearly between MAF and several other lymphokine activities (IL-2, IL-3, colony-stimulating factor) on the basis of differential rates of secretion or production at a clonal level. Partial biochemical analysis of MAF activity produced by a murine thymoma-derived cell line (EL-4) indicated that 50–75% of the activity was inseparable from γ-IFN (Meltzer et al., 1982). However, this study also identified a unique chromatographic species of 23,000 molecular weight that displayed MAF activity in the absence of concomitant antiviral activity. This result suggested that MAF activity might be effected by either γ-IFN-like or unique molecular species. Although another study indicated that a different tumor cell line (denoted BFS) produced γ-IFN but not MAF (Benjamin *et al.*, 1982), reanalysis of culture supernatants showed both activities to be present (R. D. Schreiber, M. S. Meltzer, W. R. Benjamin, and J. J. Farrar, unpublished observations). Using another approach, several laboratories constructed murine T-cell hybridomas and examined their ability to produce lymphokines. One T-cell hybridoma (24/G1) was identified that produced MAF and γ-IFN in concentrations 10–25 times higher than were present in conventional splenic cell culture supernatants (Schreiber *et al.*, 1982). Concurrently, two other MAF-producing T-cell hybridomas were reported (Ratliff *et al.*, 1982a,b; Erickson *et al.*, 1982). However, the MAF-containing supernatants from these latter hybridomas were found to be devoid of antiviral activity. The MAF produced by these hybridomas may therefore be the non-IFN type (see Section IV).

The final identification of γ-IFN as an MAF came in 1983. In this year the murine γ-IFN gene was cloned by Gray and Goeddel (1983) and the first recombinant murine γ-IFN was produced by introduction of the gene into the transformed monkey cell line COS-7. When tested for biologic activity, culture supernatants of the transfected COS-7 cells were found to display MAF activity, while supernatants of control cells did not (Pace *et al.*, 1983b; Varesio *et al.*, 1983; Schultz and Kleinschmidt, 1983). The MAF activity of these supernatants was neutralized by rabbit anti-γ-IFN and was acid labile. The amount of MAF activity relative to antiviral activity produced by the transfected cells was comparable to supernatants of normal splenic cells (Schultz and Kleinschmidt, 1983) or of the T-cell hybridoma 24/G1 (Pace *et al.*, 1983a). When titrated on an antiviral activity basis, highly purified recombinant γ-IFN produced in *Escherichia coli* was found to display MAF activity equal to that produced by hybridoma 24/G1, and rat antiserum against the purified protein neutralized all the hybridoma-derived MAF activity (R. D. Schreiber and P. W. Gray, unpublished observations). While these experiments were ongoing, considerable independent biochemical and biosynthetic data was accumulated that conclusively identified γ-IFN as the major and possibly only molecular species responsible for the hybridoma-derived MAF activity (Pace *et al.*, 1983a; Schreiber *et al.*, 1983). These data are discussed in detail in Section III.

Thus the accumulated evidence from a number of laboratories using a number of experimental approaches strongly supports the concept that γ-IFN can function as an MAF to prime macrophages for nonspecific tumoricidal activity. This conclusion does not rule out the possibility that other factors, unrelated to γ-IFN can act as MAF as well. However, before the existence of such unique factors can be accepted, it will be necessary to rigorously establish their molecular identity and independence from γ-IFN.

III. MAF PRODUCED BY A T-CELL HYBRIDOMA:
IDENTIFICATION AS γ-IFN

A. Introduction

This section presents several pieces of experimental evidence that indicate that the MAF activity produced by the murine T-cell hybridoma 24/G1 is attributable to γ-IFN. These experiments have been made possible because of recent developments in the fields of cell biology, biochemistry, and molecular biology. Over the past few years sensitive assays have been developed to quantitate MAF and IFN reproducibly. Somatic cell hybridization techniques have allowed for the construction of stable T-cell hybridomas capable of expressing normal T-cell functions, including the production and secretion of lymphokines. The availability of new biochemical chromatographic media and the adaptation of the high-performance liquid chromatography technique to proteins have provided new pathways to protein purification. Most recently, the isolation and translation of the murine γ-IFN gene has made accessible substantial amounts of this lympokine in pure form and has prompted a number of studies into its immunoregulatory functions. The concurrent appearance of these developments have thus established the appropriate environment to investigate the molecular identity of MAF.

B. Measurement of MAF and IFN

MAF was quantitated with a bioassay that measured the induction of macrophage cytocidal activity toward radiolabeled tumor cells (Schreiber et al., 1981, 1982). The assay is based on the observation that two signals are required for the development of nonspecific tumoricidal activity in macrophages (Weinberg et al., 1978; Meltzer, 1981a; Pace and Russell, 1981). By incorporating excess amounts of a second or triggering signal into the reaction mixture, such as lipopolysaccharide (LPS) or heat-killed Listeria monocytogenes (HKLM), the level of tumoricidal activity induced was found to be solely proportional on the quantity of MAF added. In this assay, adherent, elicited macrophage populations are incubated with dilutions of MAF in the presence of HKLM and then exposed to ^{51}Cr-labeled P815 mastocytoma cells at an effector cell:target cell ratio of 10:1. After incubation for 18-24 hr at 37°C, the amount of ^{51}Cr released into the supernatant is determined. One unit of MAF is defined as that amount that produces 50% of the maximal specific ^{51}Cr release.

The antiviral activity of murine interferon was quantitated using a cytopathic effect assay (Schreiber et al., 1982). This assay detects all three types of IFN (α, β, or γ). Murine L cells are cultured with dilutions of an IFN source and then challenged with purified vesicular stomatitus virus. After incubation for an

additional 24 hr, the cultures are examined microscopically to assess the cytopathic effect of the virus. If sufficient IFN was present in the initial incubation,
the L cells are protected from viral infection and the cell monolayer remains
intact. If IFN was absent or limiting in the initial reaction mixture, the cells are
permissive to viral infection and cell death ensues. One unit of IFN is defined as
the reciprocal of the dilution that produces 50% protection of the L-cell monolayer. This value is standardized to the international murine β-IFN standard and
is expressed as the number of international units per milliliter (IU/ml).

C. Production of Murine T-Cell Hybridomas That Secrete MAF

One major obstacle that impeded previous attempts at determining the
molecular identity of MAF was the lack of a suitable large-scale source of the
lymphokine with which a biochemical purification could be attempted. We reasoned that identification of a continuous cell line capable of producing large
quantities of MAF would solve this problem. This was accomplished by con-
structing murine T-cell hybridomas by fusion of alloantigen-activated T cells
with the hypoxanthine–aminopterin–thymidine (HAT)-sensitive T-lymphoma
cell line denoted BW5147. Figure 1 outlines the protocol used to generate these
hybridomas. This particular protocol was chosen because it had successfully
produced T-cell hybridomas capable of secreting high amounts of allogeneic effect factor or IL-2 (Katz et al., 1980; Altman et al., 1982). Alloantigen-activated
T cells were generated by injection of host-derived thymocytes and irradiated
semiallogeneic spleen cells into an irradiated host. Seven days later, the alloactivated T cells were recovered from the spleens of the recipients and reactivated
by short-term (18- to 24-hr) culture with fresh irradiated stimulator cells. The resulting T-cell blasts were enriched by sedimentation on Ficoll cushions and then
fused with the BW5147 cell line using polyethylene glycol. Hybrids were selected by culture in HAT-containing medium. Cells in growth-positive wells were
expanded, and stimulated with mitogen; the resulting culture supernatants were
tested for production of MAF and IFN.

After one such fusion, 32 parental T-cell hybrids were obtained. None of the
cultures produced MAF or IFN constitutively. Upon stimulation with Con A,
10 hybrids secreted 3–12 times more MAF activity than was present in conventional lymphokine-rich supernatants, and 11 hybrids produced 3–16 times more
antiviral activity (Fig. 2). The conventional supernatants were produced by mitogen stimulation of normal murine splenic cells. None of the hybridomas produced
either MAF or IFN in the absence of the other. Although production of the two
lymphokine activities correlated for most of the hybridomas, two hybridomas
secreted a disproportionately high amount of MAF activity and one produced an
unusually high amount of antiviral activity.

One hybridoma (from a different fusion) was cloned by limiting dilution.

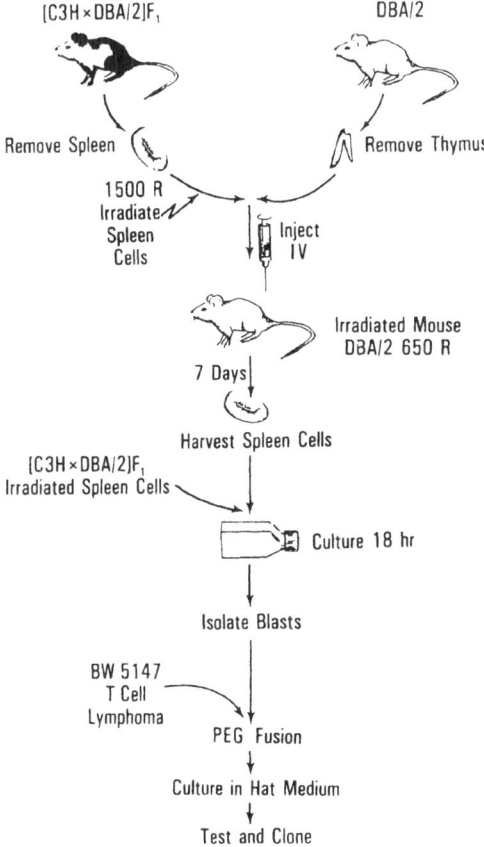

Figure 1. Production of murine T-cell hybridomas. IV, intravenous; PEG, polyethylene glycol; R, rads.

Twenty-seven clones were obtained, none of which produced MAF or IFN constitutively. Upon stimulation, seven clones produced high amounts of MAF activity. One clone (24/G1) was of particular interest, since it produced 23 times more MAF activity (55,000 units/ml) than was present in conventional MAF preparations (2,400 units/ml). Although at an early point in our studies with 24/G1 we were unable to detect antiviral activity in its supernatants, studies performed after the clone had stabilized indicated that it produced both MAF and antiviral activities.

The possibility was considered that the differences observed between the hybridoma-derived and conventional supernatants reflected the presence of inhibitory factors in the splenic cell preparation (such as prostaglandins) or

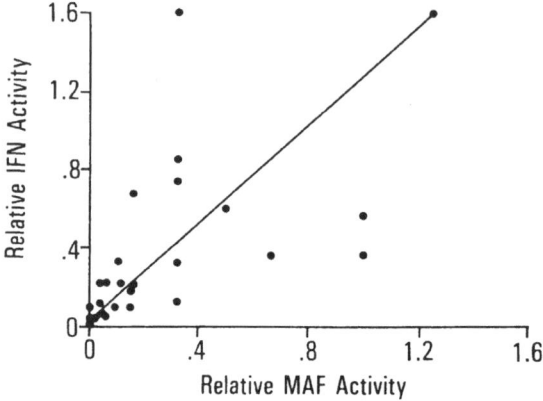

Figure 2. Correlation between MAF and IFN activities produced by various T-cell hybridomas after mitogen stimulation. MAF activity and IFN antiviral activity were titrated relative to a standard supernatant of mitogen stimulated splenic cells. This standard was assigned a value of 0.1 for each activity. IFN, interferon; MAF, macrophage-activating factor. (From Schreiber *et al.*, 1983.)

unrecognized enhancing factors in the hybridoma supernatant. To test this hypothesis, the two supernatants were mixed in varying ratios and then titrated in the quantitative MAF assay (Schreiber *et al.*, 1982). The observed MAF activity of the mixtures was found to be in close agreement with the calculated theoretical values based on the mixture composition (Table I). Thus, the higher amounts of MAF activity observed in the hybridoma supernatant compared with the splenic cell supernatant appeared to reflect actual differences in lymphokine concentrations.

Table I. Mixing of Hybridoma (24/G1) and Conventional (8080) MAF Preparations: Lack of Enhancement or Inhibition[a,b]

MAF ratio 24/G1 : 8080	MAF activity (units/ml)	
	Measured	Theoretical
1 : 0	51,200	—
0 : 1	3,200	—
3 : 1	35,800	39,200
1 : 1	25,600	27,200
1 : 3	16,000	15,200

[a]From Schreiber *et al.* (1982).
[b]*Abbreviation:* MAF, macrophage-activating factor.

These results indicated that a stable murine T-cell hybridoma (24/G1) had been identified that could produce MAF in amounts far exceeding those that could be obtained by conventional techniques. This particular hybridoma clone was then used for the remainder of the study.

D. Comparison of Physicochemical and Functional Properties of Hybridoma-Derived and Conventional MAF

A series of experiments were performed to compare the hybridoma-derived MAF with that produced by normal splenic cells, as summarized in Table II. The activities in both preparations showed identical temperature and pH sensitivities and exhibited similar molecular weights when subjected to molecular sieve chromatography on Sephadex G100. Both MAF preparations, when added at high concentrations, could elicit a tumoricidal response without the addition of a second signal. However, both activities were enhanced in the presence of a second signal. The magnitude of this enhancement was quantitatively similar for the preparations: 42-fold for the hybridoma MAF and 50-fold for the splenic MAF (Schreiber et al., 1982).

The hybridoma supernatant was also found to induce a number of other functional and biochemical changes in macrophages (Johnson et al., 1983) associated with macrophage activation and previously shown to be affected by conventional supernatants. Two such changes, appear to be important for the tumoricidal response: (1) the acquisition of a selective binding capacity for tumor cells (Marino and Adams, 1980a,b; Adams and Marino, 1981), and (2) the induction of an enzyme (cytolytic protease) that can express cytocidal

Table II. Comparison of Biochemical and Functional Properties of Hybridoma-Derived MAF and Conventional MAF Preparations[a,b]

Property	T-cell hybridoma MAF	Conventional splenic MAF
pH sensitivity	85% inactivated at $\leqslant$ pH 4.0	85% inactivated at $\leqslant$ pH 4.0
Temperature sensitivity	Inactivated at 65°C (1 hr)	Inactivated at 65°C (1 hr)
Enhancement of activity by second signal	42-fold	50-fold
Molecular weight	55,000	50,000
Induction of tumor cell binding	Yes	Yes
Induction of cytolytic protease	Yes	Variable

[a]*Abbreviations:* Con A, concanavalin A; MAF, macrophage-activating factor.
[b]MAF prepared by Con A stimulation of T-cell hybridoma clone 24/G1 or normal murine splenic cell suspensions.

activity (Adams, 1980). On the basis of these physicochemical and functional criteria, the MAF produced by the hybridoma clone 24/G1 appeared to be identical with MAF produced by normal murine splenic cells.

E. Concomitant Biosynthesis of MAF and Antiviral Activities

In an attempt to dissociate MAF from γ-IFN on a biosynthetic basis, the hybridoma clone 24/G1 was subcloned by limiting dilution. As was the case for the parental T-cell hybridomas, no subclone could be identified which secreted one of the activities in the absence of the other. The amount of MAF and antiviral activities produced by the subclones was found to correlate ($r = 0.95$) (Pace *et al.*, 1983a). In other experiments the clone was stimulated under a variety of conditions that included different types and concentrations of mitogens and different concentrations of fetal calf serum in the stimulation medium. Once again, a correlation was found between the secretion of MAF and antiviral activities even when changes in experimental conditions resulted in a 20-fold difference in the amount of each activity produced (Fig. 3) (Schreiber *et al.*, 1983). When analyzed on a kinetic basis, the appearance of MAF and antiviral activities in culture supernatants after mitogen stimulation of the hybridoma culture were found to be indistinguishable. Both activities were first detectable after 6–8 hr of stimulation and were maximally expressed after 12 and 22 hr. A concomitant decrease in both activities was observed between 36 and 48 hr after stimulation.

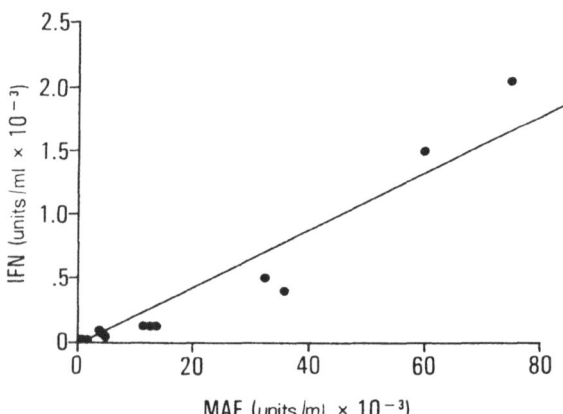

Figure 3. Correlation between MAF and IFN activities produced by the T-cell hybridoma clone 24/G1. Cultures of the hybridoma clone were stimulated under a variety of culture conditions including (1) different concanavalin A concentrations (2, 5, 10, 20, and 40 μg/ml), (2) different fetal calf serum concentrations (0%, 0.5%, and 5%), and (3) different times of stimulation (0, 12, 24, 36, 48 hr). Abbreviations as in Fig. 1. (From Scheriber *et al.*, 1983.)

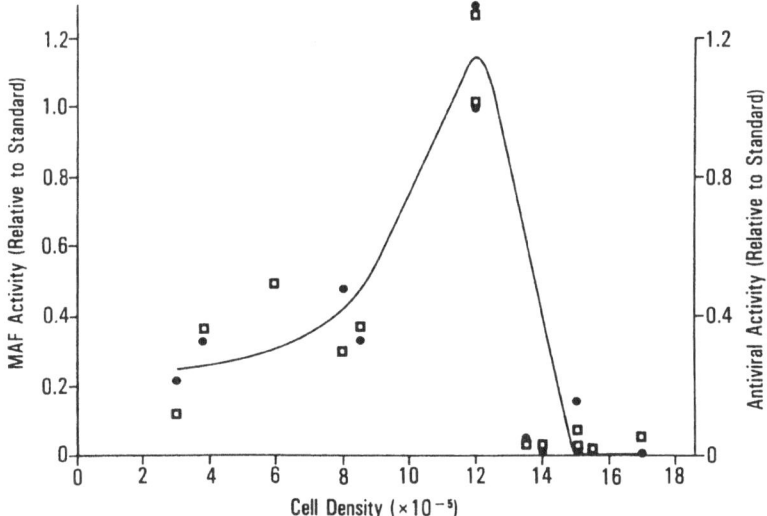

Figure 4. Influence of hybridoma cell culture density on production of MAF and IFN activities. Cultures of clone 24/G1 were grown to different cell densities and stimulated with 10 μg/ml concanavalin A. MAF (●) and antiviral (□) activities were titrated relative to a standard hybridoma supernatant. This was assigned a value of 1.0. Abbreviations as in Fig. 1. (From Schreiber *et al.*, 1983.)

In another series of experiments, the effect of cell density on MAF and anti-viral activity production was studied (Fig. 4). Although these studies revealed an unusual behavior of this particular hybridoma clone, they still could not dissoci-ate MAF and γ-IFN on a biosynthetic basis (Schreiber *et al.*, 1983). MAF and antiviral activities secreted by clone 24/G1 increased in proportion to the cell density of the culture up to a level of 1.2×10^6 cells/ml. At each cell density tested, the relative amounts of the two activities correlated with each other. However, at higher cell densities, production of either activity precipitously de-clined despite the fact that the cell culture was capable of reaching densities of 2.4×10^6 cells/ml. This unusual cell dose–response profile could not be altered even when different mitogen concentrations were used or when the stimulation times were changed.

F. Production of MAF and Antiviral Activities by a Cloned Murine NK Cell Line

In the past, MAF and γ-IFN were thought to be primarily produced by T lymphocytes. Recently we have found that an IL-2-dependent murine natural killer (NK) cell line, denoted B6NK, produced both MAF and antiviral activities (R. D. Schreiber, J. Warner, and G. Dennert, unpublished observations). Earlier

Table III. Production of γ-IFN and MAF by
Cloned Murine NK Cells[a]

Culture supernatant	IFN (IU/ml)	MAF (units/ml)
Unstimulated B6NK	40	2,750
Stimulated B6NK	96	8,000
Stimulated B6NK, 56°C, 60 min	48	4,050
Stimulated B6NK, dialyzed pH 7	96	12,000
Stimulated B6NK, dialyzed pH 4	6	450
Stimulated normal spleen cells	40	2,000
Stimulated 24/G1	364	24,000

[a]Abbreviations: IFN, interferon; MAF, macrophage-activating factor; NK, natural killer.

studies had established the NK cell character of this cell line on the basis of surface marker analysis, target cell specificity and reconstitution of the bone marrow rejection response in NK-deficient mice (Dennert et al., 1981; Warner and Dennert, 1982). When stimulated with Con A, B6NK cells secreted MAF and antiviral activities in amounts intermediate between those expressed in stimulated supernatants of the T-cell hybridoma and normal splenic cells (Table III). The antiviral activity was identified as γ-IFN because of its acid lability (pH 4.0) and relative stability to incubation at 56°C. In addition to this result, B6NK cells secreted both activities in the absence of mitogen stimulation, although in lower amounts. [Recently Handa et al. (1983) also observed the secretion of γ-IFN from cloned NK cells grown in IL-2-containing medium.] This result was considered unusual because, of the more than 150 T-cell hybridomas, hybridoma clones, or T-cell lines tested in this laboratory, none has ever been found to secrete MAF or γ-IFN in the absence of mitogen stimulation. In fact, the literature cites only one report of a T-cell hybridoma that secretes an MAF-like activity constitutively (Erickson et al., 1982). Thus, the observation that this unusual cell source can produce both MAF and antiviral activities in the absence of mitogen stimulation represents another indication that the two lymphokines are related.

G. Immunochemical Comparison of MAF and γ-IFN

Attempts were made to discriminate between MAF and γ-IFN activities using antisera to murine γ-IFN produced in two different laboratories. One antiserum (Osborn et al., 1980) neutralized both the antiviral activity (120 units) and MAF activity (10,500 units) in a hybridoma culture supernatant (Pace et al., 1983a). Nonimmune sera and antiserum to α-IFN and β-IFN were devoid of significant inhibitory activity. In another study, a second anti-γ-IFN serum

Table IV. Inhibition of MAF activity by Polyvalent Rabbit
Anti-Murine γ-IFN[a]

Culture supernatant	Inhibitor	MAF activity		IFN activity	
		U/ml	% inhibition	IU/ml	% inhibition
24/G1 hybridoma	Medium	35,000	–	192	–
	NRS	25,000	29	128	33
	Anti-γ-IFN	0 (<800)	100	0 (<32)	100
Normal splenic cell	Medium	4,050	–	ND	–
	NRS	2,700	33	ND	–
	Anti-γ-IFN	0 (<800)	100	ND	–

[a]Abbreviations: IFN, interferon; MAF, macrophage-activating factor; NRS, normal rabbit serum.

(Havell and Spitalny, 1983) neutralized both the antiviral and MAF activities in either hybridoma culture supernatants or stimulated splenic cell supernatants (Table IV) (R. D. Schreiber, G. L. Spitalny, and E. A. Havell, unpublished data). Normal serum produced only minimal amounts of inhibition and even this inhibition was quantitatively identical for MAF and antiviral activities. Although these experiments are open to criticism because of the polyclonal nature of the antiserum and the partial purity of the antigen, they are nevertheless strengthened because two independently produced antisera displayed identical inhibition profiles.

H. Comparison of Temperature and pH Sensitivities of 24/G1 MAF and γ-IFN

The temperature and pH sensitivities were determined for the MAF and antiviral activities present in stimulated supernatants of the hybridoma (Fig. 5). Both activities were stable to incubation for 1 hr at 4° or 37°C and were completely destroyed at 65°C. At 52° or 56°C, a partial (25-50%) inactivation of each activity was observed (Schreiber et al., 1983). When the kinetics of inactivation at 56°C were determined, both activities were found to be concomitantly destroyed in a time-dependent manner. Fifty percent inactivation of either the MAF or antiviral activity was observed after 90 min at 56°C (Pace et al., 1983a). Both activities in the hybridoma supernatant also displayed identical sensitivities to acid pH. No significant inactivation occurred after treatment at pH 7.0 or 5.0 for 18 hr at 4°C. However, when the supernatant was brought to ≤pH 4.0, a 75% loss of both the MAF and antiviral activities was observed (Schreiber et al., 1983).

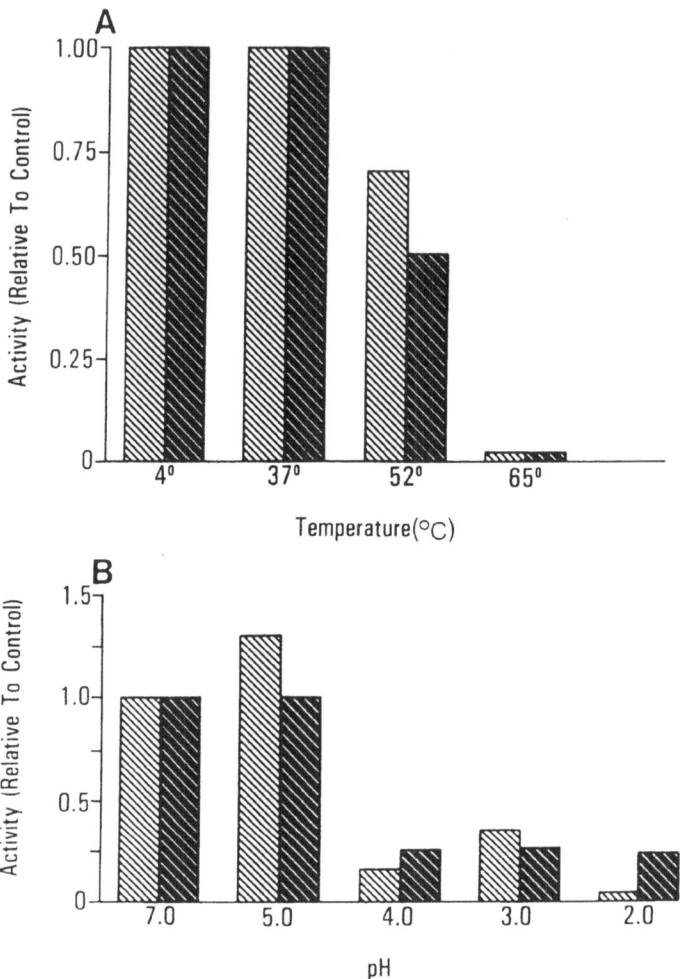

Figure 5. Comparison of heat and pH sensitivities of the MAF and antiviral activities produced by T-cell hybridoma clone 24/G1. (A) Heat sensitivity was determined by incubation of samples of a stimulated culture supernatant for 1 hr at the designated temperatures. (B) Sensitivity to pH was investigated by dialysis of samples for 18 hr at 4°C against 1 liter of physiologic saline buffered to the designated pH values. Samples were then neutralized by dialysis against 0.1 M phosphate buffer pH. 7.4 for 4 hr and then against phosphate-buffered saline. (Light crossed bar) MAF; (dark-crossed bar) IFN. Abbreviations as in Fig. 1. (From Schreiber et al., 1983.)

I. Affinity Chromatography of MAF and γ-IFN

It has been reported that the murine interferons can bind reversibly to columns containing certain polynucleotides (DeMaeyer-Guignard et al., 1977; Wietzerbin et al., 1978). This binding appeared to be somewhat selective because most of the other proteins in the IFN-containing samples did not adsorb to the column. To examine whether MAF and IFN shared the capacity to bind polynucleotides, hybridoma- or splenic cell-derived lymphokine preparations were subjected to chromatography on a polyinosine–Sepharose column (Table V). Ninety percent of the total protein present in the culture supernatants eluted from the column unretarded. In contrast to the behavior of the total protein, most of the antiviral and MAF activity bound to the column and was recovered by elution with buffers of high ionic strength (Schreiber et al., 1983). Recovery of the activities was essentially quantitative. On control columns of unsubstituted Sepharose, little adsorption of MAF or antiviral activity was observed. By quantitating the amount of antiviral and MAF activity in each column fraction, the elution profiles of the two activities were found to be quantitatively identical (Pace et al., 1983a).

Another method of affinity chromatography was then investigated, based on the observation of Kniep et al. (1981) that MAF activity bound somewhat selectively to a particular dye ligand incorporated onto a crosslinked agarose support (Matrex Gel Red A). Hybridoma culture supernatants were incubated with various amounts of the gel and the resulting supernatants titrated for residual MAF and antiviral activities (Schreiber et al., 1983). Both activities were bound by the gel in an identical dose-dependent manner (Fig. 6). Thirty microliters of

Table V. Binding of Hybridoma-Derived γ-IFN and MAF to Polynucleotide Columns[a]

Column	IFN (%)	MAF (%)
Poly (I) Sepharose		
Breakthrough[c]	12	23
Desorbed[c]	88	77
Recovery	113	150
Sepharose		
Breakthrough[c]	88	69
Desorbed[c]	12	31
Recovery	68	161

[a]From Schreiber et al. (1983).
[b]Abbreviations: IFN, interferon; MAF, macrophage-activating factor.
[c]Expressed as the percentage of the recovered activities.

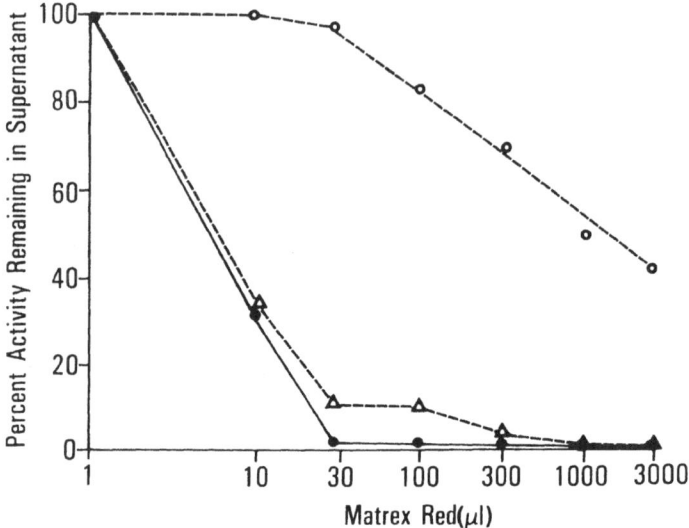

Figure 6. Dose-dependent adsorption of MAF and antiviral activity by Matrex Gel Red A. Varying amounts of packed Matrex Gel Red A were incubated 1 hr at 4°C with 4-ml samples of a stimulated clone 24/G1 supernatant. (●) MAF; (△) interferon; (○) total protein. Abbreviations as in Fig. 1. (From Schreiber *et al.*, 1983.)

gel was sufficient to remove all the MAF and antiviral activity from 4 ml of a hybridoma supernatant. This behavior was distinct from most of the other proteins present in the culture supernatant. No significant protein adsorption could be detected with 30 μl of gel, and 2400 μl of the gel adsorbed only 50% of the total protein.

The information obtained from the analytic-scale adsorption experiments was then used in preparative-scale dye ligand chromatography experiments (Schreiber *et al.*, 1983). Eight hundred and fifty milliliters of the hybridoma supernatant was concentrated and applied to a 5-ml column of Matrex Gel Red A (Fig. 7). Ninety percent of the applied protein eluted from the column unretarded, and an additional 3% was desorbed by treatment with 0.5 M sodium chloride. The unretarded and sodium chloride pools contained only minimal amounts of MAF activity (4.4% and 4.1%, respectively) or antiviral activity (1.5% and 0.7%, respectively). Elution of the column with an ammonium sulfate gradient resulted in the concomitant elution of the MAF and the antiviral activities. Virtually all the MAF activity, but only 0.07% of the total protein originally applied to the column, was recovered in the ammonium sulfate pool. This constituted a 1500-fold purification of the activity. Although the elution behavior of the antiviral activity was qualitatively similar to that for the MAF activity, only 35–50% of the applied antiviral activity was recovered.

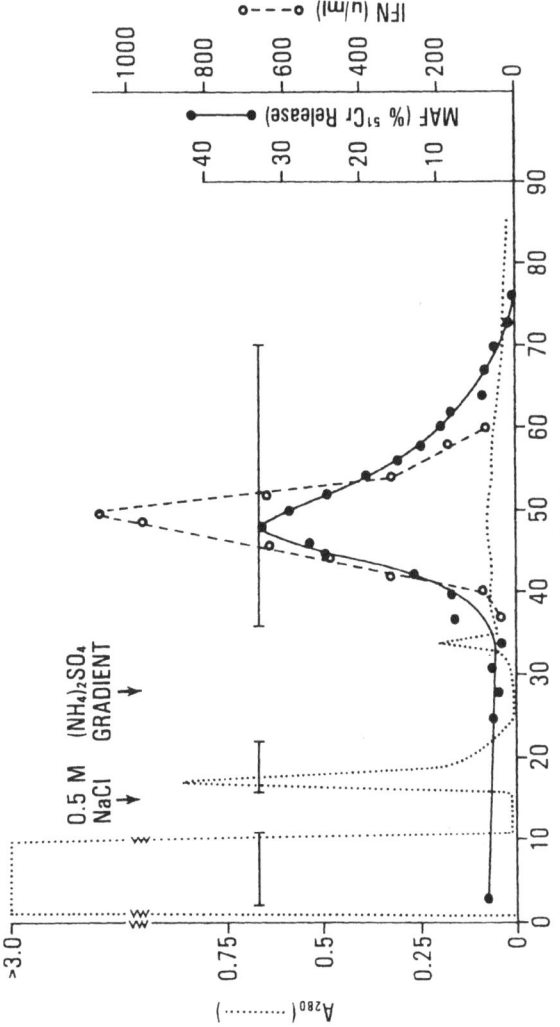

Figure 7. Dye-ligand chromatography of 24/G1 supernatant. Eight hundred and fifty ml of a stimulated hybridoma supernatant was concentrated 10-fold and subjected to chromatography on a 1.5 × 7.0-cm column of Matrex Gel Red A at 4°C. Horizontal bars represent column pools. Abbreviations as in Fig. 1. (From Schreiber *et al.*, 1983.)

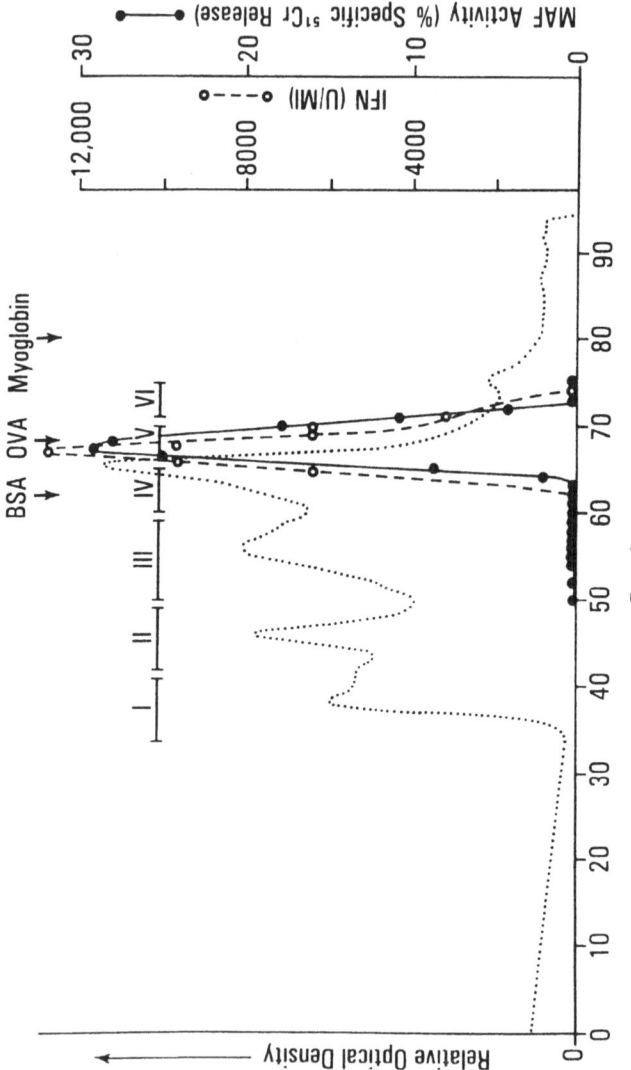

Figure 8. High-performace liquid chromatography gel filtration of Matrex Red purified MAF and IFN. Horizontal bars represent column pools. BSA, Bovine serum albumin; OVA, ovalbumin. Other abbreviations as in Fig. 1 (From Schreiber *et al.*, 1983.)

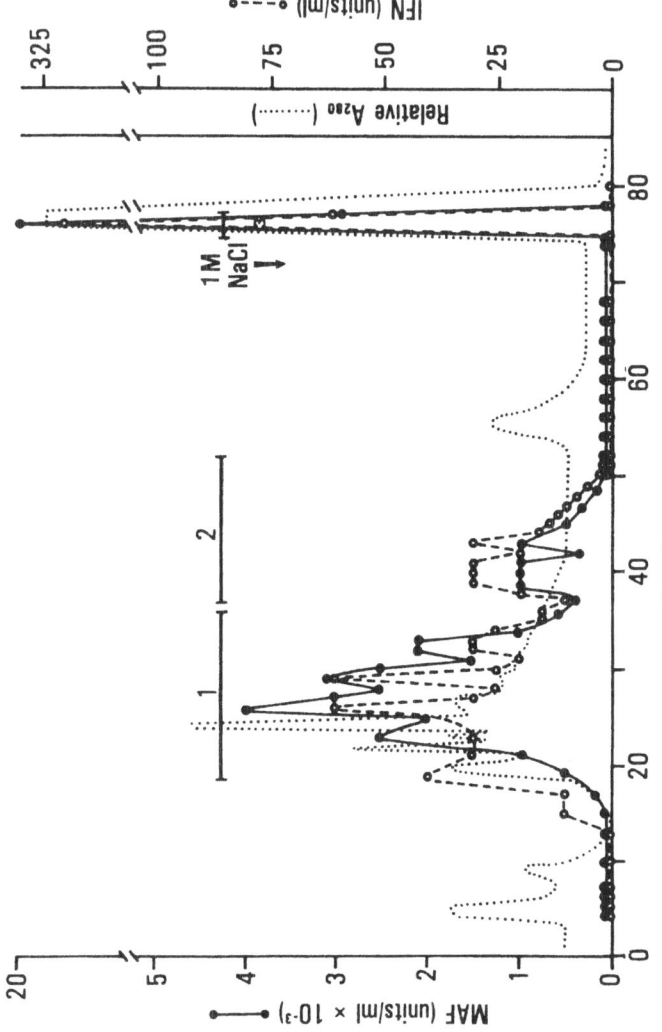

Figure 9. High-performance liquid chromatography chromatofocusing of Matrex Red purified MAF and IFN. Seventy-five μl of the concentrated ammonium sulfate pool from the Matrex Red column was diluted to 2 ml with 0.025 M bis-Tris-HCl, pH 7.1 and injected into a 7.5 × 300-mm Mono P column at 23°C. Elution was performed at 0.5 ml/min with Polybuffer 74 diluted 1:10 pH 5.0. Fraction size was 1 ml. At the end of the pH gradient, 2 ml of buffer containing 1 M NaCl was injected into the column, and elution was contained for an additional 15 min. Abbreviations as in Fig. 1. (From Schreiber et al., 1983.)

J. Copurification of Hybridoma-Derived MAF and γ-IFN Using High-Resolution Chromatography

The ammonium sulfate pool from the Matrex Gel Red A column was concentrated and a portion subjected to gel filtration on a high-performance liquid chromatography system (HPLC). MAF activity eluted as a single symmetrical peak and displayed an apparent molecular weight of 50,000 (Fig. 8). HPLC gel filtration was able to separate MAF successfully from the bulk of the contaminating proteins in the Matrex Gel Red A pool. Most of the latter exhibited molecular weights higher than that of MAF. However, this technique failed to resolve MAF from IFN. The elution profile of the antiviral activity was virtually superimposable on that of the MAF activity. Forty-three percent of the MAF activity, 67% of the antiviral activity, and 10% of the applied protein was recovered in pool *v* from this column. This constituted an additional 5- to 10-fold purification of the two lymphokine activities (15,000-fold overall) without resolving one from the other.

The Matrex Gel Red A pool was also subjected to HPLC chromatofocusing. This procedure involves the use of a special gel and buffer system. Under the proper conditions, proteins bind to the column and are eluted at their isoelectric points. When combined with HPLC technology, this technique displays very high-resolving capabilities. A portion of the Matrex Gel Red A pool was applied to a chromatofocusing column, and elution was accomplished by generation of an isoelectric gradient between the values of pH 7.0 and 5.0. When the limit pH was reached the column was eluted with buffer containing 1 M sodium chloride. MAF activity eluted over a relatively broad range in a complex pattern consisting of six chromatographically distinct species (Fig. 9). A mean isoelectric point of 5.8 was calculated for the bulk of the eluted MAF in pool 1. MAF activity was also eluted with the high-salt-containing buffer at the end of the gradient. The elution profile of the antiviral activity was virtually identical to that of the MAF activity, both qualitatively as well as quantitatively. When pools were made from the fractions and titrated in the respective biologic assays, nearly identical recoveries were obtained for each activity in each pool. The overall recovery of the MAF activity (76%) agreed with the recovery of the antiviral activity (85%). These results thus indicated that while charge heterogeneity existed for the molecular species that acted as MAF, the identical heterogeneity existed for the molecules that displayed antiviral activity. Since the recovery of each of the two activities was nearly quantitative, it is difficult to conceive of any possibility other than that of the molecular identity of MAF and γ-IFN to explain such a result.

K. Comparison of Hybridoma MAF with Recombinant γ-IFN

The gene for murine γ-IFN was recently cloned, isolated, and sequenced by Gray and Goeddel (1983). Expression of the gene was accomplished initially by transfection of the appropriate cDNA into the monkey cell line COS-7. Culture supernatants of the transfected cells displayed 6000 units of antiviral activity/ml, while medium from nontransfected cells was devoid of IFN activity. The product of the transfected cells resembled natural γ-IFN by a variety of criteria. It displayed the appropriate species specificity, pH, and temperature sensitivities and was neutralized by rabbit anti-γ-IFN serum.

When compared with the hybridoma supernatant, transfected COS-7 supernatants displayed the appropriate amount of MAF activity (Pace *et al.*, 1983a). Like the antiviral activity in the monkey cell supernatant, the MAF activity was sensitive to acid pH and was neutralized by rabbit antiserum to γ-IFN (Pace *et al.*, 1983a). In other experiments (R. D. Schreiber and P. W. Gray, unpublished data), purified recombinant γ-IFN produced in *E. coli* was also found to display MAF activity comparable to the hybridoma MAF on an antiviral activity basis (Fig. 10). These studies have shown that 0.11 IU of hybridoma-derived IFN was required to produce one unit of MAF activity as compared with 0.24 IU of the purified recombinant γ-IFN. Based on the known specific activity of the purified γ-IFN, this indicates that 4–8 pg of γ-IFN was required to produce one unit of MAF activity in the MAF assay. These results also indicated that the hybridoma clone 24/G1 secreted 10–20 ng γ-IFN/ml.

Two rats immunized with the purified recombinant γ-IFN both responded with an antibody response as detected by means of a γ-IFN-specific enzyme-

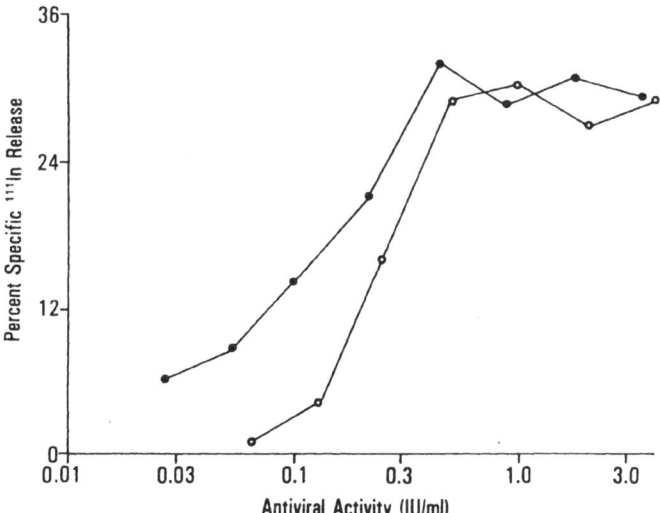

Figure 10. Comparison of MAF activities exhibited by purified recombinant murine γ-IFN (●) produced in *Escherichia coli* and hybridoma 24/G1 supernatant-purified IFN (○) had a specific activity of 30.6×10^6 IU/mg. Abbreviations as in Fig. 1. (From R. D. Schreiber and P. W. Gray, unpublished observations.)

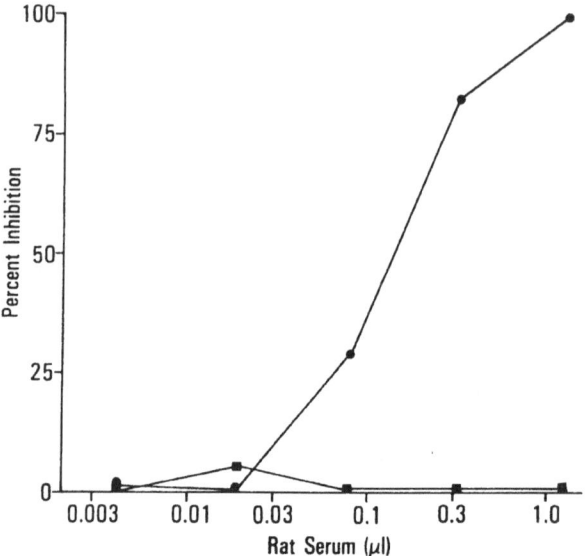

Figure 11. Inhibition of hybridoma-derived MAF activity by rat antiserum to purified recombinant murine γ-IFN. (●) Rat α-IFN-γ serum; (■) normal rat serum. Abbreviations as in Fig. 1. (From R. D. Schreiber and P.W. Gray, unpublished observations.)

linked immunoabsorbent assay (ELISA) assay. The antisera raised to the recombinant γ-IFN was capable of completely inhibiting the MAF activity in the hybridoma supernatant (Fig. 11). Inhibition was dose dependent and 1.25 μl of immune serum was found to neutralize 1.125 IU of hybridoma-derived γ-IFN (R. D. Schreiber and P. W. Gray, unpublished data). Normal rat serum was devoid of inhibitory activity.

L. Conclusions

A total of 17 pieces of evidence have been presented to support the conclusion that the MAF activity produced by the murine T-cell hybridoma 24/G1 is attributable to γ-IFN. Two pieces of evidence are conclusive in themselves: (1) the HPLC chromatofocusing experiment in which seven chromatographically distinct species were identified that displayed proportionately identical amounts of MAF and antiviral activities, and (2) the finding that purified recombinant γ-IFN could express MAF activity quantitatively and antisera raised to the purified protein could neutralize all the MAF activity in hybridoma supernatants. With the other 15 pieces of supporting evidence derived from biosynthetic, immunochemical, and biochemical experiments, this study strongly suggests that murine γ-IFN can act as an MAF. This study is thus in agreement with other studies performed on lymphokine preparations derived from normal splenic cells (Roberts and Vasil, 1982; Kleinschmidt and Schultz, 1982; Meltzer *et al.*, 1982). This conclusion must be restricted, however, to the limited definition of macrophage activation used in these studies, i.e., the development of *in vitro* nonspecific tumoricidal activity in murine macrophages. It does not address the possibility that other MAF activities exist that are distinct from γ-IFN. These possibilities are discussed in Section IV.

IV. SUPERNATANTS THAT DISPLAY MAF ACTIVITY BUT NOT ANTIVIRAL ACTIVITY

During the past few years, several laboratories have reported the production of lymphokine-containing supernatants that displayed MAF activity but not antiviral activity. Three of these studies used continuous cell lines, consisting of two T-cell hybridomas (Ratliff *et al.*, 1982a,b; Erickson *et al.*, 1982) and one thymoma cell line (Meltzer *et al.*, 1982), while another study used normal murine splenic cell cultures (Kniep *et al.*, 1981). The pertinent features of each of these studies are summarized in this section.

Kniep *et al.* (1981) used supernatants produced by Con A stimulation of C57BL splenic cells under serum-free conditions. When these supernatants were subjected to affinity chromatography on polynucleotide columns, 80% of the MAF activity was eluted unretarded and was thereby separated from the antiviral activity. The MAF activity also displayed a number of physicochemical characteristics that differed from those reported for γ-IFN (Younger and Salvin, 1973) or for conventional MAF (Leonard *et al.*, 1978; Roberts and Vasil, 1982). It exhibited a lower apparent molecular weight (30,000) and a higher isoelectric point (7.4 and 8.4) and was resistant to trypsin inactivation. In addition, while this factor displayed γ-IFN-like heat and acid pH sensitivies, it was uncharacteristically stable to alkaline pH. No immunochemical characterization was attempted in this study.

Meltzer *et al.* (1982) have also identified a low-molecular-weight MAF that did not display antiviral activity. This factor was produced by a C57BL-derived thymoma line denoted EL-4 $_{Farrar}$ and constituted approximately 25–50% of the MAF activity in stimulated culture supernatants. The activity exhibited a molecular weight of 23,000, was stable to acid pH, and was not neutralized by antisera to γ-IFN. The remainder of the MAF activity in the EL-4 supernatants was indistinguishable from γ-IFN.

Two T-cell hybridomas have also been reported that produce MAF activity in the absence of antiviral activity. However, only limited physicochemical characterization of these activities has been attempted to date. Ratliff *et al.*

(1982a,b) constructed T-cell hybridomas by fusion of alloantigen-activated splenocytes with the BW5147 cell line. One of the resulting hybridoma clones (F133) produced an activity that could induce macrophage-dependent killing of P815 target cells in a short-term assay analogous to the one presented in Section III.B. No antiviral activity was detectable in F133 supernatants. However, the amount of MAF activity produced by this clone was relatively low. Since the tumoricidal activity assay is generally more sensitive than the antiviral activity assay, the IFN activity in the supernatant may have gone undetected. Unfortunately, this hybridoma has become unstable, and no further biochemical characterizations have been performed. A second set of T-cell hybridomas were constructed by Erickson *et al.* (1982) using mitogen-stimulated C57BL splenic cells as the normal T-cell partner. One of the resulting hybridomas was particularly unusual, since it secreted MAF activity constitutively. IFN activity could not be detected in 1:5 diluted culture supernatants of this hybridoma. It is not yet clear what quantities of activity were produced by the hybridoma nor what the physicochemical properties were of the active species.

The results of these studies indicate that factors other than γ-IFN may function as MAFs for induction of nonspecific tumoricidal activity. It is possible, in some instances, that only by the use of a particular assay system could the factor(s) be detected. Such may have been the case for the study by Erickson *et al.* (1982), who used a long-term cytotoxicity assay that was significantly different from the short-term assays used in most of the other studies. It should also be noted that, of the four studies discussed here, three involved cells derived from the C57BL strains of mice. Thus, the exprssion of a low-molecular-weight MAF devoid of antiviral activity may be under genetic control. While this factor may certainly represent a unique molecule, the possibility must also be considered that it represents an abnormal or altered form of γ-IFN that can express normal tumoricidal inducing activity but reduced antiviral activity. Recently, a low-molecular-weight MAF was detected in certain partially purified preparations of the 24/G1 hybridoma (R. D. Schreiber, C. A. Nacy, and M. S. Meltzer, unpublished observation). This activity displayed a molecular weight of 25,000 and constituted 20% of the total activity in the preparation. However, in contrast to the above-mentioned studies, this MAF also displayed the appropriate amount of antiviral activity and was neutralized by the rat antiserum to recombinant γ-IFN. Whether a relationship exsits between this molecular species and the other MAF activities discussed in this section is currently under investigation.

V. CONCLUSIONS

A significant amount of data has been accumulated over the past few years that indicate that γ-IFN can function as an MAF for tumor cytotoxicity. This conclusion does not preclude the possibility that other molecular species may function as an MAF as well. However, identification of these other factors must await a detailed analysis of their biochemical identity.

The identification of γ-IFN as one type of MAF should provide the basis for new avenues of experimentation. Studies can now be performed to investigate its mechanism of action, including the identification of the topographic sites on the molecule that induce antiviral and/or MAF activities and the identification of the macrophage surface structures to which γ-IFN binds. The relationship of γ-IFN to other MAF activities including those that lead to an enhancement of intracellular killing of microbial pathogens may be examined. The biochemical changes in macrophages produced by γ-IFN can now be determined unambiguously. These studies and others like them will eventually define macrophage activation and regulation on a molecular basis and will help further elucidate the role of the macrophage in host defense.

ACKNOWLEDGMENTS

This chapter represents publication number 3097IMM from the Research Institute of Scripps Clinic. This work was supported in part by USPHS Grant numbers AI-17354 and CA-34120 and by grants from Eli Lilly Research Laboratories and the Elsa U. Pardee Foundation. I am grateful for the expert technical assistance of Lori Hicks and Mary Brothers, for the secretarial assistance of Patti Lank, and for the editorial assistance of Dr. Antonio Celada and Lori Hicks. I also wish to thank my colleagues in these studies for their valuable contributions and Genentech, Inc. for providing the recombinant γ-IFN. I would like to express my special appreciation to Dr. Hans J. Müller-Eberhard, Dr. Emil R. Unanue, and Dr. David H. Katz for their encouragement and advice in this project.

VI. REFERENCES

Adams, D. O., 1980, Effector mechanisms of cytolytically-activated macrophages. I. Secretion of neutral proteases and effect of protease inhibitors, *J. Immunol.* **124**:286–292.

Adams, D. O., and Marino, P. A., 1981, Evidence for a multistep mechanism of cytolysis by BCG-activated macrophages: The interrelationship between the capacity for cytolysis, target binding, and secretion of cytolytic factor, *J. Immunol.* **126**:981–987.

Altman, A., Sferruzza, A., Weiner, R. G., and Katz, D. H., 1982, Constitutive and mitogen-induced production of T cell growth factor by stable T cell hybridoma lines, *J. Immunol.* **128**:1365–1371.

Benjamin, W. R., Steeg, P. S., and Farrar, J. J., 1982, Production of immune interferon by an interleukin 2-independent murine T cell line, *Proc. Natl. Acad. Sci. USA* **79**:5379–5383.

Cleveland, R. P., Meltzer, M. S. and Zbar, B., 1974, Tumor cytotoxicity in vitro by macrophages from mice infected with *Mycobacterium bovis*, strain BCG, *J. Natl. Cancer Inst.* **52**:1887–1894.

David, J. R., and Remold, H. G., 1979, The activation of macrophages by lymphokines, in: *Biology of the Lynphokines* (S. Cohen, E. Picic, and J. J. Oppenheim, eds.), p. 121–139, Academic Press, New York.

DeMaeyer-Guignard, J., Thang, M. N., and DeMaeyer, E., 1977, Binding of mouse interferon to polynucleotides, *Proc. Natl. Acad. Sci. USA* **74**:3787–3790.

Dennert, G., Yamagata, S., and Yogeeswarn, G., 1981, cloned cell lines with natural killer activity: Specificity, function and cell surface markers, *J. Exp. Med.* **153**:545–556.

Den Otter, W., 1981, The effect of activated macrophages on tumor growth in vitro and in vivo, *Lymphokines* **3**:389–422.

Erickson, K. L., Cicurel, L., Gruys, E., and Fidler, I. J., 1982, Murine T-cell hybridomas that produce lymphokine with macrophage-activating factor activity as a constitutive product, *Cell. Immunol.* **72**:195–201.

Evans, R., and Alexander, P., 1970, Cooperation of immune lymphoid cells with macrophages in tumor immunity, *Nature* **228**:620–622.

Evans, R., and Alexander, P., 1972, Role of macrophages in tumor immunity. I. Cooperation between macrophages and lymphoid cells in syngeneic tumor immunity, *Immunology* **23**:615–626.

Fidler, I. J., and Raz, A., 1981, The induction of tumoricidal capacities in mouse and rat macrophages by lymphokines, *Lymphokines* **3**:345–364.

Fidler, I. J., Darnell, J. H., and Budmen, M. B., 1976, In vitro activation of mouse macrophages by rat lymphocyte mediators, *J. Immunol.* **117**:666–673.

Gray, P. W., and Goeddel, D. V., 1983, Cloning and expression of murine immune interferon cDNA, *Proc. Natl. Acad. Sci. USA* **80**:5842–5846.

Handa, K., Suzuki, R., Matsui, H., Shimizu, Y., and Kumagai, K., 1983, Natural killer (NK) cells as a responder to interleukin 2 (IL2) II. IL2-induced interferon γ production, *J. Immunol.* **130**:988–992.

Havell, E. A., and Spitalny, G. L., 1983, Production and characterization of anti-murine interferon γ antisera, *J. Interferon Res.* **3**:191–198.

Hibbs, J. B., Jr., 1974, Discrimination between neoplastic and non-neoplastic cells in vitro by activated macrophages, *J. Natl. Cancer Inst.* **53**:1487–1492.

Hibbs, J. B., Lambert, L. H., Jr., and Remington, J. S., 1972, Possible role of macrophage mediated nonspecific cytotoxicity in tumor resistance, *Nature (New Biol.)* **235**:48–50.

Johnson, W. J., Marino, P. A., Schreiber, R. D., and Adams, D. O., 1983, Sequential activation of murine mononuclear phagocytes for tumor cytolysis: Differential expression of markers by macrophages in several stages of development, *J. Immunol.* **131**:1038–1043.

Katz, D. H., Bechtold, T. E., and Altman, A., 1980, Construction of T cell hybridomas secreting allogeneic effect factor (AEF), *J. Exp. Med.* **152**:956–968.

Keller, R., 1981, The cytostatic and cytocidal effects of macrophages: Are they really specific for tumor cells? *Lymphokines* **3**:283–292.

Kelso, A., and MacDonald, H. R., 1982, Precursor frequency analysis of lymphokine-secreting alloreactive T lymphocytes, *J. Exp. Med.* **156**:1366–1379.

Kelso, A., Glasebrook, A. L., Kanagawa, O., and Brunner, K. T., 1982, Production of macrophage-activating factor by T lymphocyte clones and correlation with other lymphokine activities, *J. Immunol.* **129**:550–556.

Kleinschmidt, W. J., and Schultz, R. M., 1982, Similarities of murine immune interferon and the lymphokine that renders macrophages cytotoxic, *J. Interferon Res.* **2**:291–299.

Kniep, E. M., Domzig, W., Lohmann-Matthes, M.-L., and Kickhöfen, B., 1981, Partial purification and chemical characterization of macrophage cytotoxicity factor (MCF,

MAF) and its separation from migration inhibitory factor (MIF), *J. Immunol.* 127: 417-422.

Leonard, E. J., Ruco, L. P., and Meltzer, M. S., 1978, Characterization of macrophage activation factor, a lymphokine that causes macrophages to become cytotoxic for human cells, *Cell. Immunol.* 41:347-357.

Lohmann-Matthes, M.-L., Ziegler, F. G., and Fischer, H., 1973, Macrophage cytotoxicity factor: A product of in vitro sensitized thymus-dependent cells, *Eur. J. Immunol.* 3:56-58.

Lohmann-Matthes, M.-L., Lang, H., Sun, D. M., Kniep, E., and Kickhöfen, B., 1981, Macrophage activation to cytotoxicity by the macrophage cytotoxicity (activating) factor (MCF, MAF), *Lymphokines* 3:365-388.

Marino, P. A., and Adams, D. O., 1980a, Interaction of Bacillus Calmette-Guerin-activated macrophages and neoplastic cells in vitro. I. Conditions of binding and its selectivity, *Cell. Immunol.* 54:11-25.

Marino, P. A., and Adams, D. O., 1980b, Interaction of Bacillus Calmette-Guerin-activated macrophages and neoplastic cells in vitro. II. The relationship of selective binding to cytolysis, *Cell. Immunol.* 54:26-35.

Meltzer, M. S., 1981a, Macrophage activation for tumor cytotoxicity: Characterization of primary and triggering signals during lymphokine activation, *J. Immunol.* 127:179-183.

Meltzer, M. S., 1981b, Tumor cytotoxicity by lymphokine-activated macrophages: Development of macrophage tumoricidal activity requires a sequence of reactions, *Lymphokines* 3:319-344.

Meltzer, M. S., Benjamin, W. R., Farrar, J. J., 1982, Macrophage activation for tumor cytotoxicity: Induction of macrophage tumoricidal activity by lymphokines from EL-4, a continuous T cell line, *J. Immunol.* 129:2802-2807.

Nathan, C. F., Remold, H. G., and David, J. R., 1973, Characteristics of a lymphocyte factor which alters macrophage functions, *J. Exp. Med.* 137:275-290.

Osborn, L. C., Georgiades, J., and Johnson, H. M., 1980, Antibody to mouse immune interferon, *IRCS Med. Sci.* 8:212.

Pace, J. L., and Russell, S. W., 1981, Activation of mouse macrophages for tumor cell killing. I. Quantitative analysis of interactions between lymphokine and lipopolysaccharide, *J. Immunol.* 126:1863-1867.

Pace, J. L., Russell, S. W., Schreiber, R, D., Altman, A., and Katz, D. H., 1983a, Macrophage activation: Priming activity from a T cell hybridoma is attributable to gamma interferon, *Proc. Natl. Acad. Sci. USA* 80:3782-3786.

Pace, J. L., Russell, S. W., Torres, B. A., Johnson, H. M., and Gray, P. W., 1983b, Recombinant mouse γ interferon induces the priming step in macrophage activation for tumor cell killing, *J. Immunol.* 130:2011-2013.

Piessens, W. F., Churchill, W. H., Jr., and David, J. R., 1975, Macrophages activated in vitro with lymphocyte mediators kill neoplastic but not normal cells, *J. Immunol.* 114:293-299.

Piessens, W. F., Churchill, W. H., Jr., and Sharma, S. D., 1981, On the killing of syngeneic tumor cells by guinea pig macrophages activated in vitro with lymphocyte mediators, *Lymphokines* 3:293-318.

Prystowsky, M. B., Ely, J. M., Beller, D. I., Eisenberg, L., Goldman, J., Goldman, M., Goldwasser, E., Ihle, J., Quintans, J., Remold, H., Vogel, S. N., and Fitch, F. W., 1982, Alloreactive cloned T cell lines. VI. Multiple lymphokine activities secreted by helper and cytolytic cloned T lymphocytes, *J. Immunol.* 129:2337-2344.

Ratliff, T. L., Thomasson, D. L., McCool, R. E., and Catalona, W. J., 1982a, Production of macrophage activation factor by a T-cell hybridoma, *Cell. Immunol.* 68:311-321.

Ratliff, T. L., Thomasson, D. L., McCool, R. E., and Catalona, W. J., 1982b, T-cell hybridoma production of macrophage activation factor (MAF). I. Separation of MAF from interferon gamma, *J. Recticuloendothel. Soc.* **31**:393-397.

Roberts, W. K., and Vasil, A., 1982, Evidence for the identity of murine gamma interferon and macrophage activating factor, *J. Interferon Res.* **2**:519-532.

Rocklin, R. E., Bendtzen, K., and Greineder, D., 1980, Mediators of immunity: lymphokines and monokines, *Adv. Immunol.* **29**:56-137.

Ruco, L. P., and Meltzer, M. S., 1977, Macrophage activation for tumor cytotoxicity: Induction of tumoricidal macrophages by supernatants of PPD stimulated bacillus Calmette-Guerin-immune spleen cell cultures, *J. Immunol.* **119**:889-896.

Ruco, L. P., and Meltzer, M. S., 1978a, Macrophage activation for tumor cytotoxicity: Tumoricidal activity by macrophages from C3H/HeJ mice requires at least two activation stimuli, *Cell. Immunol.* **41**:35-51.

Ruco, L. P., and Meltzer, M. S., 1978b, Macrophage activation for tumor cytotoxicity: Development of macrophage cytotoxic activity requires completion of a sequence of short-lived intermediary reactions, *J. Immunol.* **121**:2035-2042.

Russell, S. W., Doe, W. F., and McIntosh, A. T., 1977, Functional characterization of a stable noncytolytic stage of macrophage activation in tumors, *J. Exp. Med.* **146**:1511-1520.

Schreiber, R. D., Ziegler, H. K., Calamai, E., and Unanue, E. R., 1981, Two signal requirement for macrophage tumoricidal activity, *Fed. Proc.* **40**:1002.

Schreiber, R. D., Altman, A., and Katz, D. H., 1982, Identification of a T cell hybridoma which produces large quantities of macrophage activating factor, *J. Exp. Med.* **156**: 677-689.

Schreiber, R. D., Pace, J. L., Russell, S. W., Altman, A., and Katz, D. H., 1983, Macrophage activating factor produced by a cell hybridoma: Physicochemical and biosynthetic resemblance to gamma interferon, *J. Immunol.* **131**:826-832.

Schultz, R. M., and Chirigos, M. A., 1978, Similarities among factors that render macrophages tumoricidal in lymphokine and interferon preparations, *Cancer Res.* **38**:1003-1007.

Schultz, R. M., and Kleinschimdt, W. J., 1983, Functional identity between murine recombinant gamma interferon (IFN$_\gamma$) and macrophage activating factor, *Nature* **305**:239-240.

Schultz, R. M., Papamatheakis, J. D., and Chirigos, M. A., 1977, Interferon: An inducer of macrophage activation by polyanions, *Science* **197**:674-676.

Varesio, L., Blasi, E., Gray, P., Herberman, R. B., and Wiltrout, R. H., 1983, Activation of cytotoxic macrophages by cloned murine γ-interferon (γ-IFN), *Fed. Proc.* **42**:1076.

Warner, J., and Dennert, G., 1982, Effects of a cloned cell line with NK activity on bone marrow transplants, tumor development and metastasis in vivo, *Nature* **300**:31-34.

Weinberg, J. B., and Hibbs, J. R., Jr., 1979, In vitro modulation of macrophage tumoricidal activity: Partial characterization of a macrophage-activating factor(s) in supernatants of NaIO4 treated peritoneal cell, *J. Reticuloendothel. Soc.* **26**:283-293.

Weinberg, J. B., Chapman, H. A., Jr., and Hibbs, J. B., Jr., 1978, Characterization of the effects of endotoxin on macrophage tumor cell killing, *J. Immunol.* **121**:72-80.

Wheelock, E. F., 1965, Interferon-like virus-inhibitor induced in human leukocytes by phytohemagglutinin, *Science* **149**:310-311.

Wietzerbin, J., Stafanos, S., Lucero, M., and Falcoff, E., 1978, Presence of a polynucleotide binding site on murine immune inteferon (T-type), *Biochem. Biophys. Res. Commun.* **85**:480-489.

Younger, J., and Salvin, S. B., 1973, Production and properties of migration inhibitory factor and interferon in the circulation of mice with delayed hypersensitivity, *J. Immunol.* **111**:1914-1922.

Chapter 10

Possible Autoregulatory Functions of the Secretory Products of Mononuclear Phagocytes

Robert J. Bonney and Philip Davies

Department Immunology and Inflammation
Merck Sharp & Dohme Research Laboratories
Rahway, New Jersey 07065

I. INTRODUCTION

The ability of mononuclear phagocytes to recognize, ingest, and either digest or remove through various portals of exit from the body a wide variety of infectious agents and toxic materials has been recognized since the seminal observations of Metchnikoff. Mononuclear phagocytes have specialized recognition mechanisms that facilitate these functions, which include (1) several classes of Fc receptors for immune complexes; (2) responsiveness to lymphokines, which serve to enhance their phagocytic and cidal functions; and (3) receptors for complement components, which are either chemotactic or facilitate phagocytosis. During the past decade it became clear that these activities, concerned primarily with the removal of noxious stimuli from the host environment, are accompanied by other specialized functions that have far-reaching effects on the cells and connective tissues present in the pericellular environment of the mononuclear phagocyte. One of these, the ability to present antigenic determinants of ingested materials to cells of the immune system in an immunogenic form (for review, see Rosenthal, 1980), is not discussed here. The other, the ability of mononuclear phagocytes to secrete a diverse range of products, preoccupied many investigators during the past decade.

The list of products is now extensive and has been reviewed elsewhere (Davies and Bonney, 1979; Gordon, 1980; Nathan *et al.*, 1982). Although a cohesive framework for the function of these products has yet to be established, it is be-

199

coming increasingly clear that they provide a vehicle for broad-based functions of mononuclear phagocytes, both in physiological situations and pathological conditions. In this chapter we discuss one aspect of the function of these secretory products, namely, the effect that they have on the mononuclear phagocyte itself. We raise the question of whether the secretory products of mononuclear phagocytes exercise an autocrine function, a phenomenon that has been more associated with primitive cellular systems in the past. Only selected secretory products are discussed to illustrate possible ways in which mononuclear phagocytes regulate their function through the action of endogenously derived substances.

II. PROSTAGLANDINS

A. Synthesis by Mononuclear Phagocytes

Many studies have shown that mononuclear phagocytes from a variety of sources synthesize and secrete large amounts of arachidonic acid oxygenation products of the cyclooxygenase pathway, particularly protaglandin E_2 (PGE_2) and prostaglandin I_2 (PGI_2). Initial studies from our own laboratory and those of others have been reviewed by Davies et al. (1980) and more recent studies by Scott et al. (1983). Early studies showed that when mouse peritoneal macrophages were exposed to a particulate stimulus, such as yeast zymosan (Humes et al., 1977; Brune et al., 1978; Bonney et al., 1978), there is a dramatic increase in PGE_2 and PGI_2 synthesis. Prostaglandins are also synthesized in response to other inflammatory stimuli, such as antigen–antibody complexes, phorbol myristate acetate (PMA), and lipopolysaccharide (LPS) (Bonney et al., 1979, 1980a; Kurland and Bockman, 1978). However, similar numbers of macrophages elicited by thioglycollate broth, bacillus Calmette-Guérin (BCG), Corynebacterium parvum, or carrageenan synthesize and secrete only small amounts of PGE_2 and PGI_2 when exposed to these same stimuli (Fig. 1) (Humes et al., 1980).

B. Effects of Prostaglandins on Mononuclear Phagocyte Function

Prostaglandins E_2 and I_2 are potent vasodilators that also synergize with other inflammatory mediators to cause edema and pain. Prostaglandins have been shown to regulate a wide variety of cellular functions. In many instances it is thought that the initial response of cells to prostaglandins involves stimulation of adenyl cyclase and changes in cyclic adenosine monophosphate

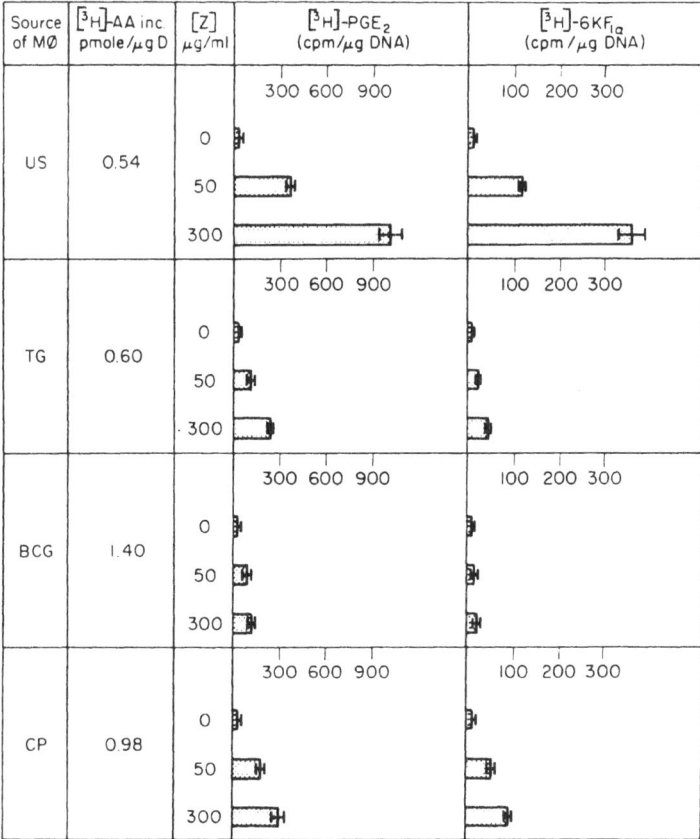

Figure 1. Prostaglandin synthesis and release from macrophages. Macrophages were isolated from the peritoneal cavity of untreated mice or from bacillus Calmette–Guérin, *Corynebacterium parvum,* and thioglycollate-treated mice. The cells were prelabeled with [^{3}H]arachidonic acid and then challenged with zymosan for 4 hr. The amount of radioactivity released into the medium that was cochromatographic with PGE$_2$ was measured. Results are the mean ± SD; N = 3.

(cAMP) levels. Such responses are discussed further as seen in mononuclear phagocyte populations.

1. Intracellular cAMP Levels

Macrophages respond to exogenous prostaglandins with elevated cellular cAMP levels (see Table I). The response of resident and thioglycollate-elicited mouse perioneal macrophages to exogenous PGE$_2$ differ markedly (Bonney

Table I. Thioglycollate-Elicited Macrophages—More
Sensitive Than Resident Macrophages to PGE_2-Induced
Increases in Cellular cAMP Levels[a]

| Additions | Time (min) | cAMP (pmol/μg DNA)[b] | |
		Resident	Elicited
None	2	0.12 ± 0.02	0.06 ± 0.02
None	30	0.10 ± 0.05	0.04 ± 0.0
PGE_2, 3 μM	1	0.13 ± 0.01	0.36 ± 0.03
PGE_2, 3 μM	2	0.11 ± 0.02	0.68 ± 0.20
PGE_2, 3 μM	30	0.11 ± 0.02	0.41 ± 0.1

[a]*Abbreviation:* cAMP, cyclic adenosine monophosphate.
[b]DNA content: resident, 23 μg/culture; elicited, 38 μg/culture.

et al., 1980*b*). Much greater increases in cAMP levels occur in elicited populations than in resident populations of macrophages. In a similar manner, PGI_2
elevates cAMP levels to a greater extent in elicited populations, whereas cholera
toxin was equally effective in both cell populations. In contrast, $PGF_{2\alpha}$ at
similar concentrations had no effect on cAMP levels in either cell population.
The phosphodiesterase inhibitor theophylline, at 5×10^{-3} M, potentiated
the effect of both PGE_2 and PGI_2 on cAMP levels in elicited macrophages
(Bonney *et al.*, 1980*b*). Gemsa *et al.* (1979) have shown that during phagocytosis the responsiveness of elicited rat peritoneal macrophages to PGE_1 is
increased with a prolonged elevation of cellular cAMP levels.

In view of these findings that exogenously added prostaglandins influence
cAMP levels in mononuclear phagocyte populations, it can be speculated that
they may also influence other functions of mononuclear phagocytes responsive to exogenous prostaglandins. The clearest indication that endogenous
prostaglandins increase cellular levels of cAMP comes from the studies of Smith
et al. (1980). These workers found that elicited rat peritoneal macrophages
showed rapid elevations in cellular cAMP levels upon challenge with A23187
or zymosan particles. This elevation was inhibited by indomethacin and was
not seen in broken cell preparations exposed to zymosan. Similar findings
have been reported by Gemsa *et al.* (1979).

2. Neutral Proteinase Secretion

a. Collagenase. The addition of PGE_2 and endotoxin, but not PGE_2 alone,
enhances the production of collagenase by mineral oil-induced guinea pig peritoneal macrophages (Wahl *et al.*, 1977). The increase was detectable with as
little as 10 nM PGE_2. Furthermore, these cultured macrophages exposed to

endotoxin secrete PGE_2 at levels sufficient to increase their own production of collagenase. Consonant with this was the inhibition of endotoxin-stimulated production of collagenase by the prostaglandin synthesis inhibitor indomethacin.

   b. Plasminogen Activator. Resident populations of mouse peritoneal macrophages release large amounts of prostaglandins (Humes *et al.*, 1980) and low amounts of plasminogen activator (Hamilton, 1981). In contrast, thioglycollate-elicited macrophages secrete low levels of prostaglandins (Humes *et al.*, 1980) and increased levels of plasminogen activator (Hamilton, 1981; Vassalli *et al.*, 1977). Therefore, it is possible that prostaglandin may regulate the secretion of this enzyme by macrophages. However, the addition of cyclooxygenase inhibitors, namely indomethacin and flufenamic acid in the 10^{-6}–10^{-5} *M* range, did not consistently effect the basal secretion or the PMA-induced secretion of plasminogen activator (Hamilton, 1981). Consequently, it is difficult to conclude that endogenous prostaglandins have any significant effect on plasminogen activator secretion.

   c. Elastase. Mouse macrophages also secrete an elastinolytic activity characterized as a metalloproteinase (Dahlgren *et al.*, 1980; Banda and Werb, 1981). The synthesis and secretion of this elastase by both resident and elicited macrophages are dramatically increased when PMA is added to the cultures. This stimulated synthesis is blocked by dexamethasone, but not by indomethacin (Dahlgren *et al.*, 1980), suggesting that endogenous prostaglandins do not play a regulatory role in the secretion of this product.

3. Regulation of Macrophage Proliferation

   Colony-stimulating factors (CSF) are growth factors required for granulocyte and macrophage production from undifferentiated hematopoietic cells (Bradley and Metcalf, 1966). One such factor, CSF-1 from L cells, is a sialic acid-containing glycoprotein of 70,000 daltons. This factor binds specifically to receptors on cells of the mononuclear phagocytic series (Guilbert and Stanley, 1980). Binding to this receptor is saturated at $\sim 5 \times 10^4$ sites/cell. CSF-1 is necessary for both survival and proliferation of bone marrow-derived monocytes, which are also capable of degrading this protein (Guilbert and Stanley, 1980; Tushinski *et al.*, 1982). CSF-1 differs from the CSF produced by monocyte/macrophage populations described by Chervenick and LoBuglio (1972). It appears that both materials stimulate prostaglandin synthesis by macrophages (Pelus *et al.*, 1979).

   Clonal macrophage and granulocyte growth from committed hematopoietic progenitor cells has been studied in semisolid agar culture by Pelus *et al.* (1979). These workers examined the effects of a variety of compounds, including CSF and PGE_1, on macrophage and granulocyte proliferation. It was found that colony-forming units that are committed to macrophage differentiation are specifically inhibited by PGE_1 at concentrations as low as 10^{-10} M. This PGE_1-mediated regulation of macrophage colony formation was also demonstrated using adherent mouse peritoneal macrophages, which are good PGE_2 producers, as feeder layers. Addition of indomethacin to these cultures selectively enhances macrophage colony formation. CSF induces macrophage prostaglandin synthesis (Pelus et al., 1980), thereby initiating the feedback loop for prostaglandin inhibition of proliferation. The growth of macrophage cell lines such as WEHI-3 is also inhibited by PGE_2 (Kurland et al., 1979). This PGE_2-mediated inhibition of myeloid colony formation is abnormal in chronic myeloid leukemia, as colony-forming unit populations from such patients are insensitive to normal feedback regulation by PGE (Broxmeyer et al., 1977; Pelus et al., 1980).

4. Macrophage-Mediated Tumor Cell Killing

Macrophages collected from animals treated with intracellular parasites (Hibbs et al., 1971) or killed bacteria (North et al., 1976) are tumoricidal in cell culture. The extent of killing and the duration of time in culture in which the macrophages remain tumoricidal can be varied by factors in the culture media. Prostaglandins have been shown to reduce tumor cell killing by interferon-activated macrophages (Schultz et al., 1978). Resident mouse peritoneal macrophages can be rendered tumoricidal by purified bacterial LPS. This tumor cell killing can be inhibited by PGE_2, but not PGI_2, at concentrations as low as 10^{-8} M (Taffet et al., 1982). This level of PGE_2 can be found in the culture media of the LPS-triggered macrophages, suggesting a feedback loop (Taffet and Russell, 1980). The rise in intracellular cAMP levels in macrophages treated with prostaglandins does not appear to be related to the interference with tumoricidal activity. As expected, the addition of the cyclooxygenase inhibitor indomethacin increases the ability of macrophages to kill tumor cells.

In a somewhat different protocol, McCarthy and Zwilling (1981) found that PGE_1, PGE_2, and $PGF_{2\alpha}$ did not affect BCG-activated macrophage killing of B16 melanoma cells. However, these workers report that these same prostaglandins at concentrations of 10^{-7}–10^{-5} M suppressed the cytostatic activity of BCG-activated macrophages and enhanced the cytostatic activity of resident macrophages. These data suggest that the effect of prostaglandins on the ability of macrophages to effect tumor growth varies according to their state of activation.

Macrophage activation in vitro as defined by increased plasminogen activator secretion and cellular levels of lysosomal enzymes is also inhibited by PGE_2 and enhanced by indomethacin (Schnyder et al., 1981).

5. Antibody-Dependent Cellular Cytotoxicity

Monocytes isolated from patients with malignant melanomas who were receiving treatment with C. parvum exhibited higher antibody-dependent cellular cytotoxicity (ADCC) than did monocytes from controls (Murray, 1982). Patient monocyte ADCC levels were augmented by the presence of 10^{-6} M indomethacin. These data suggest that PGE_2 also acts via a feedback loop to limit monocyte ADCC.

6. Complement

Mononuclear phagocytes synthesize most of the components of the alternative and classic components of the complement cascade (Whaley, 1980). The production of these components has been shown to vary according to the intracellular cAMP levels of these cells (Lappin and Whaley, 1981). Therefore, the role of prostaglandin and prostaglandin synthetase inhibitors in the regulation of complement component synthesis by monocytes was examined by Lappin and Whaley (1982). Arachidonic acid, PGE_2, PGD_2, and PGI_2, but not $PGF_{2\alpha}$, when added to human monocyte cultures, inhibited the production of the second component of complement, C2. Furthermore, addition of indomethacin increased the synthesis of C2, C4, C3, factor B, properdin, $\beta1H$, and C3b inactivator by the monocyte. The basal synthesis and the increase induced by indomethacin was blocked by cycloheximide, puromycin, and mitomycin C. Other cyclooxygenase and cyclooxygenase/lipoxygenase inhibitors, such as flufenamic acid, aspirin, phenylbutazone, benoxaprofen, BW 755C, and ETYA, showed similar effects. The enhancing effects of these agents on complement component synthesis parallels their potency as cyclooxygenase inhibitors. Finally, the effects of the inhibitors were reversed by the addition of prostaglandins. It is therefore concluded that prostaglandins produced by monocytes could be endogenous regulators of complement component synthesis.

7. Ia Expression

The expression of I region-associated antigen (Ia) by macrophages can be increased by interaction with a lymphokine termed macrophage Ia-recruiting factor (Scher et al., 1980). This lymphokine is generated in adult mice injected with live Listeria monocytogenes or immune T cells challenged with the organism in vitro. However, neonatal mice failed to respond to these stimuli (Snyder et al., 1982a). This failure was linked to suppressor molecules released by cells of the mononuclear phagocyte lineage. The suppressor molecules appear to be prostaglandins (Snyder et al., 1982a) thereby suggesting autoregulation of Ia surface expression.

Adult macrophages obtained from control and Listeria-infected mice also altered their expression of Ia when cultured in the presence of prostaglandins or prostaglandin synthesis inhibitors (Snyder et al., 1982b). PGE_2 was found to be a potent inhibitor (10^{-7}–10^{-11} M) of macrophage Ia expression. Furthermore, the intraperitoneal injection of PGE_1 inhibits the lymphokine-induced increase of peritoneal exudate cells bearing Ia. Indomethacin augments the effects of the lymphokine and will increase the basal level as well. These observations have recently been extended by Steeg et al. (1982). Macrophage Ia expression induced by lymphokine can be inhibited by endotoxin. The endotoxin effects are though to be mediated by prostaglandin induction of intracellular cAMP levels. Furthermore, the lymphokine used in this study was shown to have the properties of immune interferon.

8. Other Effects

a. Morphology, Phagocytosis, and Enzyme Content. PGE_1 $(2.5 \times 10^{-6} M)$ added to mouse bone marrow macrophages resulted in drastic changes in morphology. Ruffling ceased and processes disappeared as the cells became flat and more spreadout. This effect was reversed by simply washing the cells (Oropeza-Rendon et al., 1979). The same study showed that zymosan phagocytosis was also inhibited by PGE_1, but not by $PGF_{2\alpha}$. The cultivation of human monocytes in autologous serum for 3 days causes a 35-fold incease in the specific activity of the membrane enzyme 5'-nucleotidase (Picker et al., 1980). Indomethacin, 10^{-7} M, blocked the rise in 5'-nucleotidase, which was countered by the simultaneous addition of PGE_2. Furthermore, when the monocyte population was fractionated into subpopulations, only the PGE_2-producing population enhanced the activity of this enzyme.

III. LEUKOTRIENES

A. Synthesis by Mononuclear Phagocytes

Macrophages have the capacity to synthesize large quantities of a slow-reacting substance of anaphylaxis or leukotriene C_4 (LTC_4) and the chemoattractant leukotriene B_4 (LTB_4). The production of LTC_4 by mouse peritoneal macrophage is stimulated by the addition of zymosan to the cultures (Rouzer et al., 1980). The calcium ionophore A23187 also stimulates LTC_4 from rat peritoneal macrophages (Feuerstein et al., 1981) and from mouse macrophages (Humes et al., 1982). However, other soluble stimulators of prostaglandin production, such as PMA and LPS, fail to induce LTC_4 secretion, suggesting independent regulation of these two arachidonic acid pathways of oxygenation (Fig. 2). Furthermore, LTB_4 synthesis is induced by zymosan but not by PMA (Humes et al., 1982). Calcium ionophore A23187 stimulates LTB_4 synthesis by rabbit alveolar macrophages (Hsueh and Sun, 1982) and by human alveolar macrophages (Fels et al., 1982). The production of LTC_4 by elicited macrophages is decreased significantly compared with production by resident peritoneal macrophages (Rouzer et al., 1982).

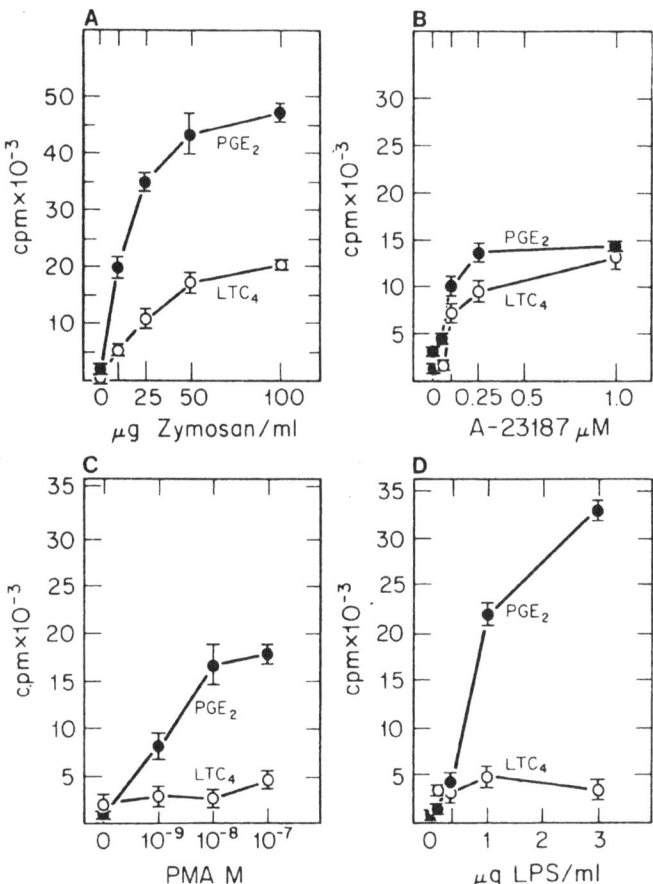

Figure 2. Independent formation of cyclooxygenase- and lipoxygenase-derived oxygenation products. Resident mouse peritoneal macrophages were prelabeled with [^{3}H]-AA, washed, and then incubated with zymosan, A-23187, phorbol myristate acetate (PMA), or lipopolysaccharide (LPS). After 3 hr, the media were removed and extracted for [^{3}H]-PGE$_2$ and [^{3}H]-LTC. The results are expressed as the mean ±SD; $N = 3$. (Reproduced with permission from Humes *et al.*, 1982.)

B. Effects on Mononuclear Phagocyte Function

LTC$_4$ is a potent bronchoconstrictor; it also can bring about increases in vascular permeability and stimulation of mucus secretion. LTB$_4$ is a potent chemoattractant for neutrophils and at higher concentrations causes these cells to degranulate. There are some indications that leukotrienes may modulate mononuclear phagocyte function.

1. Prostaglandin Production

The production of PGE_2 and 6-keto-$PGF_{1\alpha}$ by rat peritoneal macrophages can be stimulated by the addition of LTC_4 to the cultures (Feuerstein et al., 1981). PGE_2 synthesis was stimulated by 10^{-7}-10^{-4} M LTC_4, whereas 6-keto-$PGF_{1\alpha}$ synthesis was stimulated by 10^{-6}-10^{-5} M LTC_4. These investigators also suggest that prostaglandin production induced by LPS is mediated through leukotrienes. The conclusion is based on the inhibition of LPS-induced PGE_2 synthesis by the lipoxygenase inhibitor, nordihydroguaiaretic acid.

2. Lysosomal Enzyme Release

β-Glucuronidase release from elicited rate peritoneal macrophages was increased by the addition of LTC_4 from 10^{-8}-10^{-6} M (Schenkelaars and Bonta, 1983). These preliminary observations do not provide clear indications of a similar role for endogenously produced leukotrienes, which will only be clarified by the use of inhibitors of leukotriene synthesis or antagonists of their action with appropriate reconstitution experiments.

IV. PLATELET-ACTIVATING FACTOR

A. Synthesis by Mononuclear Phagocytes

Macrophages isolated from peritoneal cavity of rats and mice have been shown to synthesize and release platelet-activating factor (PAF) (1-0-alkyl-2-acetyl-sn-glyceral-3-phosphorylcholine) (Mencia-Huerta et al., 1982). PAF secretion can be increased by the calcium ionophore A-23187 (Mencia-Huerta and Benveniste, 1979). PAF is one of the most potent activators of platelets known, causing change in shape as well as aggregation and mediator release at concentrations of 10^{-11}-10^{-10} M. The synthesis of PAF in macrophages occurs via an acetyl transferase that incorporates acetyl-CoA into 2-lyso-PAF-acether (Mencia-Huerta et al., 1982).

B. Effects on Mononuclear Phagocyte Function

We have recently studied the effects of PAF on mouse peritoneal macrophages in culture. At concentrations of 10^{-6}-10^{-4} M PAF stimulates a modest increase in the synthesis and release of PGE_2 and PGI_2 without an effect on the synthesis of LTC_4 (Table II). This finding is in contrast to the marked stimulation of prostaglandin and leukotriene synthesis obtained with zymosan.

Table II. Effect of Platelet-Activating Factor on PGE$_2$
and LTC$_4$ Secretion by Macrophages[a,b]

Additions to culture	Concentration (nM)	Relative change fold stimulation	
		PGE$_2$	LTC$_4$
Without cytochalasin B			
None	–	1.0	1.0
Zymosan	50 μg/ml	17.1	2.9
PAF	10	2.1	0.9
PAF	3	2.7	0.9
PAF	1	1.9	0.9
Lyso-PAF	10	1.0	0.8
With cytochalasin B			
None	–	1.0	1.0
Zymosan	50 μg/ml	12.7	3.3
PAF	10	2.2	0.8
PAF	3	1.6	0.8
PAF	1	0.9	0.4
Lyso-PAF	10	1.0	0.7

[a]*Abbreviations:* LTC$_4$, leukotriene C$_4$; PAF, platelet-activating factor; PGE$_2$, prostaglandin E$_2$; TLC, thin-layer chromatography.
[b]Resident mouse peritoneal macrophages were incubated with [^{3}H]-archidonic acid for 18 hr. PAF and zymosan (50 μg/ml) were added in 1 ml Dulbecco's MEM in the presence or absence of cytochalasin B (2 μg/ml). After 3 hr, the media were removed and extracted for [^{3}H]-PGE$_2$ and [^{3}H]-LTC$_4$, which were separated by TLC and counted.

V. COMPLEMENT

A. Synthesis of Components of the Complement System by Mononuclear Phagocytes

Components of the complement system are synthesized by at least three cell types, i.e., hepatic parenchymal cells, gastrointestinal epithelial cells, and mononuclear phagocytes (Colten, 1976). A recent study by Whaley (1980) has confirmed and extended earlier studies showing that peripheral blood monocytes synthesize several the components of the classic pathway and all components of the alternate pathway, as well as the regulatory proteins C3b inactivator, β1H globulin, and C1 inhibitor. C2, factors B and D, properidin, and C3b inactivator were identified on the basis of their functional activity. C5 was present in some but not all cultures. Table III summarizes a number of studies that have established that mononuclear phagocytes obtained from various sources produce one or more products of the complement system.

Table III. Synthesis of Components of the Complement System
by Mononuclear Phagocytes

Complement component	Mononuclear phagocyte source	Reference
C1q	Monkey peritoneum	Stecher (1970)
	Monkey lung	Stecher (1970)
C1	Human and guinea pig peritoneum	Mueller *et al.* (1978)
	Human peritoneal macrophages	Stecher (1970)
C2	Guinea pig peritoneal	Wyatt *et al.* (1972)
	Human peripheral blood	Einstein *et al.* (1975, 1976)
		Lai A Fat and Van Furth (1975)
		Whaley (1980)
	Human alveolar	Littman and Ruddy (1977)
		Ackerman *et al.* (1978)
C3	Human peripheral blood	Lai A Fat and Van Furth (1975)
		Einstein *et al.* (1976)
		Whaley (1980)
	Mouse peripheral blood	McClelland and Van Furth (1976)
	Mouse, rabbit, monkey, and human peritoneum	Stecher (1970)
	Guinea pig peritoneal	Bentley *et al.* (1978)
C4	Monkey, human, and rat peritoneum	Stecher (1970)
	Guinea pig peritoneum	Littleton *et al.* (1970)
		Wyatt *et al.* (1972)
		Colten and Borsos (1974)
	Human peripheral blood	Whaley (1980)
		Einstein *et al.* (1976)
C5	Mouse peritoneum	Ooi and Colten (1979)
	Human peripheral blood	Whaley (1980)
Factor B	Guinea pig peritoneum	Bentley *et al.* (1978)
	Mouse peritoneum	Bentley *et al.* (1977)
	Human peripheral blood	Whaley (1980)
Factor D	Human peripheral blood	Whaley (1980)
	Guinea pig peritoneal	Bentley *et al.* (1978)
Properdin	Guinea pig peritoneal	Bentley *et al.* (1978)
	Human peripheral blood	Whaley (1980)

B. Responses of Mononuclear Phagocytes to Complement Components and Their Cleavage Products

The activation of the complement system with the resultant formation of several cleavage products possessing potent biologic activities, some of which are exerted on phagocytic cells, raises the possibility that complement components produced by mononuclear phagocytes trigger autoregulatory responses by these cells.

1. Factor B Acitivity

Gotze *et al.* (1979) have established that mouse peritoneal macrohages can be induced to spread by the active b fragment of factor B (Bb) of the alternative pathway. Bb was purified from activated serum by ion exchange and molecular sieve chromatography as well as by polyacrylamide gel electrophoresis (PAGE). The capacity of the activated serum to cause macrophage spreading was found to associate closely with Bb during the purification procedures; 1.6 ug of purified Bb was found to induce spreading of 50% of glass-adhered resident mouse peritoneal macrophages. The significance of Bb-induced spreading of macrophages may be considerable at sites of inflammation. Gotze *et al.* (1979) point out that mononuclear phagocytes synthesize all the complement components necessary to generate Bb and that appropriate stimulation of the cells by inflammatory mediators such as lymphokines may accelerate this process. Indeed, cultures of human peripheral blood monocytes in a lymphokine-rich medium accelerate their synthesis of C2 (Littman and Ruddy, 1977). The spreading of monocytes by Bb may facilitate their retention and function of sites of inflammation.

2. C3 Activity

In seeking a common property among a variety of agents that induce chronic inflammation *in vivo* and lysosomal hydrolase secretion from macrophages in culture, our attention was drawn to the fact that many of these agents activate complement by the alternative pathway. This is true of carrageenans and dextran sulfates (Burger *et al.*, 1975), zymosan, moldy hay dust containing *Micropolysporum faeni*, and streptococcal cell walls (for review, see Davies and Bonney, 1980). An early event in such activation is the cleavage of C3 to a smaller fragment C3a and a larger fragment C3b. The highly purified C3a component, when incubated with mouse peritoneal macrophages in culture, was found to release both lysosomal hydrolases and loctate dehydrogenase, thus indicating cell death (Schorlemmer *et al.*, 1976). In contrast, when highly purified C3b was incubated with the macrophages, selective concentration and time-dependent release of several glycosidases, but not of lactate dehydrogenase, was observed. Similar results have been obtained with guinea pig peritoneal macrophages using highly purified guinea pig C3a and C3b (Schorlemmer and Allison, 1976). A purified preparation of guinea pig C3b stimulated the selective release of lysosomal hydrolases by starch-elicited guinea pig peritoneal macrophages in a concentration-dependent manner over the range of 10-40 ug/ml of culture medium. Enzyme release began after a considerable lag period of 24 hr after addition of C3b and continued to increase thereafter for ≥72 hr of culture. Under these experimental conditions, C3b also produced significant, dose-dependent increases in cellular levels of lactate dehydrogenase, indicating the continued viability of the cells. The preincubation of C3b with an anti-C3b Fab preparation completely inhibited the stimulation of acid hydrolase release when the complex was added to the culture. Partial inhibition of enzyme release was observed with anti-C3 IgG, but preincubation with an unrelated IgG had no inhibitory effect.

On the basis of their observations that exogenous C3b stimulates the selective release of lysosomal acid hydrolases from macrophages, Schorlemmer and Allison (1976) and Schorlemmer *et al.* (1977) have suggested that endogenously synthesized components of the alternative pathway of complement play a critical role in this process. They base this idea on the following experimental findings:

1. Many of the substances that stimulate the selective release of lysosomal acid hydrolases activate the alternative pathway of complement as well.

2. Macrophages synthesize both factor B (Bentley *et al.*, 1977; Whaley, 1980) and C3 (Stecher, 1970; Lai A Fat and Van Furth, 1975; Bentley *et al.*, 1977; Brade *et al.*, 1977; Whaley, 1980) as well as enzyme activity corresponding to that of factor D. On the basis of such experimental findings, Schorlemmer *et al.* (1977) have hypothesized that various agents eliciting the selective release of lysosomal enzymes from mouse peritoneal macrophages either provide or create a suitable environment for the formation of a C3bBb complex, producing an amplification system for both the synthesis and activation of C3. This activity provides C3b, which stimulates the selective release of lysosomal enzymes from macrophages as has been observed experimentally (Schlorlemmer *et al.*, 1976).

Rutherford and Schenkein (1983) have shown that fragments of C3, i.e., C3b, C3bi, and C3c, but not C3d or C3 itself, stimulate the synthesis of immunoreactive PGE by human peripheral blood monocytes. Maximum stimulation with C3b occurred at a concentration of 25 ug/ml over a period of 18-24 hr. In view of the inhibitory effects of prostaglandins on the synthesis of complement components reported by Lappin and Whaley (1982), the activity of the C3 fragments may represent a negative feedback mechanism for regulating synthesis of complement components by monocytes.

3. C5a and C5a$_{des \, arg}$ Activity

The chemotactic activity of C5a and C5a $_{des \, arg}$ for mononuclear phagocytes is now well established (Gallin and Gallin, 1977). Recently, McCarthy and Henson (1979) described a number of other effects of C5a and C5 $_{des \, arg}$ on macrophage function. C5a and C5a $_{des \, arg}$ purified from human serum were shown to increase the secretion of lysosomal glycosidases from rabbit alveolar macrophages, so that after 72-hr exposure the levels in treated cultures exceeded those in control cultures by 5-fold. It appeared that this accumulation of lysosomal enzyme in the medium was not the result of the depletion of intracellular enzyme, since there was no concomitant decrease in intracellular lysosomal enzyme. This response of alveolar macrophages to C5a and $C5a_{des\ arg}$ was accompanied by a number of other changes. Pinocytic activity was increased as measured by uptake of horseradish peroxidase as was the secretion of a proteolytic enzyme active at neutral pH. Also, alveolar macrophages treated with $C5a_{des\ arg}$ produced a factor chemotactic for rabbit neutrophils desensitized with C5a. It thus appears that the cleavage fragments of C5 have a number of effects on macrophages in addition to chemotaxis. The secretion of hydrolytic enzymes by the macrophages may contribute to tissue damage in the local environment of the cell, while the generation of neutrophil chemotactic factors can lead to an influx of these cells to site of inflammation.

Recent observations by Goodman *et al.* (1982*a*) have indicated that C5a stimulates the primary humoral antibody response of splenocytes. Also, C5a has been shown by these investigators to induce the secretion of interleukin 1-like material by mouse peritoneal macrophages and the $P388D_1$ cell line (Goodman *et al.*, 1982*b*). Both types of macrophages bear a high-affinity receptor for C5a (Chenoweth *et al.*, 1982). In contrast, the P388 cell line, which lacks the high-affinity C5a receptor, fails to produce IL-1 in response to C5a, an example of a feedback effect on the macrophage of a biologically active fragment of one of its secretory products. C5a production is presumably limited to environments in which complement activation is being triggered via the classic or alternate pathways.

VI. FIBRONECTIN

A. Synthesis of Fibronectin by Mononuclear Phagocytes

Fibronectin is a high-molecular-weight glycoprotein found in a soluble form in plasma and other body fluids, and in an insoluble form in connective tissue and also in association with basement membranes. Fibronectin interacts with various cell types, with extracellular molecules, and even with itself. The involvement of fibronectin in the control of cell morphology has been surmised from its presence in abundant amounts in the extracellular matrix of nontransformed cells in contrast to its absence from the surface of transformed cells.

The molecule has been characterized as a 440-kd disulfide-linked dimer. Structural studies on fragments of the molecule generated by proteolytic digestion show them to possess a series of distinct functional properties (Ruoslahti *et al.*, 1982). These properties include separate domains that bind gelatin, collagen, fibrin, fibrinogen, staphylococci, and actin. In addition, there is another domain to which cells can bind.

One of the first demonstrations of fibronectin synthesis by mononuclear phagocytes was that of Johansson *et al.* (1979) who showed secretion of metabolically labeled authentic material by mouse peritoneal macrophages.

Alitalo *et al.* (1980) demonstrated that human peripheral blood monocytes maintained in culture secrete considerable amounts of a material identified as fibronectin by immunoprecipitation and gel electrophoresis to reveal a 220-kd molecule. Immunofluorescent staining with a specific antiserum to fibronectin demonstrated a cytoplasmic localization for fibronectin in cells permitted to differentiate in culture for several days, with an absence of pericellular localization characteristic of fibroblasts. Although not detected in the culture medium of young monocytes, up to 150 ng/ml was released into the medium of cells after 3–6 days of culture.

Rennard *et al.* (1981) have shown that human alveolar macrophages synthesize and release fibronectin in a time-dependent manner at a rate of ~0.5 ng/10^6 cells/hr. This rate is greatly enhanced in alveolar macrophages from patients with idiopathic pulmonary fibrosis and sarcoidosis. It should be noted that these enhanced levels of production ranging from 5 to 10 ng/10^6 cells/hr are still low compared with rates of 500–2000 ng/10^6 cells/hr reported for fibroblasts (Rennard *et al.*, 1980; Mosher *et al.*, 1977).

Villiger *et al.* (1981) have provided further evidence of fibronectin synthesis by human pulmonary alveolar macrophages. Pulse-chase experiments showed rapid secretion of newly synthesized fibronectin with no evidence of pericellular association, as has been commonly observed in fibroblasts. Fibronectin was observed, however, at the cell surface at binding sites for gelatin-coated latex beads.

B. Effects of Fibronectin and Its Cleavage Products on Mononuclear Phagocytes

Bevilacqua *et al.* (1981) have demonstrated a trypsin-sensitive binding site for fibronectin on the surface of human monocytes. In the presence of fibronectin, monocytes adhered to gelatin-coated surfaces in a concentration-dependent manner with a requirement for Mg^{2+} ions. In addition, it was found that fibronectin promoted the adherence, but not ingestion, of gelatin-coated latex particles or tanned erythrocytes to monocytes. Fibronectin did, however, facilitate the ingestion of immunoglobulin-coated erythrocytes via Fc receptors as well as enhancing C3 receptor expression as indicated by increased binding of complement-coated erythrocytes.

In contrast, Gudewicz *et al.* (1980) have provided good evidence of fibronectin-dependent phagocytosis through a trypsin labile binding site in caseinateelicited rat peritoneal macrophages. Studies were made with ^{125}I-labeled gelatin-coated latex beads. Interiorization was confirmed by electron microscopy and shown to require metabolic energy and to be dependent on intact intracellular skeletal elements.

Norris et al. (1982) have found that fragments of fibronectin are chemotactic for human peripheral monocytes. Such activity was found in purified fragments generated by the endogenous proteinase activity of the human plasma. Fragments of 90–220 kd showed both chemotactic and chemokinetic activity at concentrations of 0.001–0.1 µg/ml. The activity is selective for monocytes, both neutrophils and lymphocytes being unresponsive. The chemotactic activity of these fragments for monocytes is effectively blocked by antifibronectin serum, but not by an antibody to C5a. The relationship between fibronectin fragments generated by cathepsin D proteolysis and which are chemotactic for fibroblasts (Postlethwaite et al., 1981) and those described by Norris et al. (1982) is not clear. It is conceivable that there are differences relevant to the function of the two cell types; such differences should be resolved by generation of fully characterized fragments from pure fibronectin by controlled proteolysis.

Much remains to be learned about fibronectin secretion by mononuclear phagocytes, but it is already apparent that this molecule can influence the function of these cells in a number of ways. The capacity of fibronectin to bind to endogenous molecules such as collagen and fibrin suggests a specific mechanism by which macrophages can recognize newly formed fibrin at sites of tissue injury as well as collagen, which is either exposed in the course of tissue damage or newly synthesized at sites of tissue repair. In this way macrophages can become specifically associated with fibrin for the purposes of fibrinolysis mediated by another of its secreted products, i.e., plasminogen activator, and the turnover of collagen, which is normally associated with tissue repair after injury. The chemotactic function of fibronectin fragments serves as a mechanism to attract further mononuclear phagocytes for the purpose of host defense, tissue debridement, and repair.

VII. PROTEINASES AND PROTEINASE INHIBITORS

A. Synthesis by Mononuclear Phagocytes

The secretion of neutral proteinases by mononuclear phagocytes has been extensively investigated. Most of these studies have been reviewed elsewhere (Davies and Bonney, 1980) and are not covered further here. Suffice it to point out that these enzymes are secreted by undefined mechanisms that may involve pathways similar to those in other cells performing endocrine functions. In most instances the rate of enzyme secretion is low in nonactivated cells but is increased when cells are activated by immunologic stimuli such as lymphokines or after ingestion of nondigestible phagocytic loads. The activity of neutral proteinases is under tight control in the extracellular environment. Plasma and, to a lesser extent, other biologic fluids contain large amounts of high-molecular-weight inhibitors of all known classes of proteolytic enzymes. The two most abundant of these inhibitors, α_1-proteinase inhibitor and α_2-macroglobulin are secreted by mononuclear phagocytes under certain conditions.

Van Furth et al. (1983) have demonstrated the synthesis of α_1-proteinase inhibitor by human peripheral blood monocytes. De novo synthesis was demonstrated by autoradiography of incorporated radioactive amino acid precursors on immunoelectrophoretic slides prepared from cell-free lysates of monocytes or tissue containing mononuclear phagocytes. The functional activity of the synthesized material was shown by treatment of lysates with porcine pancreatic elastase and the demonstration of a change in electrophoretic mobility commensurate with a formation of a complex between α_1-proteinase inhibitor and elastase. It is also noteworthy that monocytes from individuals of the Pi MS and Pi ZZ phenotypes synthesized proportionately less α_1-proteinase inhibitor.

White et al. (1981a) have used similar techniques to detect the incorporation of [^{35}S]methionine into α_1-macroglobulin. These include human peripheral

A number of investigators have shown that mononuclear phagocytes from a variety of sources secrete α_2-macroglobulin. These include human peripheral blood monocytes (Hovi et al., 1977), human alveolar macrophages (White et al., 1980), and murine peritoneal macrophages (White et al., 1981b).

B. Effects of α_2-Macroglobulin–Proteinase Complexes on Mononuclear Phagocyte Function

Kaplan and Nielsen (1979a,b) have demonstrated the presence of receptors for α_2-macroglobulin–trypsin complexes on rabbit alveolar macrophages and have shown an ability to endocytose the complexes with the receptor being recycled (Kaplan, 1980). The function of neutral proteinases in the extracellular environment is believed to be tightly controlled by endogenous inhibitors such as α_1-proteinase inhibitor, active against serine proteinase, and α_2-macroglobulin, active against all known classes of proteinases. The balance between the proteinases and their inhibitors will therefore be a critical determinant of their activity. Excess inhibitor should ensure rapid removal of enzyme as a complex by mechanisms described by Kaplan (1980). Alternatively, excess enzyme should have access to its extracellular substrates. We know very little of the dynamics of the interactions of the proteinases and proteinase inhibitors secreted by macrophages. Banda et al. (1980) have shown that elastase purified from the culture medium of mouse peritoneal macrophages inactivates human α_1-proteinase inhibitor by limited proteolysis, indicating a novel way by which the macrophage can alter the proteinase-proteinase inhibitor balance. No clear conclusions regarding the overall implications of the interaction of these two classes of macrophage secretory products can be made until we know more about the specific mechanisms by which individual enzymes interact with each inhibitor.

A recent study by Johnson *et al.* (1982) suggests that α_1-macroglobulin-protease complexes may have profound effects on macrophage function. These workers found that BCG-elicited mouse peritoneal macrophages, which have a low basal level of secretion of three neutral proteinases, i.e., plasminogen activator, neutral caseinase, and a cytotoxic proteinase, can be triggered to secrete high levels of these enzymes by maleylated bovine serum albumin. This triggering is though to occur via specific receptors. It was not accompanied by release of the lysosomal acid hydrolase acid phosphatase nor was the phenomenon elicited by manose-bovine serum albumin complexes or lactoferrin, which bind to macrophages via separate receptors. The triggering of neutral proteinase secretion was accompanied by the increased expression of tumor cell cytolytic activity. In marked contrast, exposure of these cells to α_2-macroglobulin-trypsin complexes resulted in a significant diminution of neutral proteinase secretion as well as cytolytic activity toward a tumor cell target. These findings suggest that specific engagement of its receptor by the proteinase-proteinase inhibitor complex not only leads to its removal, but also has a profound subsequent effect on the cell's secretory and functional capacities.

VIII. CONCLUDING REMARKS

From the selected examples reviewed in this chapter, it can be suggested that the secretory products of mononuclear phagocytes may have important autoregulatory functions that facilitate the control of the cell in the exercise of its host-defense functions. Its secretory products can serve either to amplify or to inhibit mononuclear phagocyte function. For example, fragments of fibronectin, C5a, and probably LTB_4 act as chemotactic stimuli for the recruitment of cells to sites of infection, inflammation and tissue damage. Upon their arrival, fibronectin and other products of complement activation enhance microbicidal and phagocytic function. Since the initial stimulus for the secretion of various products by the cells is the target of the cidal and phagocytic activity, its removal will serve to limit the generation of further secretory products, which have an inhibitory influence on mononuclear phagocyte function, e.g., prostaglandins inhibit stem cell proliferation as well as secretion of other products. Finally, two products, such as proteases and their inhibitors, may regulate each other's activity in the pericellular environment while the complexes of the two, such as α_2-macroglobulin-trypsin, have profound effects on the function of the mononuclear phagocytes.

We emphasize that the constructs we have reviewed in this chapter are based on simple *ex vivo* systems and might not actually be operational *in vivo*. It does appear logical however, that the mononuclear phagocyte should be able to respond to certain of its own secretory products in order to facilitate a balanced interaction with its pericellular environment while mediating essential host-defense functions.

ACKNOWLEDGMENTS

We are grateful to Carolyn Kradjel for her excellent secretarial work.

IX. REFERENCES

Ackerman, S. K., Friend, P. S., Hoidal, J. R., and Douglas, S. D., 1978, Production of C2 by human alveolar macrophages, *Immunology* 35:369-372.

Alitalo, K., Hovi, T., and Vaheri, A., 1980, Fibronectin is produced by human macrophages, *J. Exp. Med.* 151:602-613.

Banda, M., Clark, E. J., and Werb, Z., 1980, Limited proteolysis by macrophage elastase inactivates human α_1-proteinase inhibitor, *J. Med. Exp.* 152:1563-1570.

Banda, M., and Werb, Z., 1981, Mouse macrophage elastase. Purification and characterization as a metalloproteinase, *Biochem. J.* 193:589-605.

Bentley, C., Hadding, U., Bitter-Suermann, D., and Brade, V., 1977, Effect of *in vivo* stimulation of mice on the secretion of factor B of the alternative complement pathway by peritoneal macrophages, *Eur. J. Immunol.* 7:188-190.

Bentley, C., Fries, W., and Brade, V., 1978, Synthesis of factors D, B and P of the alternative pathway of complement activation, as well as of C3 by guinea pig peritoneal macrophages *in vitro*, *Immunology* 35:971-980.

Bevilacqua, M. P., Amrani, D., Mossesson, M. W., and Bianco, C., 1981, Receptors for cold insoluble globulin (plasma fibronectin) on human monocytes, *J. Exp. Med.* 153:42-60.

Bonney, R. J., Wightman, P. D., Davies, P., Sadowski, S., Kuehl, F. A., Jr., and Humes, J. L., 1978, Regulation of prostaglandin synthesis and of the release of lysosomal hydrolases by mouse peritoneal macrophages, *Biochem. J.*, 176:433-442.

Bonney, R. J., Naruns, P., Davies, P., and Humes, J. L., 1979, Antigen-antibody complexes stimulate the synthesis and release of prostaglandins by mouse peritoneal macrophages, *Prostaglandins* 18:605-616.

Bonney, R. J., Wightman, P. D., Dahlgren, M. E., Davies, P., Kuehl, F. A., Jr., and Humes, J. L., 1980a, Release of inflammatory mediators by macrophages in response to phorbol myristate acetate: Effect of RNA and protein synthesis inhibitors, *Biochim. Biophys. Acta* 633:410-421.

Bonney, R. J., Burger, S., Kuehl, F. A., Jr., and Humes, J., 1980b, Prostaglandin E_2 and prostacyclin elevate cAMP levels in elicited but not resident populations of macrophages, *Adv. Prostaglandin Thromboxane Res.* 8:1691-1693.

Brade, V., Hall, R. E. and Colten, H., 1977, Biosynthesis of pro-C3, a precursor of the third component of complement, *J. Exp. Med.* 146:759-765.

Bradley, J. R., and Metcalf, D., 1966, The growth of mouse bone marrow cells *in vitro, Aust. J. Exp. Bio. Med. Sci.* 44:287-293.

Broxmeyer, H. E., Mendelsohn, N., and Moore, M. A. S., 1977, Abnormal granulocyte feedback regulation of colony stimulating activity-producing cells from patients with chronic myelogenous leukemia, *Leukemia Res.* 1:3-12.

Brune, K., Glatt, M., Kalin, H., and Peskar, H., 1978, Pharmacological control of prostoglandin and thromboxane release from macrophages, *Nature* 274:261-263.

Burger, R., Hadding, U., Schorlemmer, H. U., Brade, V., and Bitter-Suerman, D., 1975, Dextran sulphate: A synthetic activator of C3 via the alternative pathway. I. Influence of molecular size and degree of sulphation on the activation potency, *Immunology* 29:549-554.

Chenoweth, D. E., Goodman, M. G., and Weigle, W. O., 1982, Demonstration of a specific receptor for human C5a anaphylatoxin on murine macrophages, *J. Exp. Med.* 156:68-78.

Chervenick, P. A., and LoBuglio, A. E., 1972, Human blood monocytes: stimulators of granulocyte and mononuclear colony formation *in vitro, Science* 178:164-166.

Colten, H., 1976, Biosynthesis of complement, *Adv. Immunol.* 22:67-118.

Colten, H. R., and Borsos, T., 1974, Biosynthesis of the second and fourth components of complement: Inhibition *in vitro* by chemical carcinogens, *J. Immunol.* 112:1107-1114.

Dahlgren, M. E., Davies, P., and Bonney, R. J., 1980, Phorbol myristate acetate induces the secretion of an elastase by populations of resident and elicited mouse peritoneal macrophages, *Biochim. Biophys. Acta* 630:338-351.

Davies, P., and Bonney, R. J., 1979, Secretory products of mononuclear phagocytes, *J. Reticuloendothel. Soc.* 26:37-47.

Davies, P., and Bonney, R. J., 1980, The secretion of hydrolytic enzymes by mononuclear phagocytes, in: *The Cell Biology of Inflammation* (G. Weissmann, ed.), pp. 497-542, Elsevier North-Holland, New York.

Davies, P., Bonney, R. J., Humes, J. L., and Kuehl, F. A., Jr., 1980, The synthesis of arachidonic acid oxygenation products by various mononuclear phagocyte populations, in: *Mononuclear Phagocytes, Functional Aspects* (R. van Furth, ed.), pp. 1317-1350, Martinus Nijhoff, The Hague.

Einstein, L. P., Alper, C. A., Bloch, K. J., Herin, J. T., Rosen, F. S., David, J. R., and Colten, H. R., 1975, Biosynthetic defect in monocytes from human beings with genetic deficiency of the second component of complement, *N. Engl. J. Med.* 292:1169-1171.

Einstein, L. P., Schneeberger, E. E., and Colten, H. R., 1976, Synthesis of the second component by long-term primary cultures of human monocytes, *J. Exp. Med.* 143:114-126.

Fels, A. O. S., Pawlowski, N. A., Cramer, E. B., King, T. K. C., Cohn, Z. A., and Scott, W. A., 1982, Human alveolar macrophages produce leukotriene B4, *Proc. Natl. Acad. Sci. USA* 79:7866-7870.

Feuerstein, N., Bash, J. A., Woody, J. N., and Ramwell, P. W., 1981, Leukotriene C stimulates prostaglandin release from rat peritoneal macrophages, *Biochem. Biophys. Res. Commun.* 100:1085-1090.

Gallin, E. K., and Gallin, J. I., 1977, Interaction of chemotactic factors with human macrophages. Induction of transmembrane potential changes, *J. Cell Biol.* 75:277-289.

Gemsa, D., Seitz, M., Kramer, W., Till, G., and Resch, K., 1979, The effects of phagocytosis, dextran sulfate, and cell damage on PGE_1 sensitivity and PGE_1 production of macrophages, *J. Immunol.* 120:1187-1194.

Goodman, M. G., Chenoweth, D. E., and Weigle, W. O., 1982a, Potentiation of the primary humoral immune response *in vitro* by C5a anaphylatoxin, *J. Immunol.* 129:70-75.

Goodman, M. G., Chenoweth, D. E., and Weigle, Wo. O., 1982*b*, Induction of interleukin 1 secretion and enhancement of humoral immunity by binding of human C5a to macrophage surface C5a receptors, *J. Exp. Med.* 156:912–917.

Gordon, S., 1980, Lysozyme and plasminogen activator: Constitutive and induced secretory products of mononuclear phagocytes, in: *Mononuclear Phagocytes, Functional Aspects* (R. Van Furth, ed.), pp. 1273–1298, Martinus Nijhoff, The Hague.

Gotze, O., Bianco, C., and Cohn, Z. A., 1979, The induction of macrophage spreading by Factor B of the properdin system, *J. Exp. Med.* 149:372–386.

Gudewicz, P. W., Molnar, J., Lai, M. Z., Beezhold, D. W., Siefring, Jr., G. E., Credo, R. B., and Lorand, L., 1980, Fibronectin mediated uptake of gelatin coated latex particles by peritoneal macrophages, *J. Cell Biol.* 87:427–433.

Guilbert, L. J., and Stanley, E. R., 1980, Specific interaction of murine colony-stimulating factor with mononuclear phagocytic cells, *J. Cell. Biol.* 85:153–159.

Hamilton, J. A., 1981, Regulation of prostaglandin and plasminogen activator production by mouse peritoneal macrophages, *J. Reticuloendothel. Soc.* 30:115–128.

Hibbs, J. R., Lambert, L. H., and Remington, J. S., 1971, Resistance to murine tumors conferred by chronic infection with intracellular protozoa, *Toxoplasma gondii* and *Resonitia jellisoni, J. Infect. Dis* 124:587–591.

Hovi, T., Mosher, D., and Vaheri, A., 1977, Cultured human monocytes synthesize and secrete α$_2$-macroglobulin, *J. Exp. Med.* 145:1580–1589.

Hsueh, W., and Sun, F. F., 1982, Leukotriene by biosynthesis by alveolar macrophages, *Bioch. Biophys. Res. Commun.* 106:1085–1091.

Humes, J. L., Bonney, R. J., Pelus, L., Dahlgren, M. E., Sadowski, S. J., Kuehl, F. A., Jr., and Davies, P., 1977, Macrophages synthesize and release prostaglandins in response to inflammatory stimuli, *Nature* 269:149–151.

Humes, J. L., Sadowski, S., Galavage, M., Kuehl, F. A., Jr., Wightman, P. D., Dahlgren, M. E., Davies, P., and Bonney, R. J., 1980, The diminished production of arachidonic acid oxygenation products by elicited peritoneal macrophages: Possible mechanisms, *J. Immunol.* 124:2110–2116.

Humes, J. L., Sadowski, S., Galavage, M., Goldenberg, M., Subers, E., Bonney, R. J., and Kuehl, F. A., Jr., 1982, Evidence for two sources of arachidonic acid for oxidative metabolism by mouse peritoneal macrophages, *J. Biol. Chem.* 257:1591–1594.

Johansson, S., Rubin, K., Hook, M., Ahlgren, T., and Seljelid, D. R., 1979, *In vitro* biosynthesis of cold insoluble gloublin (fibronectin) by mouse peritoneal macrophages, *FEBS Lett.* 105:313–316.

Johnson, W. J., Pizzo, S. V., Imber, M. J., and Adams, D. O., 1982, Receptors for maleylated proteins regulate secretion of neutral proteases by murine macrophages, *Science* 218:574–576.

Kaplan, J , 1980, Evidence for reutilization of surface receptors for α-macroglobulin protease complexes in rabbit alveolar macrophages, *Cell* 19:197–205.

Kaplan, J., and Nielsen, M. L., 1979*a*, Analysis of macrophage surface receptors. 1. Binding of α-macroglobulin–protease complexes to rabbit alveolar macrophages, *J. Biol. Chem.* 254:7323–7328.

Kaplan, J., and Nielsen, M. L., 1979*b*, Analysis of macrophage surface receptors. II. Internalization of α-macroglobulin–trypsin complexes by rabbit alveolar macrophages, *J. Biol. Chem.* 254:7329–7335.

Kurland, J. J., and Bockman, R., 1978, Prostaglandin E production by human blood monocytes and mouse peritoneal macrophages, *J. Exp. Med.* 147:952–957.

Kurland, J. I., Pelus, L. M., Ralph, P., Bockman, R. S., and Moore, M. A. S., 1979, Induction of prostaglandin E synthesis in normal and neoplastic macrophages: Role for colony-stimulating factor(s) distinct from effects on myeloid progenitor cell proliferation, *Proc. Natl. Acad. Sci. USA* 76:2326–2330.

Lai A Fat, R. F. M., and Van Furth, R., 1975, *In vitro* synthesis of some complement components (C1q, C3 and C4) by lymphoid tissues and circulating leucocytes in man, *Immunology* 28:359-368.

Lappin, D., and Whaley, K., 1981, Cyclic AMP-mediated modulation of the production of the second component of human complement by monocytes, *Int. Arch. Allergy Appl. Immunol.* 65:85-90.

Lappin, D. F., and Whaley, K., 1982, Prostaglandins and prostaglandin synthetase inhibitors regulate the synthesis of complement components by human monocytes, *Clin. Exp. Immunol.* 49:623-630.

Littleton, C., Kessler, D., and Burkholder, T. M., 1970, Cellular basis for the synthesis of the fourth component of guinea pig complement as determined by a hemolytic plaque technique, *Immunology* 18:693-704.

Littman, B. H., and Ruddy, S., 1977, Production of the second component of complement by human monocytes: Stimulation by antigen-activated lymphocytes or lymphokines, *J. Exp. Med.* 145:1344-1352.

McCarthy, K., and Henson, P. M., 1979, Induction of lysosomal enzyme secretion by alveolar macrophages in response to purified complement fragments C5a and C5a$_{des\ arg}$, *J. Immunol.* 123:2511-2517.

McCarthy, M. E., and Zwilling, B. S., 1981, Differential effects of prostaglandins on antitumor activity of normal and BCG-activated cells, *Immunology* 60:91-99.

McClelland, D. B. L., and Van Furth, R., 1976, *In vitro* synthesis of $\beta_1 C/\beta_1 A$ globulin (the C3 component of complement) by tissues and leucocytes of mice, *Immunology* 31:855-861.

Mencia-Huerta, J-M, and Benveniste, J., 1979, Platelet-activating factor and macrophages I. Evidence for the release from rat and mouse peritoneal macrophages and not from mastocytes, *Eur. J. Immunol.* 9:409-415.

Mencia-Huerta, J.-M., Roubin, R., Morgot, J-L, and Benveniste, J., 1982, Biosynthesis of platelet-activating factor, III. Formation of PAF-acether synthetic substrates by stimulated murine macrophages, *J. Immunol.* 129:804-808.

Mosher, D. F., Saksela, O., Keski-Oja, J., and Vaheri, A., 1977, Distribution of a major surface-associated glycoprotein, fibronectin, in cultures of adherent cell, *J. Supramol. Struct.* 6.551-557.

Mueller, W., Anausje-Abel, M., and Loos, M., 1978, Biosynthesis of the first component of complement by human and guinea pig peritoneal macrophages: Evidence for independent production of C1 subunits, *J. Immunol.* 121:1578-1584.

Murray, J. L., 1982, Prostaglandin E$_2$ modulation of human monocyte antibody-dependent cell-mediated cytotoxicity against human red cells, *Cell Immunol.* 71:196-201.

Nathan, C. F., Murray, H. U., and Cohn, Z. A., 1982, The macrophage as an effector cell, *N. Engl. J. Med.* 303:622-626.

Norris, D. A., Clark, R. A. F., Swigart, L. M., Huff, J. C., Weston, W. M., and Howell, S. E., 1982, Fibronectin fragment(s) are chemotactic for human peripheral blood monocytes, *J. Immunol.* 129:1612-1618.

North, R. J., Kirstein, D. P., and Remington, J. S., 1976, Subversion of host defense mechanisms by murine tumors. I. A circulating factor that suppresses macrophage-mediated resistance to infection, *J. Exp. Med.* 143:559-573.

Ooi, Y. M., and Colten, H. M., 1979, Genetic defect in secretion of complement C5 in mice, *Nature* 282:207-208.

Oropeza-Rendon, R. L., Speth, V., Heller, G., Weber, K., and Fisher, H., 1979, Prostaglandin E$_1$ reversibly induces morphological changes in macrophages and inhibits phagocytosis, *Exp. Cell Res.* 119:365-371.

Pelus, L. M., Broxmeyer, H. E., Kurland, J. I., and Moore, M. A. S., 1979, Regulation of macrophage and granulocyte proliferation, *J. Exp. Med.* 150:277-292.

Pelus, L. M., Broxmeyer, H. E., Clarkson, B. D., and Moore, M. A. S., 1980, Abnormal responsiveness of granulocyte–macrophage committed colony forming cells from patients with chronic myeloid leukemia to inhibition by prostaglandin E, *Cancer Res.* 40:2512–2515.

Picker, L. J., Raff, H. V., Goldyne, M. E., and Stobo, J. D., 1980, Metabolic heterogenity among human monocytes and its modulation by PGE_2, *J. Immunol.* 124:2557–2562.

Postlethwaite, A. E., Keski-Oja, J., Balian, G., and Kang, A. H., 1981, Induction of fibroblast chemotaxis by fibronectin. Localization of the chemotactic region to a 140,000-molecular weight non-gelatin-binding fragment, *J. Exp. Med.* 153:494–499.

Rennard, S. I., Berg, R., Martin, G. R., Foidart, J. M., and Gehron-Robey, P., 1980, Ensyme-linked immunoassay (ELISA) for connective tissue components, *Anal. Biochem.* 104:205–214.

Rennard, S. I., Hunninghake, G. W., Bitterman, P. B., and Crystal, R. G., 1981, Production of fibronectin by the human alveolar macrophage: Mechanism for the recruitment of fibroblasts to sites of tissue injury in interstitial lung diseases, *Proc. Natl. Acad. Sci. USA* 78:7147–7151.

Rosenthal, A. S., 1980, Regulation of the immune response–role of the macrophage, *N. Engl. J. Med.* 303:1153–1156.

Rouzer, C. M., Scott, W. A., Conn, Z. A., Blackburn, P., and Manning, J., 1980, Mouse peritoneal macrophages release leukotriene C in response to a phagocytic stimulus, *Proc. Natl. Acad. Sci. USA* 77:4928–4932.

Rouzer, C. M., Scott, W. A., Hamill, A. L., and Cohn, Z. A., 1982, Synthesis of leukotrienes and other arachidonic acid metabolites by mouse peritoneal macrophages, *J. Exp. Med.* 155:720–733.

Ruoslahti, E., Pierschbacher, M., Hayman, E. G., and Engvalle, E., 1982, Fibronectin: A molecule with remarkable structural and functional diversity, *Trends Biochem. Sci.* 7:188–190.

Rutherford, B., and Schenken, H. A., 1983, C3 cleavage products stimulate release of prostaglandins by human mononuclear phagocytes *in vitro*, *J. Immunol.* 130:874–877.

Schenkelaars, E. J., and Bonta, I. L., 1983, Effect of leukotriene C_4 on the release of secretory products by elicited populations of rat peritoneal macrophages, *Eur. J. Pharmacol.* 86:477–480.

Scher, M. G., Beller, D. I., and Unanue, E. R., 1980, Demonstration of a soluble mediator that induces exudates rich in Ia-positive macrophages, *J. Exp. Med.* 152:1684–1693.

Schnyder, J., Dewald, B., and Baggiolini, M., 1981, Effects of cyclooxygenase inhibitors and prostaglandin E_2 on macrophage activation *in vitro*, *Prostaglandins* 22:411–421.

Schorlemmer, H. U., and Allison, A. C., 1976, Effects of activated complement components on enzyme secretion by macrophages, *Immunology* 31:181–186.

Schorlemmer, H. U. Davies, P., and Allison, A. C., 1976, Ability of activated complement components to induce lysosomal enzyme release from macrophages, *Nature* 261:48–49.

Schorlemmer, H. U., Bitter-Suerman, D., and Allison, A. C., 1977, Complement activation by the alternative pathway and macrophage enzyme secretion in the pathogenesis of chronic inflammation, *Immunology* 32:929–940.

Schultz, R. M., Pavlidis, N. A., Styles, W. A., and Chirigos, M. A., 1978, Regulation of macrophage tumorcidal function: A role for prostaglandins of the E series, *Science* 202:320–321.

Scott, W. A., Rouzer, C. A., and Cohn, Z. A., 1983, Leukotriene C release by macrophages, *Fed. Proc.* 42:129–133.

Smith, R. L., Hunt, N. H., Merritt, J. E., Evans, T., and Weidemann, M. J., 1980, Cyclic nucleotide metabolism and reactive oxygen production by macrophages, *Biochem. Biophys. Res. Commun.* 96:1079–1087.

Snyder, D. S., Lu, C. Y., and Unanue, E. R., 1982a, Control of macrophage Ia expression in neonatal mice role of a splenic suppressor cell, *J. Immunol.* 128:1458–1465.

Snyder, D. S., Beller, D. I., and Unanue, E. R., 1982b, Prostaglandins modulate macrophage Ia expression, *Nature* 299:163–165.

Stecher, V., 1970, Synthesis of proteins by mononuclear phagocytes, in: *Mononuclear Phagocytes* (R. Van Furth, ed.), pp. 133–150, F. A. Davis, Philadelphia.

Steeg, P. S., Johnson, H. M., and Oppenheim, J. J., 1982, Regulation of murine macrophage Ia antigen expression by an immune interferon-like lymphokine: Effect of endotoxin, *J. Immunol.* 129:2402–2406.

Taffet, S. M., and Russell, S. W., 1980, Macrophage mediated tumor cell killing: Regulation of expression of cytolytic activity by prostaglandin E., *J. Immunol.* 126:424–427.

Taffet, S. M., Eurell, T. E., and Russell, S. W., 1982, Regulation of macrophage-mediated tumor cell killing by prostagladins: comparison of PGE_2 and PGI_2, *Prostaglandins* 24:763–774.

Tushinski, R. J., Oliver, I. T., Guilbert, L. J., Tyhan, P. W., Warner, J. R., and Stanley, E. R., 1982, Survival of mononuclear phagocytes depends on a lineage-specific growth factor that the differentiated cells selectively destroy, *Cell* 28:71–81.

Van Furth, R., Kramps, J. A., and Diesselhoff-Den Dulk, M. M. C., 1983, Synthesis of α_1-antitrypsin by human monocytes, *Clin. Exp. Immunol.* 51:551–557.

Vassalli, J-D., Hamilton, J., and Reich, E., 1977, Macrophage plasminogen activator: Induction by concanavalin A and phorbol myristate acetate, *Cell* 11:695–705.

Villiger, B., Kelley, D. G., Engleman, W., Kuhn, C. III, and McDonald, J. A., 1981, Human alveolar macrophage fibronectin: Synthesis, secretion, and ultrastructural localization during gelatin-coated latex particle binding, *J. Cell. Biol.* 90:711–720.

Wahl, L. M., Olsen, C. E., Sandberg, A. L., and Mergenhagen, S. E., 1977, Prostaglandin regulation of macrophage collagenase production, *Proc. Natl. Acad. Sci. USA* 74:4955–4958.

Whaley, K., 1980, Biosynthesis of the complement components and the regulatory proteins of the alternative complement pathway by human peripheral blood monocytes, *J. Exp. Med.* 151:501–516.

White, R., Janoff, A., and Godfrey, H. P., 1980, Secretion of alpha-2-macroglobulin by human alveolar macrophages, *Lung* 158:9–14.

White, R., Leed, D., Habicht, G. S., and Janoff, A., 1981a, Secretion of alpha₁-proteinase inhibitor by cultured rat alveolar macrophages, *Am. Rev. Respir. Dis.* 123:447–449. ·

White, R., Habicht, G., S., Godfrey, H. P., Janoff, A., Barton, E., and Fox, C., 1981b, Secretion of elastase and alpha-2-macroglobulin by cultured murine peritoneal macrophages: studies on their interaction, *J. Lab. Clin. Med.* 97:718–729.

Wyatt, H. V., Colten, H. R., and Borsos, T., 1972, Production of the second and fourth components of guinea pig complement by single peritoneal cells: Evidence that one cell may produce both components, *J. Immunol.* 108:1609–1614.

Chapter 11

Surveillance Role of Various Leukocytes in Preventing the Outgrowth of Potentially Malignant Cells

James Urban and Hans Schreiber

LaRabida-University of Chicago Institute
Department of Pathology and Committee on Immunology
The University of Chicago
Chicago, Illinois 60649

I. INTRODUCTION

Several different types of leukocytes can destroy cancer cells *in vitro* (Klein *et al.*, 1960; Hibbs *et al.*, 1972; Herberman *et al.*, 1975; Kiessling *et al.*, 1975). Rather little, however, is known about the relative importance of these different leukocytes in the normal host, where they may exert a surveillance function and prevent the outgrowth of potentially malignant cells. We have studied the relative efficiency of the different leukocytes in restraining malignant growth *in vivo* by comparing the effects of the leukocytes on regressor tumors and on progressively growing tumor variants that have escaped the immunity of the host. This type of approach is based on the premise that if a leukocyte operates effectively *in vivo* in restraining the growth of a tumor, a tumor cell must become resistant to it before it can grow progressively. A study of such phenotypic changes in tumor variants should therefore give insight into the relative importance and hierarchy of the different naturally occurring immune defense cells. This type of analysis is analogous to that performed by the microbiologist who deduces the mechanism of action of an antibiotic from the type of change found in a bacterium that has become resistant to the drug.

The most stringent conditions for this type of analysis would involve the study of variant tumors that have escaped a highly effective natural host resistance. Such a model is given by the murine fibrosarcomas induced experimentally by ultraviolet (UV) radiation. These tumors are exceedingly immunogenic and are usually rejected when transplanted into normal syngeneic hosts after initial

225

growth during the first 10 days (Kripke, 1981). For example, in one of our experiments (Urban *et al.*, 1982*a*) the UV-induced regressor (RE) tumor 1591, designated 1591-RE for convenience, was completely rejected by all but 1 of 300 syngeneic C3H mice challenged with the tumor, despite the fact that none of the mice had been previously immunized and that multiple viable fragments had been used for challenge. (Such tumor fragments always grew progressively in athymic nude mice.) On all the rare occasions in which progressively growing tumors, or progressors (PRO), designated 1591-PRO, were observed, the tumors were reisolated and reinjected into additional normal mice to determine whether their growth capacity had been heritably altered. In every instance, it was found that these tumors had acquired a different heritably stable growth behavior in that they would now grow progressively in more than 80% of normal young mice injected with the tumor (Urban *et al.*, 1982*a*). It was also found that the frequency of generation of such variants could be increased by serially passing the parental 1591-RE tumor in immunodeficient mice before transplantation into normal mice (Urban *et al.*, 1982*a*). Presumably, this occurred because the likelihood of generating a progressor variant is a direct function of the number of generations of tumor cells (Nowell, 1976).

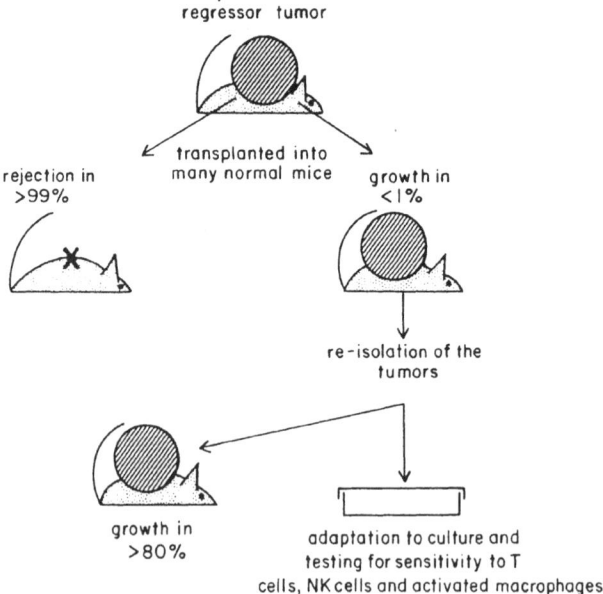

Figure 1. Protocol for the analysis of progressor variants, 1591-PRO, in terms of their growth potential in normal mice and their susceptibility to various immune effector cells. NK, Natural killer.

For the experiments outlined in Sections II and III, we used a number of the 1591-PRO tumors that had been isolated independently and that showed a heritable progressor phenotype. In all instances, cell lines had been readily established *in vitro* from the original variant progressor tumor, immediately grown up to mass culture, and frozen in aliquots for later use (Urban *et al.*, 1982*a,b*; Urban and Schreiber, 1983). These variant tumor cell lines and the parental 1591-RE tumor cells were then tested *in vitro* for sensitivity to cytolytic T cells, natural killer (NK) cells, and activated macrophages (Fig. 1). This way, we could determine whether the variants had become resistant to one or several classes of leukocytes that supposedly operate *in vivo*.

II. TUMOR-SPECIFIC T CELLS

Injection of the 1591-RE tumor into normal mice elicits a strong and highly 1591-specific T-cell response. For example, cytolytic T cells can be generated from spleen cells of these animals in a mixed lymphocyte–tumor cell culture (MLTC). These T cells lyse 1591-RE tumor cells selectively but lyse none of the numerous other currently available (non-1591) UV-induced regressor tumor cell lines that are antigenically different. We were interested in determining whether the progressor tumor variants could still be lysed by such 1591-RE-specific T cells. Figure 2 shows that the variant 1591-PRO tumors had become resistant to 1591-RE-specific T cells; in fact, the 1591-PRO tumors were as insensitive as control regressor fibrosarcomas, which were also UV-induced but unrelated to 1591-RE.

The *in vitro* generation of 1591-RE-specific cytolytic T cells depends on reexposure to the relevant antigen. We could therefore also test whether the 1591-PRO tumors expressed the 1591-RE-specific antigen on their surface by determining whether they could restimulate 1591-RE-specific T cells *in vitro*. We found that the progressor cells failed to restimulate primed 1591-RE-specific T cells in culture, which suggested further that the progressor tumors had lost the regressor-specific tumor antigen (Urban *et al.*, 1982*a*). Thus, it appears that the expression of the 1591-RE-specific antigen recognized by tumor-specific cytolytic T cells is essential for the resistance of the unimmunized host to the 1591-RE tumor. These results are complementary to our earlier studies in which anti-idiotypic immune suppression was shown to suppress the capability of mice to reject a primary 1591-RE tumor challenge selectively and to respond to 1591-RE cells *in vitro* (Flood *et al.*, 1980,1981). In these studies anti-idiotypic immunity suppressed both the proliferative as well as the cytolytic responses to 1591-RE, suggesting that helper and cytolytic subclasses of 1591-RE-specific T lymphocytes, either alone or together, may be important for normal mice to reject the 1591-RE tumor.

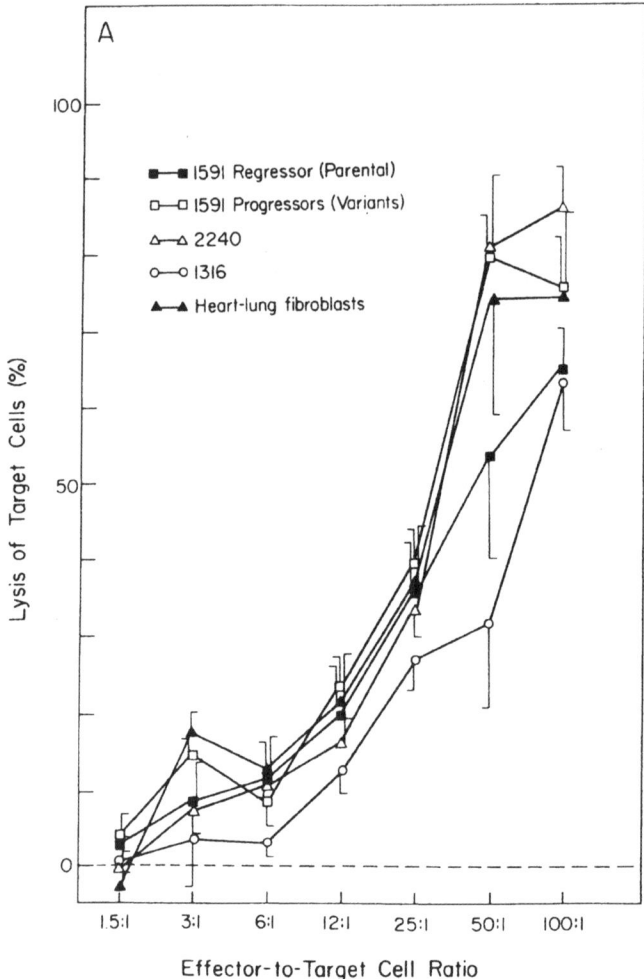

Figure 2. (A) Cross-reactive cytolytic immunity of animals injected with progressor tumor cells.

III. NATURAL KILLER CELLS

1591-RE-specific T cells can be generated directly *in vivo* in the peritoneal cavities of tumor-injected mice. It was therefore of interest to compare the capacity of 1591-RE and 1591-PRO tumors to induce such a response. Figure 3 shows that a single injection of the 1591-RE tumor induced highly specific cytolytic reactivity against the regressor tumor, which did not lyse 1591-PRO tumors or

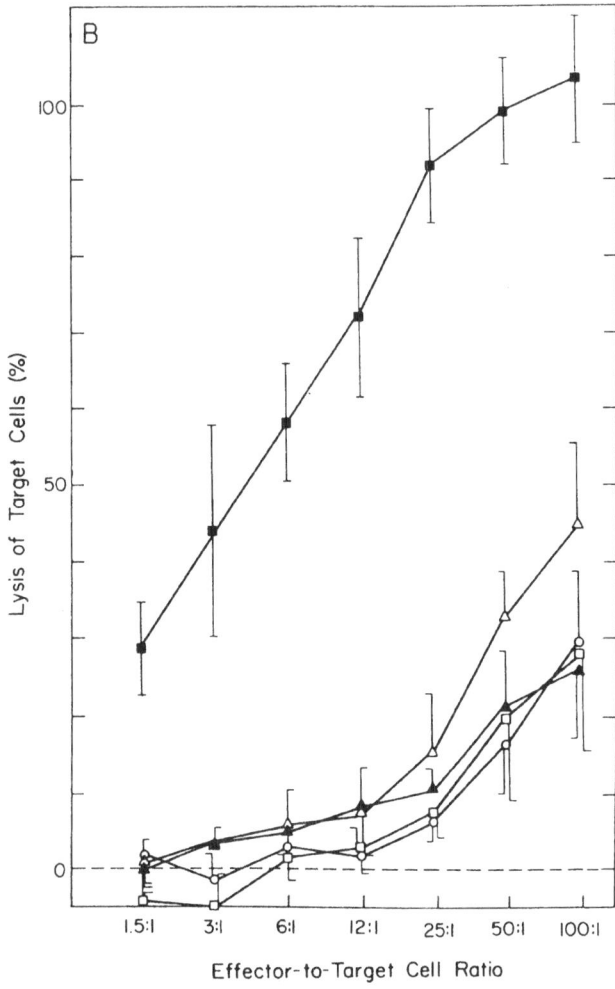

Figure 2. (*Continued*) (B) Specific cytolytic immunity of animals injected with regressor tumor cells. Spleen cells were removed 20–40 days after tumor challenge, restimulated for 6 days in culture with the same cell line used for *in vivo* injection, and then used as effectors in a 6-hr ^{51}Cr-release assay. Thirteen animals injected with progressor tumor cells and 11 animals injected with regressor tumor cells were analyzed individually in nine separate experiments; a total of five progressor cell lines were tested separately as targets in these experiments. The results of these experiments were similar and were therefore pooled. Vertical bars indicate the (SE) standard error of the results for individual animals. (From Urban *et al.,* 1982*a*.)

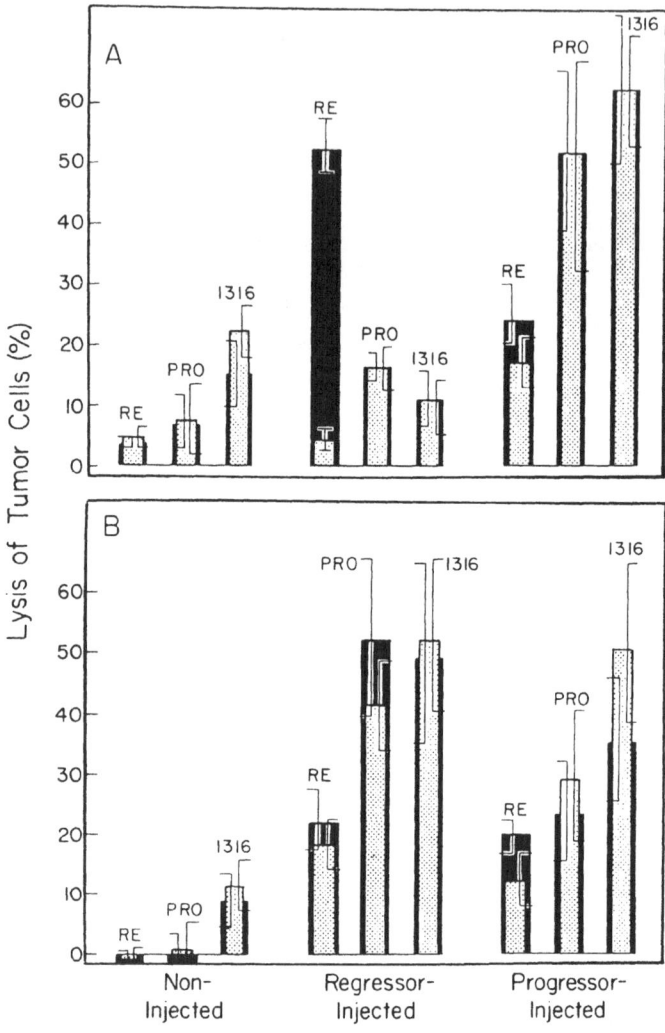

Figure 3. (A) In normal C3H mice, induction of cross-reactive cytolytic immunity by intra-peritoneal injection of progressor tumor cells 8 days previously and induction of tumor-specific cytolytic immunity by a similar injection of regressor tumor cells. Vertical bars indicate the percentage lysis of target cells–1591-RE (parental) tumor cells, 1591-PRO (variant) tumor cells, and 1316 RE tumor cells–in a 6-hr ^{51}Cr-release assay after treatment of the effector cells with anti-Lyt-2 and complement (ϖ) or after treatment with complement alone ($\blacksquare$). Progressor targets are totally insensitive to Lyt-2^{+} regressor specific T cells, and regressor targets are less sensitive than progressor targets to Lyt-2^{-} lymphocytes. Vertical

Table I. Differentiation Markers on Natural Killer Lymphocytes and Increase of Natural Killer Activity after Injection of the Host with 1591 Progressor Tumor Cells[a,b]

Pretreatment of host[c]	Source of effector cells	Pretreatment of effector cells[d]	Percentage specific lysis of target cell[e]			
			1591-PRO cells		YAC cells	
			6 hr	16 hr	6 hr	16 hr
Noninjected	Peritoneal cavity	C' alone	10	17	38	53
		C' + anti-Lyt 2	10	19	30	62
		C' + anti-NK 1.2	2	11	<0	<0
	Spleen	C' alone	10	15	66	74
		C' + anti-Lyt 2	11	16	66	81
		C' + anti-NK 1.2	7	2	6	7
1591-PRO-injected	Peritoneal cavity	C' alone	45	83	85	94
		C' + anti-Lyt 2	37	82	81	90
		C' + anti-NK 1.2	7	20	28	17
	Spleen	C' alone	20	56	93	100
		C' + anti-Lyt 2	19	51	87	86
		C' + anti-NK 1.2	6	16	15	21

[a]From Urban et al. (1982a).
[b]Abbreviations: NK, natural killer; PRO, progressor.
[c]C3H mice were injected intraperitoneally with 10^7 live 1591-PRO.2 progressor tumor cells. Spleen or peritoneal exudate cells were removed and tested 8 days after injection.
[d]Effector cells were depleted of adherent cells and then treated with antisera and/or complement before testing in a ^{51}Cr-release assay.
[e]Effector cells were tested in a 6-hr and 16-hr ^{51}Cr-release assay using 1591-PRO.2 progressor or YAC tumor cells as target cells at a 200:1 effector:target ratio.

an unrelated UV-induced regressor tumor. Apparently, this reactivity was mediated by cytolytic T cells, as it was almost entirely eliminated by pretreatment of the effectors with a monoclonal Lyt-2-specific antibody and complement. Interestingly, 1591-PRO tumor cells were capable of inducing a similar level of cytolytic reactivity, but this reactivity was different because it lysed unrelated UV-induced tumors nonspecifically and was not eliminated by pretreatment with the

bars show the standard error (SE) of values obtained from the individual analysis of animals injected with regressor tumor cells (11 animals), injected with progressor tumor cells (eight animals), or noninjected (five animals), obtained in four independent experiments. The effector:target ratio was 200:1. (B) In nude C3H mice, induction of cross-reactive cytolytic immunity by injection of either progressor or regressor tumor cells, individually tested in three independent experiments (one animal per injection per experiment). Target cells: PRO, progressor; RE, regressor. Other details as in (A). (From Urban et al., 1982a.)

anti-Lyt-2 antibody and complement. A similar nonspecific cytolytic reactivity could also be induced in nude mice by injection of the 1591-RE as well as by injection of the 1591-PRO tumor. This finding suggested that the cytolytic effect might not be attributable to T cells, but rather to NK cells. These 1591-PRO-induced cytolytic effector cells were therefore tested for cytolysis of the NK-sensitive YAC tumor target cells [a tissue culture of a Moloney virus-induced lymphoma of A/Sn origin (Cikes *et al.*, 1973)] and for susceptibility to an NK-specific antiserum and complement. As shown in Table I, these 1591-PRO-induced 1591-PRO-reactive leukocytes not only lysed 1591-PRO target cells, but YAC tumor cells, as well, and their activity was almost totally removed by pre-treatment with anti-NK-1.2 antibody and complement. Such cytolytic activity was also present, although at a considerably lower level, in mice not previously injected with the 1591-PRO tumor. Taken together, these results indicate that no selection against NK activity occurred during the evolution of progressor variants from the parental 1591-RE tumor. In fact, tumor cells of the different progressor variants induced NK activity more effectively than did the parental regressor tumor cells. We do not know whether the increased capacity of our progressor variants to induce and to be killed by NK cells in an obligatory event occurring simultaneously with the loss of a strong tumor-specific antigen. If tumor cells commonly show an increased susceptibility to NK cells after the loss of a strong tumor antigen, NK cells may have a protective role against local or metastatic growth of tumors that escape tumor-specific immunity. NK cells have been reported to protect against the outgrowth of UV-induced progressor tumor cells that have entered the vascular circulation (Hanna and Fidler, 1980; Hanna and Burton, 1981). Nevertheless, we have found that the progressor variants grow faster in nude than in normal animals (Urban *et al.*, 1982a), although nude mice have equal or higher levels of NK activity.

IV. ACTIVATED MACROPHAGES

Macrophages (Mϕ) can show highly selective cytotoxicity toward malignant cells *in vitro* (Hibbs *et al.*, 1972), and there is some evidence that they may de-stroy neoplastic cells *in vivo* (for review see Adams and Snyderman, 1979). We therefore used the above-mentioned approach of comparative analysis of regres-sor tumors and their progressor tumor variants in order to determine the role of the Mϕ in the resistance of normal mice to 1591-RE tumor cells *in vivo* (Urban and Schreiber, 1983). Mϕ were activated by priming purified thioglycollate-elicited peritoneal exudate cells with lymphokine-containing supernatant for 5 hr *in vitro* (Meltzer, 1981). These activated Mϕ were used as effector cells in a

16-hr ^{51}Cr-release assay to test for possible differences in Mϕ sensitivity between the 1591-RE parental tumor and the host-selected 1591-PRO variants. Figure 4B shows that all of eight host-selected 1591-PRO tumors showed a considerably decreased sensitivity to activated Mϕ, as compared with the 1591-RE tumor (Fig. 4A). Interestingly, sensitivity to activated Mϕ was not decreased if progressively growing tumors were reisolated from mice incapable of mounting tumor-specific T-cell-mediated immunity. Thus, tumors isolated from nude mice (1591-NU tumors), from mice in which the 1591-specific T cells had been suppressed selectively by the induction of anti-idiotypic immunity (1591-ID tumors) (Flood et al., 1980), or from mice that carried a progressively growing tumor and that simultaneously failed to mount a tumor-specific T-cell response (1591-PB tumors) were as sensitive as the parental 1591-RE tumor to activated Mϕ, indicating that tumor-specific T-cell reactivity might be required in vivo before Mϕ could select for resistant variants. Alternatively, it was possible that loss of Mϕ sensitivity was not related to selection pressure exerted by Mϕ in vivo, but rather to a physical linkage of the Mϕ target site with the tumor antigen. In order to determine whether these two target sites were independent, parental 1591-RE tumor cells were exposed repeatedly to purified activated Mϕ in vitro, and Fig. 4C shows that such tumor cells became highly resistant to activated Mϕ (1591-Mϕ tumors). This resistance was heritably stable, since passage of the 1591-Mϕ.1 tumor through a nude mouse (1591-Mϕ.1.NU tumor) did not result in a return of sensitivity to activated Mϕ (Fig. 4C). As shown in Figure 5C, however, these same 1591-Mϕ variants fully retained their sensitivity to 1591-RE-specific cytolytic T cells and thus retained the 1591-RE-specific antigen, in contrast to the host-selected 1591-PRO variants that had also lost the T-cell-recognized antigen (Fig. 5B). These data show that the Mϕ target site could be lost independently of the tumor antigen in vitro, despite the fact that the T-cell-recognized and Mϕ-recognized target sites were regularly lost together after selection in vivo by the host.

In order to explore further how changes in Mϕ sensitivity and T-cell sensitivity influenced tumor rejection, we compared the Mϕ-resistant variants with the parental 1591-RE tumor in terms of their growth behavior in normal and nude mice. As shown in Fig. 6, each of two Mϕ-resistant variants injected into normal mice grew significantly faster initially than the parental tumor that was rapidly rejected. This difference in growth kinetics was not due to a difference in general growth potential of the different tumor cells in vivo, since no differences in growth kinetics were observed in nude mice. Despite an early rapid growth phase in normal animals, the Mϕ-resistant variants were finally also rejected, possibly because of recognition of the tumor-specific antigen on their surface by cytolytic T cells. These results, taken together, indicated that Mϕ play a role in restraining tumor growth during the early stages of tumor development, at a time before an effective cytolytic T-cell-mediated tumor immunity may have been established.

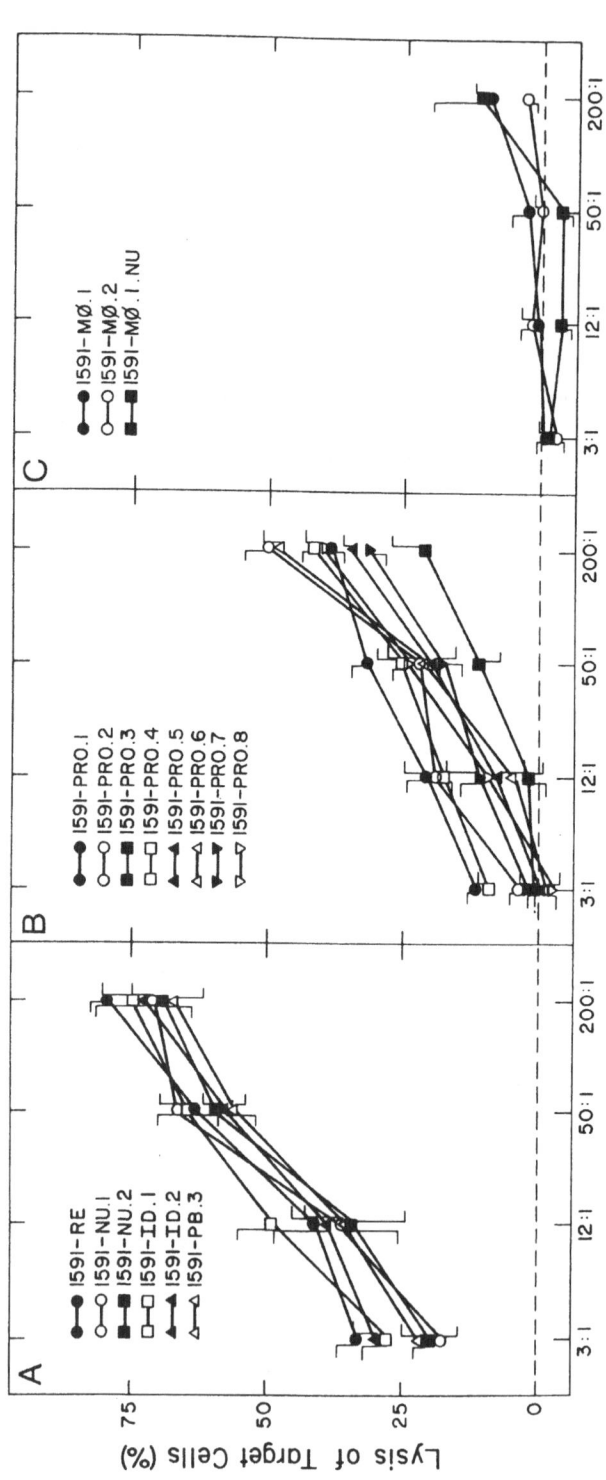

Figure 4. (A–C) Decreased sensitivity to Mφ-mediated lysis in host-selected progressor tumor variants (1591-PRO) isolated after one passage of the parental 1591 regressor tumor (1591-RE) through mice possessing immunocompetent T cells. The 1591-RE tumor was implanted subcutaneously into mice. One month later, the progressing tumors were adapted to growth *in vitro*. The resulting cell lines were compared for Mφ sensitivity in a 16-hr ^{51}Cr-release assay. Mφ, Macrophage. Target cells; ID, idiotypically suppressed; NU, nude ; PB, progressor bearing; PRO, progressor. (From Urban and Schreiber, 1983.)

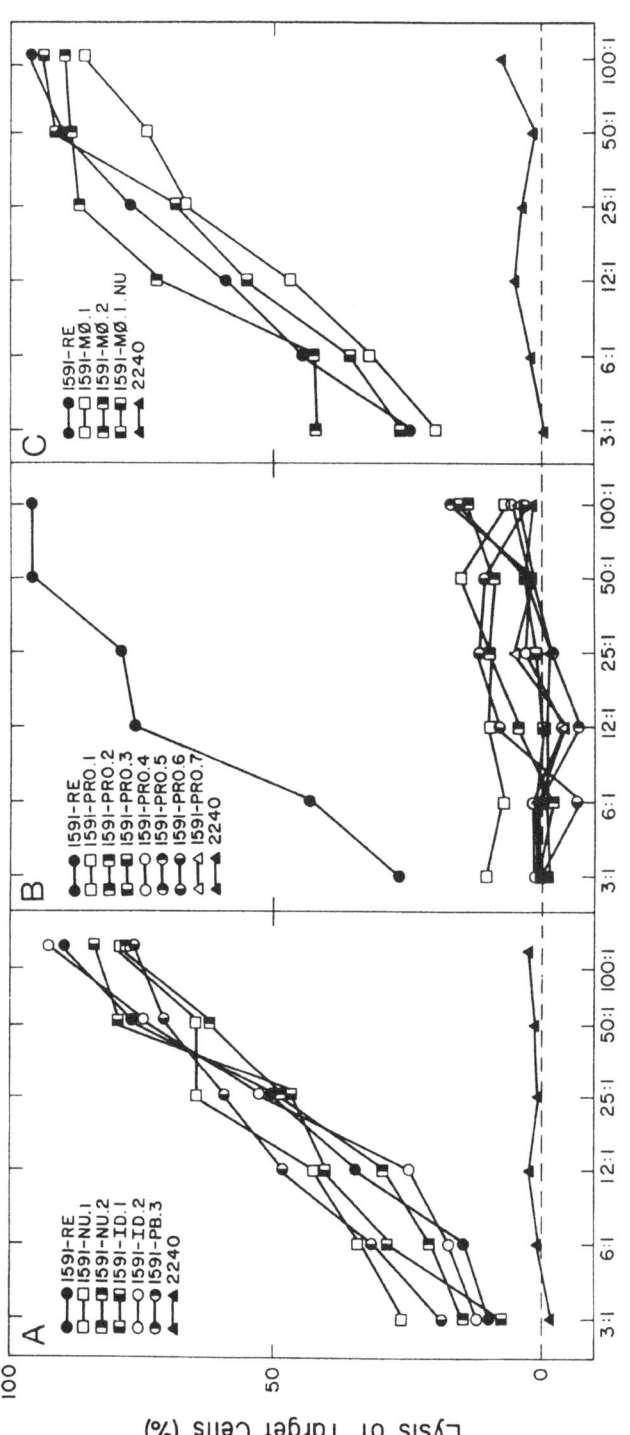

Figure 5. Retention of the 1591-RE tumor-specific target antigen by the Mɸ-resistant 1591-Mɸ tumor cells. The specific T cells were generated in mixed lymphocyte-tumor cell cultures using spleen cells from 1591-RE-immunized mice. All experiments shown in (A), (B), and (C) were done independently of each other and were therefore independently controlled with 1591-RE and 2240-RE target cells used in each of the experiments. Abbreviations as in Fig. 4. (From Urban and Schreiber, 1983.)

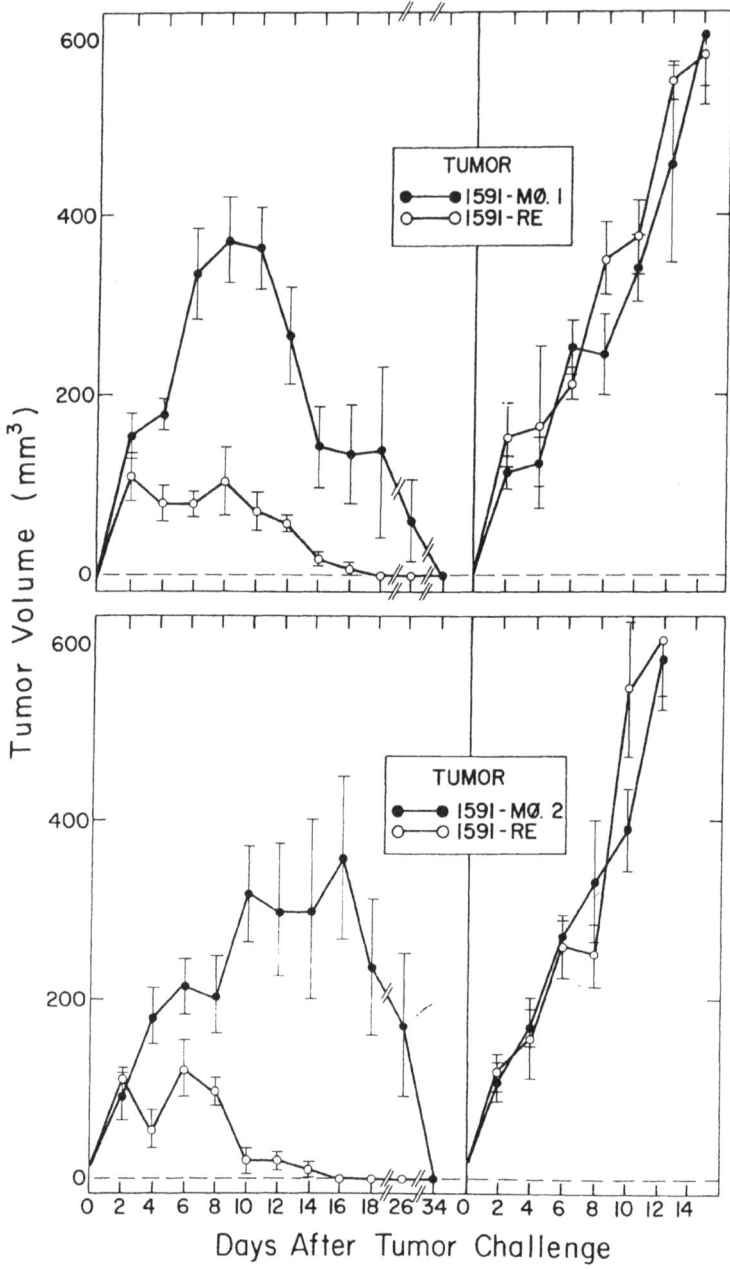

Figure 6. Differences in growth kinetics of the parental 1591-RE tumor and the variant 1591-Mφ tumors in normal but not in nude mice. The tumors were implanted subcutaneously in the shaved inguinal fossae of groups of 10 normal or 4 nude mice each, two 1-mm^3 fragments per mouse. The mean tumor size per mouse was measured with a caliper every 2 days

V. COOPERATION BETWEEN T CELLS AND MACROPHAGES

We have shown that selection for Mϕ-resistant variants only occurred in hosts capable of mounting a tumor-specific T-cell-mediated immune response. This suggested that the activation of Mϕ to become tumoricidal might require T cells that recognize the tumor-specific antigen on the tumor cell surface. Consequently, one should expect that 1591 tumor cell variants lacking the T-cell-recognized tumor antigen should not evoke a Mϕ-mediated cytotoxic response. This hypothesis was tested using tumor cell lines obtained by exposing 1591-RE parental tumor cells *in vitro* to cytolytic T-cell clones directed against the 1591-RE-specific target antigen (Wortzel *et al.*, 1983). Such variants lost this antigen and acquired progressive growth behavior in normal mice, but retained normal sensitivity to activated Mϕ (Br-1 tumor) (Fig. 7). This normal level of sensitivity to activated Mϕ was not altered by passage through normal mice (Br-1-NORM tumor). These results suggested that T cells needed to be attracted to the tumor by a strong tumor-specific antigen in order to induce an effective Mϕ-mediated immune-defense cytotoxic response capable of exerting a selective pressure on the tumor.

Although the precise frequency of Mϕ-resistant variants in the parental population was not determined in our experiments, it appeared to be a rare event. For example, in repeated experiments only 3–10 clones of tumor cells survived an exposure to activated Mϕ in a plate initially seeded with $>10^5$ tumor cells. This finding suggested that the frequency of Mϕ-resistant variants was in the range of <1 in 10^4 cells and was therefore similar to that of T-cell-resistant variants (Wortzel *et al.*, 1983). Having established that loss of the Mϕ target site is independent from that of the T-cell antigen, one would expect that the frequency of variants losing both target structures would be less than 1 in 10^8 cells. This low frequency means that such variants are unlikely to occur *in vivo* unless the selections occur sequentially in the host. The possibility of such sequential selection was also suggested by the growth kinetics of the Mϕ-resistant T-cell-susceptible variants in· normal mice described above. These data had indicated that Mϕ-mediated suppression of tumor cell growth probably preceded direct T-cell-mediated destruction of the tumor. We therefore designed an experiment to explore the possibility that the Mϕ-recognized target site was lost before the T-cell-recognized antigen. It was impossible to examine this question under conditions in which variants arose only sporadically, so conditions were chosen under which tumor-escape variants could be observed with much greater frequency. Such conditions exist in mice that are partially immunosuppressed by UV expo-

thereafter and the tumor volume was computed as the average product of three perpendicular tumor diameters. Vertical bars represent the standard error (SE) for the individual analysis of the number of mice indicated. Abbreviations as in Fig. 4. (Form Urban and Schreiber, 1983.)

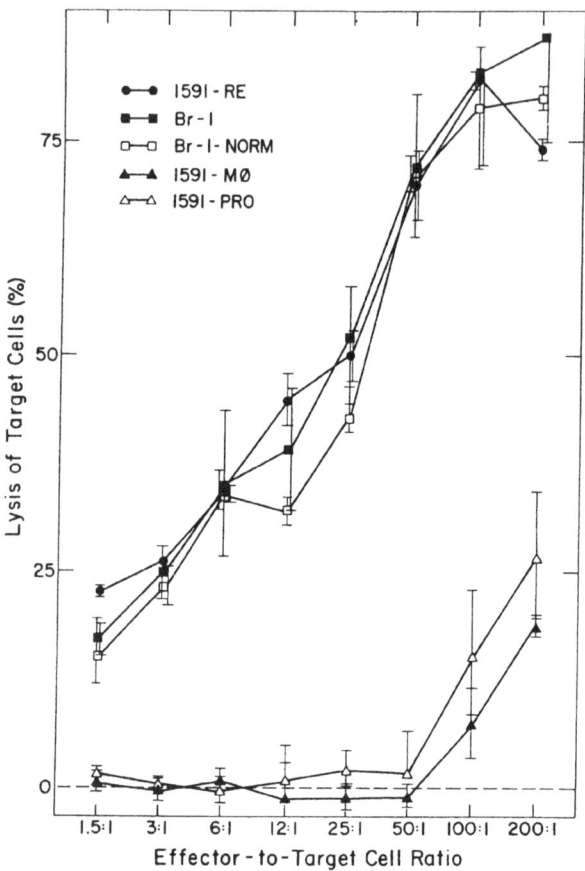

Figure 7. A 1591 tumor variant lacking the 1591 regressor-specific antigen (Br-1) does not lose sensitivity to activated Mφ during passage through normal mice (Br-1-NORM). The Br-1 variant tumor was derived by exposing parental 1591-RE tumor cells to T cells specific for the 1591-RE tumor *in vitro* (Wortzel *et al.*, 1983). The Br-1 tumor was then injected into normal immunocompetent C3H mice, two 1-mm^3 fragments per mouse, and the resulting progressively growing tumor was reisolated, adapted to tissue culture, and then tested for sensitivity to activated Mφ in a 16-hr ^{51}Cr-release assay. The Mφ-sensitive 1591-RE and Br-1 tumors and the Mφ-resistant 1591- Mφ and 1591-PRO tumors were tested in the same experiments for comparison. The data represent the mean ± SE for two separate experiments. Abbreviations as in Fig. 4. (From R. D. Wortzel, J. L. Urban, and H. Schreiber, unpublished results.)

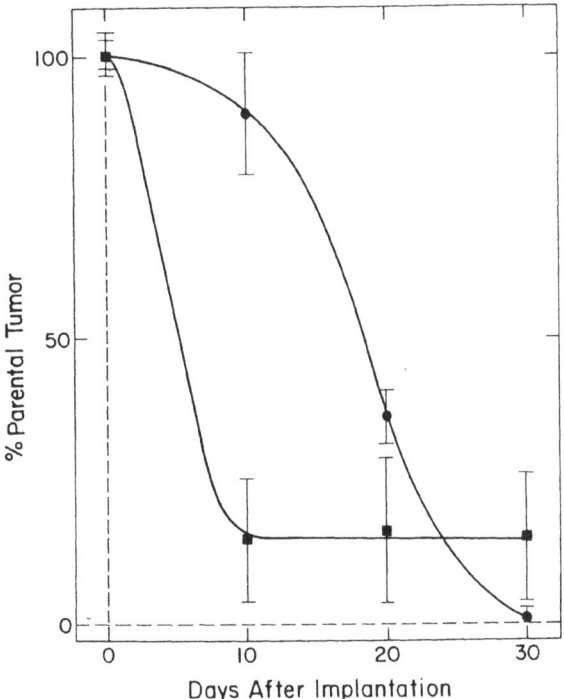

Figure 8. Sequential loss of sensitivity to activated Mϕ and T cells during immunoselection *in vivo*. Ultraviolet (UV)-treated mice were injected with 5×10^7 1591-RE tumor cells; at various times thereafter, the progressing tumors were excised and adapted to tissue culture. The cell lines were tested for sensitivity to activated 1591-RE-specific T cells (■) in a 6-hr ^{51}Cr-release assay and for sensitivity to Mϕ (•) in a 16-hr ^{51}Cr-release assay. Relative sensitivity was the reciprocal of the number of T cells or Mϕ required to lyse 50% of 1×10^4 tumor target cells and was expressed as a percentage of the sensitivity of the parental 1591-RE tumor. Vertical bars indicate the SE for the analysis of four independently isolated tumor cell lines.

sure; these mice retain an inadequate but significant level of tumor-specific immunity. This host environment seems to favor the frequent (>80%) outgrowth of antigen-loss variants (Urban *et al.*, 1982*b*) and allowed us to compare the kinetics of the loss of the Mϕ target site with that of the tumor-specific antigen. Thus, UV-treated mice were challenged with 1591-RE tumor fragments. At various times after tumor challenge, the developing tumors were reisolated and analyzed for sensitivity to Mϕ and tumor-specific T cells. As shown in Figure 8, loss of Mϕ sensitivity preceded the loss of the 1591-RE tumor-specific antigen by several days.

VI. SUMMARY

We have found that when potentially malignant cells escape the immune defenses of the normal host they become resistant to activated Mϕ and to tumor-specific T cells, yet retain their sensitivity to NK cells. This observation seems to imply that NK cells do not play a role in the resistance of the normal immuno-competent host against this tumor, since they were neither necessary nor sufficient for the rejection of this tumor, and since progressor variants both induced and were as susceptible to NK cells as was the parental tumor. In contrast, the regular selection for resistance to activated Mϕ and regressor-specific T cells strongly argues for the relevance of these two types of leukocytes in the resistance of the normal host against the tumor. Apparently Mϕ and T cells acted together in a concerted manner, since T cells were required for the exertion of selective pressure by activated Mϕ. Thus, if the host did not mount a tumor-specific T-cell response, because of suppression of the T cells or because of absence of the tumor antigen, selection for Mϕ-resistant variants was not observed. However, we also found that the T-cell-recognized antigen was lost by the progressor variants after the loss of the Mϕ target site. This finding suggests that T cells were required for the early stages of Mϕ-mediated tumor rejection, although they did not exert a direct selective pressure upon the tumor cells at this time.

It is attractive to think that Mϕ-activating factors produced by T cells were needed for the *in vivo* effect of Mϕ. Such factors can activate Mϕ *in vitro* (Meltzer, 1981; Pace and Russell, 1981) and are known to be produced *in vitro* by the Lyt-1 positive subset of T cells. This subset of T cells has been implied to have a significant importance in the transplantation resistance to tumors and allografts (for review, see Loveland and McKenzie, 1982). Thus, Mϕ appeared to suppress tumor growth until a sufficient number of cytolytic T cells had been activated to eliminate the tumor. A restraint of tumor growth at an early stage may be important in reducing the chance of later development of progressor tumor variants, since the frequency of such variants is thought to be a direct function of the number of generations of tumor cells (Nowell, 1976). Thus, our data suggest that T cells and Mϕ can act together in the normal host in preventing the outgrowth of overt cancer from potentially malignant cells.

ACKNOWLEDGMENTS

We thank Dr. D. A. Rowley for critically reviewing this manuscript. This study was supported by grants ROI-CA-27326-01, CA-22677, and P01CA-19266 from the National Institutes of Health, by Research Career Development Award

CA-00432 from the U.S. Public Health Service, and by grant PHS 5-7-32-GM0-7281-04 from the National Institute of General Medical Sciences and by the National Research Service Award 5T32-AI-07090.

VII. REFERENCES

Adams, D. O., and Snyderman, R., 1979, Do macrophages destroy nascent tumors? *J. Natl. Cancer Inst.* **62:**1341–1345.

Cikes, M., Friberg, S. Jr., and Klein, G., 1973, Progressive loss of H-2 antigens with comcommitant increase of cell surface antigen(s) determined by Moloney leukemia virus in cultered murine lymphomas, *J. Natl. Cancer Inst.* **50:**347–362.

Flood, P. M., Kripke, M. L., Rowley, D. A., and Schreiber, H., 1980, Suppression of tumor rejection by autologous anti-idiotypic immunity, *Proc. Natl. Acad. Sci. USA* **77:**2209–2213.

Flood, P. M., Urban, J. L., Kripke, M. L., and Schreiber, H., 1981, Loss of tumor-specific and idiotype-specific immunity with age, *J. Exp. Med.* **154:**275–290.

Hanna, N., and Burton, R. D., 1981, Definite evidence that natural killer cells inhibit experimental tumor metastasis *in vivo, J. Immunol.* **127:**1754–1758.

Hanna, N., and Fidler, I. J., 1980, Role of natural killer cells in the destruction of circulating tumor emboli, *J. Natl. Cancer Inst.* **65:**801–809.

Herberman, R. B., Nunn, M. E., and Lavrin, D. H., 1975, Natural cytotoxic reactivity of mouse lymphoid cells against syngeneic and allogeneic tumors. I. Distribution of reactivity and phenotype, *Int. J. Cancer* **16:**216–229.

Hibbs, J. B., Jr., Lambert, L. H., Jr., and Remington, J. S., 1972, Control of carcinogenesis: A possible role for the activated macrophage, *Science* **177:**998–1000.

Kiessling, R., Klein, E., and Wigzell, H., 1975, Natural killer cells in the mouse. I. Cytotoxic cells with specificity for mouse leukemia cells. Specificity and distribution according to genotype, *Eur. J. Immunol.* **5:**112–117.

Klein, G., Sjögren, H. O., Klein, E., and Hellström, K. E., 1960, Demonstration of resistance against methylcholanthrene-induced sarcomas in the primary autochthonous host, *Cancer Res.* **20:**1561–1572.

Kripke, M., 1981, Immunoglobulin mechanisms in UV radiation carcinogenesis, *Adv. Cancer Res.* **34:**69–106.

Loveland, B. E., and McKenzie, I. F. C., 1982, Which T cells cause graft rejection? *Transplantation (Baltimore)* **33:**217–221.

Meltzer, M. S., 1981, Macrophage activation for tumor cytotoxicity: Characterization of priming and trigger signals during lymphokine activation, *J. Immunol.* **127:**179–183.

Nowell, P. C., 1976, The clonal evolution of tumor cell populations. Acquired genetic lability permits stepwise selection of variant sublines and underlies tumor progression, *Science* **194:**23–28.

Pace, J. L., and Russell, S. W., 1981, Activation of mouse macrophages for tumor cell killing. I. Quantitative analysis of interactions between lymphokine and lipopolysaccharide, *J. Immunol.* **126:**1863–1867.

Urban, J. L., and Schreiber, H., 1983, Selection of macrophage-resistant progressor tumor variants by the normal host. Requirement for concomitant T cell-mediated immunity, *J. Exp. Med.* **157:**642–656.

Urban, J. L., Burton, B. C., Holland, J. M., Kripke, M. L., and Schreiber, H., 1982a, Mechanisms of syngeneic tumor rejection: Susceptibility of host-selected progressor variants to various immunological effector cells, *J. Exp. Med.* 155:557–573.

Urban, J. L., Holland, J. M., Kripke, M. L., and Schreiber, H., 1982b, Immunoselection of tumor cell variants by mice suppressed with ultraviolet radiation, *J. Exp. Med.* 156: 1025–1041.

Wortzel, R. D., Urban, J. L., Philipps, C., Fitch, F. W., and Schreiber, H., 1983a, Independent immunodominant and immunorecessive tumor-specific antigens on a malignant tumor. Antigenic dissection with cytolytic T cell clones, *J. Immunol.* 130:2461–2466.

Wortzel, R. D., Philipps, C., and Schreiber, H., 1983b, Multiple tumor-specific antigens on a single malignant cell, *Nature* 304:165–167.

Chapter 12

Models of Adoptive T-Cell-Mediated Regression of Established Tumors

Robert J. North

Trudeau Institute, Inc.
Saranac Lake, New York 12983

I. INTRODUCTION

Most of the chapters in this volume deal with the physiology and protective functions of macrophages as revealed by *in vitro* assays. Evidence that activated macrophages can express tumoristatic or tumoricidal function *in vitro* has been accumulating for 15 years or so and has resulted in the suggestion (Adams and Snyderman, 1979) that macrophages play a role in protecting against neoplastic colonization. It might be inferred, on the basis of *in vitro* evidence, moreover, that macrophages have the potential to destroy an established growing tumor, provided ways could be found to cause these cells to acquire tumoricidal function in the tumor bed. There is evidence (Evans, 1972; Russel and McIntosh, 1977) that progressive tumors can contain surprisingly large numbers of macrophages. It also has been shown (Russel and McIntosh, 1977) that Moloney sarcomas undergoing spontaneous regression in syngeneic mice contain macrophages that, on isolation, can lyse tumor cells *in vitro*. However, because of the reductionistic nature of the evidence obtained, no number of results obtained with *in vitro* assays can permit the conclusion that macrophages destroy tumors *in vivo*. The same criticism can be leveled against evidence showing that tumor-sensitized cytolytic T cells and natural killer (NK) cells lyse tumor cells *in vitro*. Sooner or later experiments will have to be designed to determine whether any one of these types of host cells, in the absence of the others, can express tumoricidal function *in vivo* and destroy an established tumor. The best models to employ in an attempt

243

to identify the ultimate effectors of tumor regression would be models of immunologically mediated regression of established tumors.

However, convincing models of immunologically mediated regression of established syngeneic tumors are far from numerous. In fact, until fairly recently, the only evidence for immunologically mediated tumor regression has come from a few convincing examples of the regression of established tumors after immunotherapy with immunoadjuvants, such as bacillus Calmette-Guérin (BCG) and *Corynebacterium parvum* (Zbar *et al.*, 1972). These are difficult models of tumor regression to analyze, however, because the generation and expression of specific T-cell-mediated antitumor immunity that results from intralesional injection of immunoadjuvants almost certainly depend on the generation and expression of an immune response to the intralesional adjuvant itself (Bast *et al.*, 1976). A cleaner model of immunologically mediated tumor regression would be one based on tumor regression caused by the passive transfer of tumor-sensitized T cells.

The purpose of this chapter is to show that models of adoptive immunotherapy of established syngeneic tumors now exist. It is also shown that the success of attempts to demonstrate that passively transferred tumor-sensitized T cells can cause the regression of established tumors depends on eliminating a tumor-induced mechanism of T-cell-mediated suppression from the tumor-bearing recipients.

II. TUMOR REGRESSION BY ADOPTIVE IMMUNIZATION WITH T CELLS FROM IMMUNIZED DONORS

The literature indicates that it has proved relatively easy to immunize mice adoptively against the growth of an implant of tumor cells with the T cells from appropriately immunized donors, particularly if the Winn neutralization assay is employed. Nevertheless, it has proved extremely difficult to demonstrate that passive transfer of the same T cells can cause the regression of a tumor once it is established and growing progressively (Rosenberg and Terry, 1977). Apparently something happens during the early growth of an immunogenic tumor to render it resistant to the antitumor function of passively transferred, tumor-sensitized T cells. Either an established tumor represents too large a burden for donor T cells to reject, or a tumor-induced mechanism is generated that blocks the antitumor function of donor T cells. The latter possibility seems likely, in view of the reported demonstrations (Rotter and Trainin, 1975; Fujimoto *et al.*, 1976; Reinisch *et al.*, 1977) that animals with progressive tumors acquire suppressor T cells. Obviously, if suppressor T cells are generated in tumor-bearing hosts, their generation would be a consequence of an immune response to the tumor. It should follow, that any procedure that depresses the capacity of a recipient animal to respond immunologically to its tumor should prevent the production of suppressor T cells and thereby enable passively transferred tumor-sensitized T cells to destroy its tumor. This line of reasoning has resulted in the development of three models of immunologically mediated destruction of established tumors by passively transferred, tumor-sensitized T cells.

It will be shown that adoptive T-cell-mediated tumor regression can be demonstrated by employing tumor-bearing recipient mice that have been (1) made T-cell-deficient by thymectony and lethal γ-irradiation and restored with bone marrow (T X B mice), (2) immunodepressed by treatment with cyclophosphamide, or (3) immunodepressed by exposure to sublethal whole-body γ-irradiation. Regardless of whether adoptive immunization is directed against a tumor implant or an established tumor, however, any attempt to immunize a recipient adoptively would be futile without a suitably immunized donor of tumor-sensitized T cells. It is therefore necessary first to describe the method used to immunize donors.

A. Immunization of Donors by a *Corynebacterium parvum*-Potentiated Response to Tumor Antigens

A favorite procedure for immunizing against an immuogenic tumor is to give repeated subcutaneous injections of tumor cells that have been heavily irradiated to prevent them from forming a tumor. Another method is to excise a tumor that has grown to a palpable size. Either immunization procedure can leave the host specifically immune to growth of a challenge implant and with lymphocytes capable of adoptively immunizing a recipient animal against growth of a tumor implant. For the adoptive immunization experiments discussed in this chapter, however, the immunization of donors involved causing the regression of a growing tumor by intratumor treatment with *C. parvum*. This model has been employed by others (Baldwin and Pimm, 1973) to demonstrate the therapeutic action of immunoadjuvants, such as BCG and *C. parvum* and involves injecting tumor cells as an admixture with *C. parvum*. In the case of the immunogenic murine tumors employed in this laboratory, the injection of tumor cells admixed with *C. parvum* results in the rapid emergence of a tumor that grows progressively for 9–10 days and that then undergoes progressive and complete regression. That tumor regression is immunologically mediated is evidenced by the findings (Dye *et al.*, 1981) that it is associated with the acquisition of immunity to a challenge implant, and with T cells capable of passively transferring systemic immunity to a tumor implant. Again, tumor regression fails to occur in T X B mice.

The possibility that *C. parvum* induces regression by augmenting an immune response to the tumor was investigated by Mills *et al.* (1981) in the case of the P815 mastocytoma. In *in vitro* experiments these workers followed the generation of T cells capable of lysing P815 tumor cells in the lymph node, draining the site of injection of 10^6 P815 cells and 100 μg of *C. parvum*. Mills *et al.* (1981) found that, whereas growth of a control tumor implant resulted in progressive tumor growth and a cytolytic T-cell response of low magnitude, growth of the tumor containing *C. parvum* resulted in a greatly enhanced cytolytic T-cell response that peaked on day 10, at the time of onset of tumor regression. After day 10, the potentiated cytolytic response rapidly decayed in concert with tumor regression. These results represent evidence that *C. parvum*-induced tumor regression results from the capacity of *C. parvum* to potentiate the production of T cells cytolytic for tumor cells. It is important to emphasize that these cytolytic T cells are rapidly lost and that they were no longer present at the time that the hosts were employed as donors for the experiments described in Sections II.B,C. By definition, therefore, the T cells that were harvested routinely for adoptive immunization were memory or helper T cells lacking an immediate capacity of their own to destroy a tumor.

B. Adoptive T-Cell-Mediated Regression of Tumors in T X B Recipients and Its T-Cell-Mediated Suppression

Based on the rationale that a T X B tumor bearer would be incapable of generating an immune response, or T cells that suppress an immune response, it was predicted that passive transfer of tumor-sensitized T cells would cause the regression of an established tumor in T X B recipients, but not in normal recipients. The experiments were performed with the Meth A fibrosarcoma syngeneic in BALB/c mice and with the P815 mastocytoma syngeneic in DBA/2 mice. Donors of sensitized T cells were immunized by causing the regression of their tumors by *C. parvum* therapy as described in section II.A. Spleen cells were harvested 3 weeks after tumor regression, and an organ equivalent of them, $\sim 1.5 \times 10^8$, was infused intravenously into T X B recipients or into immunocompetent recipients bearing 4-day intrafootpad tumors. The experiments revealed (Berendt and North, 1980; Dye and North, 1981) that, whereas passive transfer of sensitized spleen cells had little if any effect on the growth of tumors in immunocompetent recipients, the same number of sensitized spleen cells caused complete regression of tumors in all T X B recipients. Additional experiments revealed that the donor spleen cells that caused the regression of tumors in T X B recipients were T cells, as evidenced by their susceptibility to treatment with anti-Thy-1.2 antibody and complement. It should be mentioned, moreover, that passive transfer of sensitized T cells did not cause regression of tumors until after a 6-day delay, during which time the tumors were found to grow. The results show that the failure of passively transferred sensitized T cells to mediate the regression of tumors in immunocompetent recipients was caused by the presence in these recipients of a thymus-dependent, tumor-induced mechanism that inhibits the antitumor function of tumor-sensitized T cells.

It was reasoned next that if such a tumor-induced mechanism of immunosuppression is present in immunocompetent tumor bearers, it should be possible to reveal its presence by demonstrating that it can be passively transferred. This possibility was investigated by determining whether the spleens of immunocompetent mice bearing established tumors contain cells that are capable, on passive transfer, of inhibiting the ability of passively transferred immune T cells to cause the regression of tumors in T X B test recipients. The results of this experiment showed (Berendt and North, 1980; Dye and North, 1981) that intravenous infusion of one organ equivalent, $\sim 1.5 \times 10^8$, of spleen cells from immunocompetent tumor-bearing donors inhibited the ability of immune T cells to cause the regression of tumors in T X B recipients. In contrast, the same number of spleen cells from normal mice possessed no such inhibitory activity. Again, the spleen cells from tumor-bearers that inhibited adoptive immunotherapy of tumors in T X B recipients were susceptible to treatment with anti-Thy-1.2 antibody and complement. It may therefore be concluded that after a certain stage, the growth on an immunogenic tumor in its immunocompetent host results in the induction of a population of T cells that function to suppress the antitumor function of passively transferred immune T cells.

It perhaps is surprising that the aforementioned experiments were not performed before this time, considering that evidence has existed for a number of years to suggest (Rotter and Trainin, 1975; Fujimoto et al., 1976; Reinisch et al., 1977) that growth of an immunogenic tumor induces the generation of suppressor T cells. The ease with which a number of different types of relatively large tumors growing in T X B mice can be made to undergo regression by passively transferred tumor-sensitized T cells is impressive. Thus far, it has proved possible in this laboratory to induce regression of three DBA/2 tumors—the P815 mastocytoma, P388 lymphoma, and L5178Y lymphoma—as well as the Meth A fibrosarcoma of BALB/c mice and the SA1 sarcoma of A/J mice. We predict that the same result will be obtained with any immunogenic tumor growing in T X B mice, provided adequately immunized mice are employed as donors of sensitized T cells. It is important to point out, in this connection, that identical results with the Meth A fibrosarcoma have been reported from another laboratory (Bonventre et al., 1982), in which athymic BALB/c nude mice were employed as tumor-bearing test recipients, and heterozygous littermates as donors of immune T cells, or suppressor T cells.

C. Successful Adoptive Immunotherapy of Established Tumors by Eliminating Suppressor T Cells with Cyclophosphamide

It was hypothesized in the Section II.B that any immunosuppressive treatment should prevent a tumor-bearing host from generating suppressor T cells, and thereby facilitate the antitumor function of passively transferred tumor-sensitized T cells. According to this hypothesis, the alkylating agent, cyclophosphamide, should facilitate the expression of adoptive antitumor immunity, because of its well-known immunosuppressive action. It is well to bring to mind, however, that cyclophosphamide has been shown to augment cell-mediated immune responses, provided it is administered before giving antigen. For example, pretreatment with this drug has been shown to result in the generation of increased levels of delayed sensitivity to certain antigens (Goto *et al.*, 1981) and in the augmented production of cytolytic T cells to allografts (Röllinghoff *et al.*, 1977) and virus-induced tumor syngrafts (Glaser, 1979). There is evidence, moreover, that this immuno-augmenting action of cyclophosphamide is a function of its ability to eliminate suppressor T cells selectively (Röllinghoff *et al*, 1977; Glaser, 1979), in that the augmenting effects are canceled by the passive transfer of T cells from normal mice. Needless to say this evidence does not demonstrate antigen-induced suppressor T cells, but rather cyclophosphamide-sensitive precursors of suppressor T cells.

It was not with a view to eliminate suppressor T cells selectively that cyclophosphamide was employed in this laboratory to facilitate the antitumor function of passively transferred T cells, but rather to bring about a temporary immunosuppression of the tumor-bearing recipient for a long enough period to enable passively transferred immune T cells to cause tumor regression. The Meth A fibrosarcoma was chosen to test the proposition that cyclophosphamide would faciliate the expression of adoptive immunity against an established tumor because, unlike the other immunogenic tumors available in this laboratory, when this tumor grows either intradermally or in a hind footpad, it is relatively resistant to the direct cytotoxic action of the drug. For example, a single 100-mg/kg dose of cyclophosphamide causes only a temporary halt in tumor growth, after which the tumor resumes growing at its normal rate. This enabled the action of passively transferred T cells to be measured against an appreciable tumor mass. The experiments incolved giving mice bearing an established Meth A tumor 100 mg/kg of cyclophosphamide intravenously 1 hr before infusing them intravenously with one organ equivalent (1.5×10^8) of spleen cells from immune donors. Because the drug has a half-life in mice of only 20 min or so, it has no effect on immune cells infused 1 hr later.

The experiments revealed (North, 1982) that, whereas cyclophosphamide alone caused a temporary halt in tumor growth and immune cells alone had no antitumor effect at all, combination therapy with cyclophosphamide and immune cells resulted in complete regression of tumors in all mice. In addition, it was found that the donor spleen cells responsible for tumor regression were Thy 1.2-positive T cells. The results indicate that cyclophosphamide facilitates adoptive immunotherapy by removing from the tumor-bearing recipient a cyclophosphamide-sensitive mechanism that serves to inhibit the antitumor function of passively transferred tumor-sensitized T cells. The existence of this tumor-induced mechanism of suppression was next revealed by demonstrating that tumor regression caused by combination therapy with cyclophosphamide and immune cells could be completely inhibited by the passive transfer of Thy-1.2-positive splenic T cells from donors with established tumors. In contrast, normal spleen cells had no such inhibitory effect. It was also revealed that these suppressor T cells could be eliminated from the spleens of tumor-bearing donors by treating the donors with 100 mg/kg of cyclophosphamide, i.e., with the same dose of the drug used to facilitate adoptive immunotherapy in tumor-bearing recipients. There would appear to be little doubt from these results that cyclophosphamide facilitates the expression of passively transferred immunity against an established tumor by eliminating a tumor-induced population of suppressor T cells. More recent experiments (North, 1984) have demonstrated that the dose of cyclophosphamide required to facilitate adoptive immunotherapy is ⩾100 mg/kg, and that the facilitating action lasts for only about 2 days.

This additional model of immunologically mediated tumor regression can be employed to analyze the effector mechanisms responsible for the destruction of an established tumor. However, better models of cyclophosphamide-facilitated adoptive immunotherapy will employ tumors that are totally resistant to the drug, because in these cases the antitumor effect can be attributed totally to passively transferred tumor-sensitized T cells. Tumors that are highly susceptible to cyclophosphamide, such as the P815 mastocytoma, are less desirable because passively transferred T cells function in these cases to prevent the regrowth of a tumor that has been destroyed almost totally by the direct cytotoxic action of the drug. Needless to say, models of cyclophosphamide-facilitated adoptive immunotherapy are more appealing therapeutically than are models that employ T X B tumor-bearing recipients.

D. Successful Adoptive Immunotherapy by γ-Irradiating the Tumor-Bearing Recipient

The majority of evidence shows that whole-body exposure to ionizing radiation is immunosuppression. Therefore, according to the results in Sections II. A–C, sublethal, whole-body γ-irradiation of tumor-bearing recipients should facilitate the expression of passively transferred T-cell-mediated immunity and result in tumor regression. This prediction was tested with four different immunogenic tumors: SA1 sarcoma, P815 mastocytoma, L5178Y lymphoma, and Meth A fibrosarcoma. The experiment involved exposing mice bearing a 4-day tumor to 500 rad of γ-irradiation 1 hr before infusing them intravenously with 1.5×10^8 spleen cells from immune donors.

The results were as predicted (North, 1984). With all four tumors (Fig. 1), combination therapy with γ-irradiation and immune spleen cells resulted in complete regression of tumors in all mice. In contrast, γ-irradiation alone caused only a temporary reduction in the rate of tumor growth, whereas immune cells alone were without effect. The results show that γ-irradiation served to eliminate a tumor-induced mechanism that suppressed the function of intravenously infused immune spleen cells. That this mechanism is a radiosensitive population of suppressor T cells was evidenced by the additional finding (North, 1984) that tumor

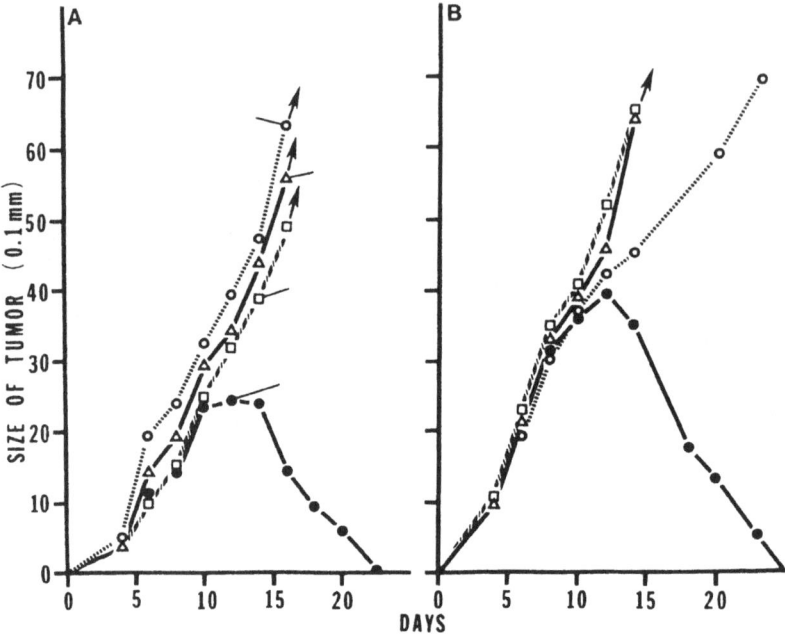

Figure 1. Evidence that 500 rad of whole-body γ-irradiation on day 4 of growth of an intra-footpad (A) Meth A fibrosarcoma or (B) SA1 sarcoma enabled 1.5 × 10^8 immune spleen cells infused 1 hr later to cause complete tumor regression after a 6-day delay. The same result was obtained with the P815 mastocytoma and L5178Y lymphoma (not shown). Means of five mice per group. ○, Immune cells; △, tumor control; □, 500-rad γ-irradiated; ● 500-rad γ-irradiated; plus immune cells.

regression caused by combination therapy with γ-irradiation and immune spleen cells was prevented by passive transfer of spleen cells from donors with established tumors. Moreover, the spleen cells that passively transferred immunity and those that suppressed the expression of this immunity were susceptible to treatment with anti-Thy-1.2 antibody and complement. Therefore, the passive transfer of immune T cells into γ-irradiated tumor-bearing recipients represents a third model of successful adoptive immunotherapy. Furthermore, it should be pointed out that immune T cells can cause tumor regression in mice that are preirradiated with either a lethal or sublethal dose of γ-irradiation (North *et al.*, 1982).

III. TUMOR REGRESSION BY PASSIVE TRANSFER OF T CELLS FROM DONORS WITH PROGRESSIVE TUMORS

The discussion in Section II indicates that mice bearing progressive immuno-genic tumors generate a state of T-cell-mediated immunosuppression. The presence of this state of immunosuppression in tumor-bearing mice was demonstrated by revealing that treatment of the mice with agents that are known to prevent a mechanism of immunosuppression from developing, or to eliminate this mechanism after it develops, were found to enable passively transferred, tumor-sensitized T cells to cause tumor regression. The presence of a mechanism of immunosuppression was further shown by its ability to be passively transferred to an appropriate test recipient with T cells from a tumor-bearing donor.

However, all the tumors employed in this laboratory, and probably immunogenic tumors in general, evoke in their hosts the generation of a state of concomitant antitumor immunity. In the case of the P815 mastocytoma (North et al., 1982) and Meth A fibrosarcoma (Berendt and North, 1980), concomitant immunity peaks at about day 9 of tumor growth and then decays. Since suppressor T cells employed in the experiments described in Section II were harvested at days 12–14 of tumor growth, it is evident that they were harvested after the onset of decay of a concomitant immunity response. Indeed it has been shown in the case of the P815 mastocytoma (North et al., 1982) that the decay of concomitant immunity is coincident with the generation of suppressor T cells capable of suppressing the expression of adoptive immunotherapy in T X B tumor-bearing test recipients. On the basis of this evidence, it was hypothesized that suppressor T cells function to down-regulate concomitant immunity before it reaches a level high enough to cause tumor regression (North et al., 1982).

The timing of the development of concomitant immunity poses a problem in interpreting the results discussed in Section II, however, because no T-cell-mediated immunosuppression should have occurred at the time that the tumor-bearers were adoptively immunized on day 4 of tumor growth. The answer to the problem rests with the knowledge that adoptive immunity is not immediately expressed. Instead, there is invariably a 6- to 8-day delay after the passive transfer of tumor-sensitized T cells before tumor regression commences. In a normal tumor-bearing recipient this delay represents enough time for suppressor T cells to emerge and to express their suppressor function. The reason for the delay in the expression of adoptive immunity has been dealt with by Mills and North (1983), who showed that the lag represents the time needed for an active cytolytic T-cell response to be generated in the recipients.

Meantime, the knowledge that the tumor-bearing mouse generates a state of T-cell-mediated concomitant immunity before it generates suppressor T cells makes it a possible donor of sensitized T cells for adoptively immunizing against an established tumor in a suitable recipient. The possibility that a mouse with a progressive tumor possesses tumor-sensitized T cells capable of causing the regression of an established tumor in a tumor-bearing recipient might seem highly unlikely, considering that it is incapable of causing the regression of its own tumor. Nevertheless, there is no good reason to doubt this possibility, provided that the recipient tumor is somewhat smaller than the donor tumor at the time that donor cells are harvested for passive transfer. Even so, the success of such an experiment would depend on using a tumor-bearing recipient that is incapable of generating suppressor T cells. On the basis of this line of reasoning, and with a knowledge of the kinetics of the generation and decay of the concomitant immunity to P815 mastocytoma (North *et al*., 1982), Meth A fibrosarcoma (Berendt and North, 1980), and the SA1 sarcoma (North and Kirstein, 1977), it was predicted that a syngeneic mouse generating peak concomitant immunity to any one of these tumors would possess enough sensitized T cells in its spleen to immunize adoptively against an established tumor growing in a γ-irradiated recipient.

It was found that one organ equivalent (1.5-2.0 × 10^8) of spleen cells from mice bearing 9-day tumors were capable, on passive transfer, of causing the complete regression of 4-day tumors growing in recipient mice given 500 rad of γ-irradiation before receiving spleen cells (Fig. 2). The spleen cells from tumor-bearing donors that passively transferred immunity were T cells, as evidenced by

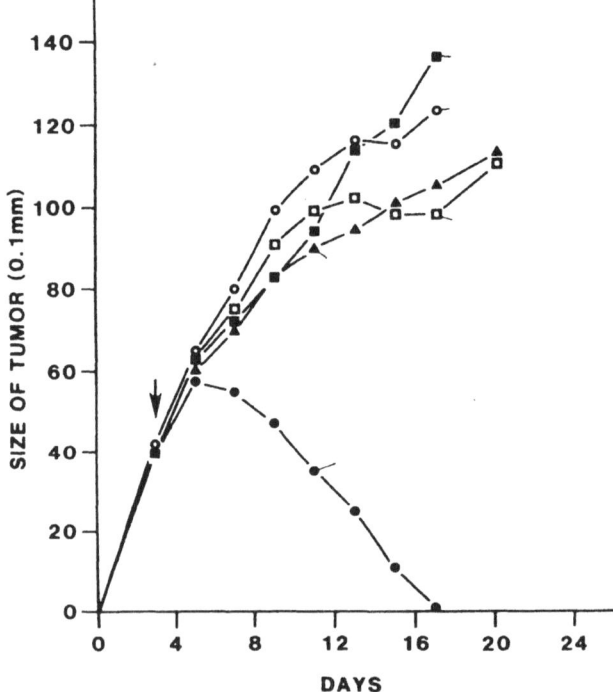

Figure 2. Paradoxical demonstration that splenic T cells (1.5 × 10^8 spleen cells) from donor mice bearing a progressing 9-day intradermal SA1 sarcoma are capable, on passive transfer, of causing the regression of an SA1 sarcoma growing in recipients given 500 rads of γ-irradiation. Donor spleen cells were infused intravenously 1 hr after exposing recipients bearing a 4-day tumor to γ-irradiation. Means of five mice per group. ○, Control; ▲, 500-rad γ-irradiated plus normal cells; □, immune cells; ■, 500-rad γ-irradiated; ●, 500-rad γ-irradiated plus immune cells.

their susceptibility to treatment with anti-Thy-1.2 antibody and complement (North and Bursuker, 1984). Obviously, these results appear paradoxical. The paradox is more apparent than real, however, when one considers that the recipient's tumor was only about one-half the size of the donor's tumor at the time of passive transfer. An obvious interpretation of the results is that the donor would have been able to reject its own tumor if its tumor on day 9 had been as small as the recipient's day 4 tumor. In other words, the donor generated too little immunity too late to destroy its own tumor.

There is another reason, however, for the suprising ease with which one organ equivalent of spleen cells from mice with a progressive immunogenic tumor can cause the regression of an established tumor in γ-irradiated recipients. The results of recent experiments in this laboratory have shown that concomitant immunity to the SA1 sarcoma and P815 mastocytoma is not suppressed by 500 rad of γ-irradiation. Consequently, the regression of tumors in γ-irradiated recipients of passively transferred immune T cells is the result of the antitumor function of the donor T cells combined with the antitumor function of the recipient's own T cells that mediate concomitant immunity. Clearly, the results obtained with γ-irradiated recipients must be interpreted with caution, as they can greatly overestimate the antitumor function of passively transferred immune T cells, regardless of whether the T cells are generated *in vivo* or *in vitro*. In fact, one organ equivalent of splenic T cells from concomitantly immune tumor-bearing donors is not capable of causing the regression of established tumors growing in T × B recipients that are incapable of generating concomitant immunity. In order for tumor regression to occur in T × B recipients a much larger number of donor spleen cells must be passively transferred. The finding that the generation of concomitant immunity is not suppressed by sublethal γ-irradiation is in keeping with the published results of others (Hellström *et al.*, 1978) which show that sublethal irradiation of mice bearing an immunogenic fibrosarcoma can cause regression of tumors in all mice and complete regression of tumors in some. The additional findings that the tumor needs to be above a certain critical size in order for irradiation to cause regression, and that regression fails to occur if the tumor bearers are infused with normal T cells immediately after irradiation, were interpreted as meaning that whole-body irradiation augments an antitumor immune response by temporarily eliminating suppressor T-cell precursors. We have confirmed these results with the SA1 sarcoma.

IV. DISCUSSION

The purpose of this chapter is to present current models of adoptive immunization against established immunogenic tumors in syngeneic mice. These models can be added to models of adoptive immunotherapy of syngeneic tumors in guinea pigs (Smith et al., 1977) and rats (Fernandez-Cruz et al., 1982). They can be used to analyze the process of immunologically mediated tumor regression and to identify the host effector cells ultimately responsible for destroying the tumor.

The results presented in Sections II and III serve to explain why it has proved so difficult in the past to demonstrate that passively transferred tumor-sensitized T cells can bring about the regression of an established tumor, in spite of the fact that the same T cells are capable of neutralizing the growth of an implant of tumor cells. The results indicate that immunogenic tumors become refractory to adoptive immunotherapy at a relatively early stage of their growth, because they evoke in their hosts the generation of a state of T-cell-mediated immunosuppression that functions to down-regulate a preceding concomitant antitumor immune response before it reaches sufficient magnitude to destroy the tumor. That growth of an immunogenic tumor evokes the generation of a state of concomitant immunity that undergoes decay is a well-documented finding (Nelson et al., 1979; Gorelik, 1983). It is suggested here and elsewhere (North et al., 1982) that concomitant antitumor immunity and its negative regulation by suppressor T cells must be taken into account in any attempt to induce tumor regression either by active immunotherapy with immunoadjuvants or by adoptive immunotherapy with tumor-sensitized T cells. It was hypothesized that any attempt to treat a tumor-bearing host by means of immunotherapy represents an attempt to superimpose an augmented immune response on an ongoing immune response that may be undergoing T-cell-mediated down regulation.

It goes without saying that, because an established immunogenic tumor in an immunodepressed recipient can be made to undergo complete regression by infusing the recipient with tumor-sensitized T cells, there can be no doubt that tumor regression is mediated by sensitized T cells. This is not to say, however, that there is direct evidence that sensitized T cells are the ultimate effectors of tumor regression. Specifically sensitized cytolytic T cells can destroy allogeneic and syngeneic tumor cells in vitro, as measured primarily by the ^{51}Cr-release assay, but there appears to be no direct evidence that these T cells perform this function in the in vivo setting. The demonstration that cytolytic T cells enter a tumor in large numbers while it is undergoing regression (Brunner et al., 1981) represents circumstantial evidence at best. Indeed, there is evidence that successful adoptive immunization against tumor allografts (Loveland et al., 1982) and tumor syngrafts (Fernandez-Cruz et al., 1982) can be achieved by the passive transfer of helper T cells, at the exclusion of cytolytic T cells. This evidence has been interpreted as meaning that tumor regression may result from the consequences of a delayed-type hypersensitivity reaction in the tumor (Loveland and McKenzie, 1982). There is certainly evidence that tumor cells can be destroyed at the site of a delayed-type hypersensitivity reaction to bacterial (Tuttle and North, 1975) and other antigens (Lagrange and Thickstun, 1979), demonstrating in turn that tumor cells can be destroyed nonspecifically.

Therefore, the macrophage remains a contender for the role of ultimate effector of tumor regression, although not necessarily exclusively. Cytolytic T cells can specifically destroy tumor cells *in vitro* and macrophages can destroy tumor cells nonspecifically *in vitro*. It is known, moreover, that macrophages do not kill tumor cells efficiently *in vitro* unless they are activated, and the evidence favors the view that macrophages are activated by sensitized lymphocytes by way of the secretion of lymphokines. Because a tumor undergoing immunologically mediated regression almost certainly contains tumor-sensitized T cells and macrophages in relatively large numbers, it contains all the ingredients necessary for the local activation of macrophages. The destruction of the tumor cells might therefore be achieved by direct lytic action of specifically sensitized cytolytic T cells, as well as by the nonspecific lytic action of macrophages.

To date, most published demonstrations of the lysis of tumor cells by macrophages *in vitro* have come from studies that have employed macrophages harvested, not from mice responding immunologically to a tumor, but from the peritoneal cavities of mice expressing cell-mediated immunity to an intraperitoneal inoculum of BCG or *C. parvum*. It is suggested that future studies of the antitumor function of macrophages would be best performed with macrophages harvested from mice responding to an immunogenic tumor. Models are now available that make this undertaking possible.

ACKNOWLEDGMENTS

This work was supported by grants CA-16642 and CA-27794 from the National Cancer Institute and by a grant-in-aid from the R. J. Reynolds Industries, Inc.

V. REFERENCES

Adams, D. O., and Snyderman, R., 1979, Do macrophages destroy nascent tumors? *J. Natl. Cancer Inst.* 62:1341.

Baldwin, R. W., and Pimm, M. V., 1973, BCG immunotherapy of a rat sarcoma, *Br. J. Cancer* 28:281.

Bast, R. C., Bast, B. S., and Rapp, H. J., 1976, Critical review of previously reported animal studies of tumor immunotherapy with non-specific immunostimulants, *J. Natl. Cancer Inst.* 277:60.

Berendt, M. J., and North, R. J., 1980, T cell-mediated immunosuppression of anti-tumor immunity. An explanation for the progressive growth of an immunogenic tumor, *J. Exp. Med.* 151:69.

Bonventre, P. F., Nickol, A. D., Ball, E. J., Michael, J. G., and Bubel, H. C., 1982, Development of protective immunity against bacterial and viral infections in tumor-bearing mice coincident with suppression of tumor immunity, *J. Reticuloendothel. Soc.* 32:25.

Brunner, K. T., MacDonald, H. R., and Cerottini, J. C., 1981, Quantitation and clonal isolation of cytolytic T lymphocyte precursors selectively infiltrating sarcoma virus-induced tumors, *J. Exp. Med.* **154**:362.

Dye, E. S., and North, R. J., 1981, T cell-mediated immunosuppression as an obstacle to adoptive immunotherapy of the P815 mastocytoma and its metastases, *J. Exp. Med.* **154**:1033.

Dye, E. S., North, R. J., and Mills, C. D., 1981, Mechanisms of anti-tumor action of *Corynebacterium parvum*. 1. Potentiated tumor-specific immunity and its therapeutic limitations, *J. Exp. Med.* **154**:609.

Evans, R., 1972, Macrophages in syngeneic animal tumors, *Transplantation* **14**:468.

Fernandez-Cruz, E., Gilman, S. C., and Feldman, J. D., 1982, Immunotherapy of a chemically-induced sarcoma in rats: Characterization of effector T cell subset and nature of suppression, *J. Immunol.* **128**:1112.

Fujimoto, S., Green, M. I., and Sehon, A. H., 1976, Regulation of immune response to tumor antigens. I. Immunosuppressor cells in tumor bearing hosts, *J. Immunol.* **116**:791.

Glaser, M., 1979, Regulation of specific cell-mediated cytotoxic response against SV40-induced tumor associated antigens by depletion of suppressor T cells with cyclophosphamide in mice, *J. Exp. Med.* **149**:774.

Gorelik, E., 1983, Concomitant tumor immunity, *Adv. Cancer Res.* **9**:71.

Goto, M., Mitsuoka, A., Sugiyama, M., and Kitano, M., 1981, Enhancement of delayed hypersensitivity reaction with varieties of anti-cancer drugs. A common biological phenomenon, *J. Exp. Med.* **154**:204.

Hellström, K. E., Kant, J. A., and Tamerius, J. D., 1978, Regression and inhibition of sarcoma growth by interference with a radiosensitive T cell population, *J. Exp. Med.* **148**:799.

Lagrange, P. H., and Thickstun, P. M., 1979, *In vivo* antitumor activity at various forms of delayed-type hypersensitivity in mice, *J. Natl. Cancer Inst.* **62**:429.

Loveland, B. E., McKenzie, I. F. C., 1982, Cells mediating graft rejection in the mouse. II. The Ly phenotypes of cells producing tumor allograft rejection, *Transplantation* **33**:174.

Loveland, B. E., Hogarth, P. M., Ceredig, R. H., and McKenzie, I. F. C., 1982, Cells mediating graft rejection in the mouse. 1. Lyt-1 cells mediate graft rejection, *J. Exp. Med.* **153**:1044.

Mills, C. D., and North, R. J., 1983, Expression of passively transferred tumor depends on generation of cytolytic T cells in recipient. Inhibition by suppressor T cells, *J. Exp. Med.* **157**:1448.

Mills, C. D., North, R. J., and Dye, E. S., 1981, Mechanisms of anti-tumor action of *Corynebacterium parvum*. II. Potentiated cytolytic T cell response and its tumor-induced suppression, *J. Exp. Med.* **154**:621.

Nelson, D. S., Nelson, M., and Hopper, K. E., 1979, Mechanisms of resistance to syngeneic methylcholanthrene-induced fibrosarcomas, *Adv. Exp. Med. Biol.* **121B**:541.

North, R. J., 1982, Cyclophosphamide-facilitated adoptive immunotherapy of an established tumor depends on elimination of tumor-induced suppressor T cells, *J. Exp. Med.* **55**:1063.

North, R. J., 1984, γ-Irradiation facilitates the expression of adoptive immunity against established tumors by eliminating suppressor T cells, *Cancer Immunol. Immunother.* (in press).

North R. J., and Bursuker, I., 1984, The generation and decay of the immune response to a progressive fibrosarcoma, *J. Exp. Med.* (in press).

North, R. J., and Kirstein, D. P., 1977, T cell-mediated concomitant immunity to syngeneic tumors. 1. Activated macrophages as the expressors of nonspecific immunity to unrelated tumors and bacterial parasites, *J. Exp. Med.* **145**:275.

North, R. J., Dye, E. S., Mills, C. D. and Chandler, J. P., 1982, Modulation of antitumor immunity. Immunobiologic Approaches, *Springer Semin. Immunopathol.* **5**:193.

Reinisch, C. L., and Andrew, S. L., and Schlossman, S. F., 1977, Suppressor cell regulation of tumor immunity: Abrogation by adult thymectomy, *Proc. Natl. Acad. Sci. USA* **74**(7): 2989.

Röllinghoff, M., Starzinski-Powitz, A., Pfizenmaier, K., and Wagner, H., 1977, Cyclophosphamide-sensitive T lymphocytes suppress *in vivo* generation of antigen-specific cytotoxic T lymphocytes, *J. Exp. Med.* **145**:455.

Rosenberg, S. A., and Terry, W. D., 1977, Passive immunotherapy of cancer in animals and man, *Adv. Cancer Res.* **25**:323.

Rotter, V., and Trainin, N., 1975, Inhibition of tumor growth in syngeneic chimeric mice mediated by a depletion of suppressor T cells, *Transplantation* **20**:68.

Russel, S. W., and McIntosh, A. T., 1977, Macrophages isolated from regressing Moloney sarcomas are more cytotoxic than those recovered from progressing tumors, *Nature* **268**:69.

Smith, H. G., Harmell, R. P., Hanna, M. G., Zwilling, B. S., Zbar, B., and Rapp, H. J., 1977, Regression of established intradermal tumors and lymph node metastases in guinea pigs after systemic transfer of immune lymphoid cells, *J. Natl. Cancer Inst.* **58**:1315.

Tuttle, R. L., and North, R. J., 1975, Mechanisms of antitumor action of *Corynebacterium parvum*: Nonspecific tumor cell destruction at site of an immunologically mediated sensitivity reaction to *C. parvum*, *J. Natl. Cancer Inst.* **55**:1043.

Zbar, B., Bernstein, I. D., Bartlett, G. L., Hanna, M. G., and Rapp, H. J., 1972, Immunotherapy of cancer: Regression of intradermal tumors and prevention of growth of lymph node metastases after intralesional injection of living *Mycobacterium bovis*, *J. Natl. Cancer Inst.* **49**:119.

Index

Activation
  alterations in secretory products by, 97, 128, 136
  definition of, 127
  genetic control of, 150-152
  induction of, 128, 140-143, 153, 171
  and kill of microorganisms, 97-101, 119-124, 149, 156-159
  and kill of tumor cells, 128, 133-134, 142-144, 149, 171, 232-236, 255
  membrane changes in, *see* Membrane alterations
  morphological changes in, 34-36
  regulation of, 141-144, 149, 204
  stages of, 134, 140, 143, 152
Antigens, surface
  CR_3, *see* Mac-1
  in differentiation, 14
  2.4G2, *see* $Fc_{\gamma 2b/\gamma 1} R$
  human, 6-7
  LFA-1, 4, 8-9
  Mac-1, 4, 5-10, 15
  Mac-2, 10-11
  Mac-3, 11-12
  in membrane recycling, 21-22
  monoclonal antibodies against, *see* Chapter 1
  murine, 2-3

Binding of tumor cells
  cell biology of, 134-136
  induction of, 134
  lymphokines and, 134
  quantification of, 128-129, 131-132
  and relationship to cytolysis, 133-134, 140, 143-144
  requirements for, 131

Binding of tumor cells (*cont.*)
  transmethylation reactions and, 135
  types of, 129-132
Binding of parasites
  characteristics and mechanisms of, 89
  quantification of, 84-86

Complement receptors
  activation for phagocytosis, 59-63
  C3b receptor, 57
  C3bi receptor, 57
  C5a receptor, 213
  and facilitation of Fc-mediated phagocytosis, 58-59, 64-65, 72, 75
  and lymphokines, 60-63
  mechanism of activation, 61
  mobility of, 61
  in particle binding, 58, 75
  types of, 75

Dendritic cells
  and antigen presentation, 12
  surface antigens of, 13

Ectoenzymes
  alkaline phosphodiesterase-I, 33
  5'-nucleotidase, 33, 73
  and prostaglandins, 206
Endocytosis, *see* Phagocytosis

Fc receptors, human
  cellular distribution of, 20-21
  heterogeneity of, 20

259